Advanced Nanomaterials for Detection of CBRN

NATO Science for Peace and Security Series

This Series presents the results of scientific meetings supported under the NATO Programme: Science for Peace and Security (SPS).

The NATO SPS Programme supports meetings in the following Key Priority areas: (1) Defence Against Terrorism; (2) Countering other Threats to Security and (3) NATO, Partner and Mediterranean Dialogue Country Priorities. The types of meeting supported are generally "Advanced Study Institutes" and "Advanced Research Workshops". The NATO SPS Series collects together the results of these meetings. The meetings are co-organized by scientists from NATO countries and scientists from NATO's "Partner" or "Mediterranean Dialogue" countries. The observations and recommendations made at the meetings, as well as the contents of the volumes in the Series, reflect those of participants and contributors only; they should not necessarily be regarded as reflecting NATO views or policy.

Advanced Study Institutes (ASI) are high-level tutorial courses to convey the latest developments in a subject to an advanced-level audience.

Advanced Research Workshops (ARW) are expert meetings where an intense but informal exchange of views at the frontiers of a subject aims at identifying directions for future action.

Following a transformation of the programme in 2006, the Series has been re-named and re-organised. Recent volumes on topics not related to security, which result from meetings supported under the programme earlier, may be found in the NATO Science Series.

The Series is published by IOS Press, Amsterdam, and Springer, Dordrecht, in conjunction with the NATO Emerging Security Challenges Division.

Sub-Series

A.	Chemistry and Biology	Springer
B.	Physics and Biophysics	Springer
C.	Environmental Security	Springer
D.	Information and Communication Security	IOS Press
E.	Human and Societal Dynamics	IOS Press

http://www.nato.int/science
http://www.springer.com
http://www.iospress.nl

Series A: Chemistry and Biology

Advanced Nanomaterials for Detection of CBRN

edited by

Janez Bonča
J. Stefan Institute
Ljubljana, Slovenia

and

Sergei Kruchinin
Bogolyubov Inst. f. Theor. Physics
Nat. Ukrainian Academy of Science
Kiev, Ukraine

Springer
Published in Cooperation with NATO Emerging Security Challenges Division

Proceedings of the NATO Advanced Research Workshop on
Advanced Nanomaterials for Detection of CBRN
Odessa, Ukraine
2–6 October 2019

ISBN 978-94-024-2029-6 (PB)
ISBN 978-94-024-2029-6 (HB)
ISBN 978-94-024-2030-2 (e-book)
https://doi.org/10.1007/978-94-024-2030-2

Published by Springer,
P.O. Box 17, 3300 AA Dordrecht, The Netherlands.

www.springer.com

Printed on acid-free paper

This Springer imprint is published by the registered company Springer Nature B.V.
The registered company address is: Van Godewijckstraat 30, 3311 GX, Dordrecht, The Netherlands.

Preface

The proceedings of the NATO ARW "Advanced nanomaterials for the detection of CBRN" emerged as a result of many presentations and discussions between participants of the workshop held in the "Mozart" Hotel, Odessa, Ukraine, in October 2–6, 2019. The Odessa workshop focused on several open problems including the current state of nanomaterials and security problems.

The program of the workshop allowed presentations and opened discussions on several modern research topics such as new nanomaterials and sensors. There was an intense discussion in the field of nanotechnology and safety systems that include advanced nanomaterials, nanosensors, nanocomposite multifunctional materials, bionanosensors, and nanoanalyzers. On the session devoted to advanced nanomaterials, the physical properties of graphene, phosphorene, multiwalled carbon nanotubes, and new composite materials were analyzed. The most promising new materials and methods for the detection of hazardous materials including explosives are: multiwalled carbon nanotubes, graphene, and optical techniques.

Participants presented new methods for the detection of chemical, biological, radiological, and nuclear (CBRN) agents with the use of chemical and biochemical sensors. The identification, protection, and decontamination are the main scientific and technological responses to the modern challenges of CBRN agents. The contemporary open problems of the physics of sensors include: the determination of sizes of nanoparticles, identification of particles, and determination of concentrations and mobilities of nanoparticles.

We are grateful to members of the International Advisory Committee, especially to F. Peeters and B. Vlahovic, for their consistent help and suggestions.

We would like to thank the NATO Science for Peace and Security Programme for the essential financial support, without which the meeting could not have taken place. We also acknowledge the generous support by the National Academy of Science of Ukraine, J. Stefan Institute (Ljubljana, Slovenia), and Faculty of Mathematics and Physics at the University of Ljubljana, Slovenia.

Ljublyana, Slovenia
Kyiv, Ukraine
March 2020

Janez Bonča
Sergei Kruchinin

MOZART
HOTEL

Сніданок
Завтрак
Breakfast

Contents

Contributors

O. Adiguzel
Department of Physics, Firat University, Elazig, Turkey

A. Ahsan
LLC Lynx Eurasia, Almaty, Kazakhstan

L. Alfonta
Department of Life Sciences and Ilse Katz Institute for Nanoscale Science and Technology, Ben-Gurion University of the Negev, Beer-Sheva, Israel

A. S. Babenko
Bioorganics Department, Belarusian State Medical University, Minsk, Belarus

O. Barmak
Khmelnitsky National University, Khmelnitsky, Ukraine

C. Bellecci
International Master Courses in Protection Against CBRNe Events, University of Rome Tor Vergata, Rome, Italy

S. Bellucci
INFN-Laboratori Nazionali di Frascati, Frascati, Italy

Y. Bondaruk
South-Ukrainian K.D. Ushynsky National Pedagogical University, Odessa, Ukraine

F. Bruno
Sogin s.p.a, Rome, Italy

R. Caldari
Sogin s.p.a, Rome, Italy

M. Carestia
Department of Industrial Engineering, University of Rome Tor Vergata, Rome, Italy
International Master Courses in Protection Against CBRNe Events, University of Rome Tor Vergata, Rome, Italy

R. F. Chakukov
Physics Department, Belarusian State University, Minsk, Belarus

A. P. Chebanenko
Odessa I.I. Mechnikov National University, Odessa, Ukraine

M. Civica
Sogin s.p.a, Rome, Italy

D. C. Critello
Department of Health Sciences, University Magna Græcia Catanzaro, Catanzaro, Italy

B. Dabrowski
Institute of Physics, Polish Academy of Sciences, Warsaw, Poland

F. d'Errico
School of Engineering, University of Pisa, Pisa, Italy

D. Di Giovanni
Department of Industrial Engineering, University of Rome Tor Vergata, Rome, Italy
International Master Courses in Protection Against CBRNe Events, University of Rome Tor Vergata, Rome, Italy
Health Safety Environmental Research Association Rome, Rome, Italy

I. Donchev
Department of Materials Engineering, Ben-Gurion University of the Negev, Beer-Sheva, Israel

I. K. Doycho
I.I. Mechnikov Odessa National University, Odessa, Ukraine

V. P. Egorova
Belarusian State Pedagogical University, Minsk, Belarus

L. M. Filevska
Odessa I.I. Mechnikov National University, Odessa, Ukraine

D. Fink
Departamento de Ciencias Ambientales, División de Ciencias Biológicas y de la Salud, Universidad Autónoma Metropolitana-Lerma, Lerma de Villada, Estado de México
Nuclear Physics Institute, ASCR, Řež, Czech Republic

A. S. Fiorillo
Department of Health Sciences, University Magna Græcia Catanzaro, Catanzaro, Italy

H. Garcia-Arrellano
Departamento de Ciencias Ambientales, División de Ciencias Biológicas y de la Salud, Universidad Autónoma Metropolitana-Lerma, Lerma de Villada, Estado de México

P. Gaudio
Department of Industrial Engineering, University of Rome Tor Vergata, Rome, Italy
International Master Courses in Protection Against CBRNe Events, University of Rome Tor Vergata, Rome, Italy

V. Ya. Gayvoronsky
Institute of Physics National Academy of Sciences of Ukraine, Kyiv, Ukraine

M. Gonchar
Drohobych Ivan Franko State Pedagogical University, Drohobych, Ukraine
Institute of Cell Biology, National Academy of Sciences of Ukraine, Lviv, Ukraine

V. S. Grinevych
Odessa I.I. Mechnikov National University, Odessa, Ukraine

H. V. Grushevskaya
Physics Department, Belarusian State University, Minsk, Belarus

V. Hnatowicz
Nuclear Physics Institute, ASCR, Řež, Czech Republic

A. Iannotti
Department of Industrial Engineering, University of Rome Tor Vergata, Rome, Italy
International Master Courses in Protection Against CBRNe Events, University of Rome Tor Vergata, Rome, Italy
International Alliance CBRN, HESAR, Rome, Italy

S. G. Ilchenko
International Center "Institute of Applied Optics" National Academy of Sciences of Ukraine, Kyiv, Ukraine

O. Ismailova
Turin Polytechnic University in Tashkent, Tashkent, Uzbekistan
Uzbek-Japan Innovation Center of Youth, Tashkent, Uzbekistan

G. Janusas
Kaunas University of Technology, Kaunas, Lithuania

T. Janusas
Kaunas University of Technology, Kaunas, Lithuania

C. Justas
Faculty of Mechanical Engineering and Design, Kaunas University of Technology, Kaunas, Lithuania

T. Kavetskyy
Drohobych Ivan Franko State Pedagogical University, Drohobych, Ukraine
The John Paul II Catholic University of Lublin, Lublin, Poland

A. Kiv
Department of Materials Engineering, Ben-Gurion University of the Negev, Beer-Sheva, Israel
South-Ukrainian K.D. Ushynsky National Pedagogical University, Odessa, Ukraine

A. E. Kiv
South-Ukrainian National Pedagogical University, Odessa, Ukraine

M. Kompitsas
NHRF-National Hellenic Research Foundation, Theoretical and Physical Chemistry Institute-TPCI, Athens, Greece

P. Koralli
NHRF-National Hellenic Research Foundation, Theoretical and Physical Chemistry Institute-TPCI, Athens, Greece

Iu. Krak
V.M. Glushkov Institute of Cybernetics, Taras Shevchenko National University, Kyiv, Ukraine

D. Kruchinin
Taras Shevchenko Kyiv National University, Kyiv, Ukraine

S. P. Kruchinin
Bogolyubov Institute for Theoretical Physics, Kyiv, Ukraine

K. Kruchynin
V.M. Glushkov Institute of Cybernetics, Taras Shevchenko National University, Kyiv, Ukraine

N. G. Krylova
Physics Department, Belarusian State University, Minsk, Belarus

Y. Kukhazh
Drohobych Ivan Franko State Pedagogical University, Drohobych, Ukraine

L. L. Li
Department of Physics and Center for Nanointegration, University of Duisburg-Essen, Duisburg, Germany
Department of Physics, University of Antwerp, Antwerpen, Belgium

I. V. Lipnevich
Physics Department, Belarusian State University, Minsk, Belarus

R. A. Lymarenko
International Center "Institute of Applied Optics" National Academy of Sciences of Ukraine, Kyiv, Ukraine

Sh. Makhkamov
Institute of Nuclear Physics Academy of Sciences, Tashkent, Uzbekistan

A. Malizia
Sogin s.p.a, Rome, Italy

A. T. Mansharipova
Kazakh Russian Medical University, Almaty, Kazakhstan

E. Manziuk
Khmelnitsky National University, Khmelnitsky, Ukraine

T. Marsagishvili
R. Agladze Institute of Inorganic Chemistry and Electrochemistry of Iv. Javakhishvili Tbilisi State University, Tbilisi, Georgia

M. Matchavariani
R. Agladze Institute of Inorganic Chemistry and Electrochemistry of Iv. Javakhishvili Tbilisi State University, Tbilisi, Georgia

R. M. Melnyk
National University of Kyiv-Mohyla Academy, Kyiv, Ukraine

V. Morozov
National Aviation University, Kyiv, Ukraine

I. Morozova
National Aviation University, Kyiv, Ukraine

G. A. Mousdis
NHRF-National Hellenic Research Foundation, Theoretical and Physical Chemistry Institute-TPCI, Athens, Greece
Electrical and Computer Engineering Department, National Technical University of Athens, Athens, Greece

V. V. Multian
Institute of Physics National Academy of Sciences of Ukraine, Kyiv, Ukraine

G. Muñoz Hernández
Departamento de Fisica, Universidad Autónoma Metropolitana-Iztapalapa, México, DF, México

V. Naginevičius
Kaunas University of Applied Sciences, Kaunas, Lithuania

W. Nawrocki
Faculty of Electronics and Telecommunications, Poznan University of Technology, Poznan, Poland

A. Normurodov
Institute of Nuclear Physics Academy of Sciences, Tashkent, Uzbekistan

A. Palevicius
Faculty of Mechanical Engineering and Design, Kaunas University of Technology, Kaunas, Lithuania

L. Palombi
International Master Courses in Protection Against CBRNe Events, University of Rome Tor Vergata, Rome, Italy
Department of Biomedicine and Prevention, University of Rome Tor Vergata, Rome, Italy

Y. Patel
Faculty of Mechanical Engineering and Design, Kaunas University of Technology, Kaunas, Lithuania

F. M. Peeters
Department of Physics, University of Antwerp, Antwerpen, Belgium

G. Petropoulou
NHRF-National Hellenic Research Foundation, Theoretical and Physical Chemistry Institute-TPCI, Athens, Greece

A. P. Polishchuk
Aerospace Faculty, National Aviation University, Kyiv, Ukraine

S. A. Pullano
Department of Health Sciences, University Magna Græcia Catanzaro, Catanzaro, Italy

A. Ramanavicius
Department of Physical Chemistry, Faculty of Chemistry and Geosciences, Vilnius University, Vilnius, Lithuania

S. P. Repetsky
Institute of High Technologies, Taras Shevchenko National University of Kyiv, Kyiv, Ukraine

C. Russo
Department of Industrial Engineering, University of Rome Tor Vergata, Rome, Italy
International Master Courses in Protection Against CBRNe Events, University of Rome Tor Vergata, Rome, Italy
Health Safety Environmental Research Association Rome, Rome, Italy

Z. Samkharadze
R. Agladze Institute of Inorganic Chemistry and Electrochemistry of Iv. Javakhishvili Tbilisi State University, Tbilisi, Georgia

O. Šauša
V.E. Institute of Physics, Slovak Academy of Sciences, Bratislava, Slovak Republic

G. Skobeeva
I.I. Mechnikov National University, Odessa, Ukraine

V. Skobeeva
Research Institute of Physics, I.I. Mechnikov National University, Odessa, Ukraine

O. Smutok
Drohobych Ivan Franko State Pedagogical University, Drohobych, Ukraine
Institute of Cell Biology, National Academy of Sciences of Ukraine, Lviv, Ukraine

V. Smyntyna
Faculty of Mathematics, Physics and Information Technologies, Department of Experimental Physics, Odesa National I.I. Mechnikov University, Odesa, Ukraine

B. A. Snopok
V.E. Lashkaryov Institute of Semiconductor Physics, National Academy of Sciences of Ukraine, Kyiv, Ukraine

O. B. Snopok
V.E. Lashkaryov Institute of Semiconductor Physics, National Academy of Sciences of Ukraine, Kyiv, Ukraine

N. Sulaymonov
Institute of Nuclear Physics Academy of Sciences, Tashkent, Uzbekistan

D. K. Sunko
Faculty of Science, Department of Physics, University of Zagreb, Zagreb, Croatia

V. B. Y. Taranenko
International Center "Institute of Applied Optics" National Academy of Sciences of Ukraine, Kyiv, Ukraine

M. Yu. Tashmetov
Institute of Nuclear Physics Academy of Sciences, Tashkent, Uzbekistan

G. Tatishvili
R. Agladze Institute of Inorganic Chemistry and Electrochemistry of Iv. Javakhishvili Tbilisi State University, Tbilisi, Georgia

A. Tereshchenko
Department of Experimental Physics, Faculty of Mathematics, Physics and Information Technologies, Odesa National I.I. Mechnikov University, Odesa, Ukraine

M. Thornton
International Master Courses in Protection Against CBRNe Events, University of Rome Tor Vergata, Rome, Italy
International Alliance CBRN, HESAR, Rome, Italy

F. Troiani
Sogin s.p.a, Rome, Italy

F. Umarova
Institute of Nuclear Physics Academy of Sciences, Tashkent, Uzbekistan
Institute of Ion-Plasma and Laser Technologies Academy of Sciences, Tashkent, Uzbekistan

S. Urbaite
Kaunas University of Technology, Kaunas, Lithuania

J. Vacik
Nuclear Physics Institute, ASCR, Řež, Czech Republic

C. Vicini
Sogin s.p.a, Rome, Italy

B. Vlahovic
North Carolina Central University, Durham, NC, USA

I. G. Vyshyvana
Institute of High Technologies, Taras Shevchenko National University of Kyiv, Kyiv, Ukraine

V. Yatsenko
Space Research Institute of NASU-NSAU, Kyiv, Ukraine

K. Zubrytska
Drohobych Ivan Franko State Pedagogical University, Drohobych, Ukraine

Part I
Advanced Nanomaterials

Chapter 1
Tuning the Electronic, Optical, and Transport Properties of Phosphorene

L. L. Li and F. M. Peeters

Abstract Phosphorene is one of the most important 2D materials, which was exfoliated from bulk black phosphorus in 2014. This 2D material features the combined properties of large band gap, high carrier mobility, and strong in-plane anisotropy, which make it well suited for future electronic and optoelectronic applications. The electronic, optical and transport properties of phosphorene tuned by gating, strain, and disorder effects are presented in this chapter. To this end, tight-binding approach, linear Kubo formula, and scattering matrix method are employed.

Keywords 2D black phosphorus · Electronic · optical and transport properties · Gating · strain and disorder effects · Tight binding method · Linear response theory · Scattering matrix method

1.1 Introduction

Since the isolation of graphene in 2004, plenty of two-dimensional (2D) materials, e.g., hexagonal boron nitride (hBN), transition metal dichalcogenides (TMDs), and black phosphorus (BP), have been discovered with a broad range of fascinating properties [1, 2]. These 2D materials exhibit a number of special properties that are not present in their bulk counterparts, e.g., high electrical mobility, widely tunable band gap, large optical transmission, and superior mechanical flexibility, which make them very promising for electronic and optoelectronic applications that

L. L. Li (✉)
Department of Physics and Center for Nanointegration, University of Duisburg-Essen, Duisburg, Germany

Department of Physics, University of Antwerp, Antwerpen, Belgium
e-mail: longlong.li@uni-due.de

F. M. Peeters
Department of Physics, University of Antwerp, Antwerpen, Belgium
e-mail: francois.peeters@uantwerpen.be

J. Bonča, S. Kruchinin (eds.), *Advanced Nanomaterials for Detection of CBRN*, NATO Science for Peace and Security Series A: Chemistry and Biology, https://doi.org/10.1007/978-94-024-2030-2_1

require atomically thin, mechanically flexible and optically transparent materials. In the context of electronics and optoelectronics, BP has drawn particular interest from the research community due to its unique structural and tunable electronic and optical properties [3, 4].

Similar to graphite, BP is a layered material [5]. Within the layers phosphorus atoms are strongly (covalently) bonded, while between the layers they are weakly coupled by van der Waals forces. However, unlike graphite (each carbon atom bonds with three neighboring atoms via sp^2 hybridization), each phosphorus atom in BP bonds with three neighboring atoms via sp^3 hybridization [6]. A single layer of BP is termed phosphorene, which was isolated by mechanical exfoliation of bulk BP in 2014 [7–9]. The lattice structures of graphene and phosphorene are sketched in Fig. 1.1a and b, respectively. As shown, graphene has a *flat* honeycomb lattice due to sp^2 hybridization of carbon atoms and a *rhombohedral* primitive unit cell with two sublattice atoms, whereas phosphorene has a *puckered* honeycomb lattice due to sp^3 hybridization of phosphorus atoms and a *rectangular* primitive unit cell with four sublattice atoms.

Apart from the difference in their lattice structure, phosphorene also differs from graphene in the band structure. As shown in Fig. 1.1c and d, graphene is a Dirac semimetal (linear energy spectrum with zero band gap in the low-energy region), whereas phosphorene is a semiconductor (quadratic energy spectrum with finite band gap in the low-energy region). It was shown by density functional theory (DFT) calculations within the GW approximation [10] that the band gap of multilayer phosphorene decreases monotonically with increasing number of phosphorene layers due to the interlayer interaction, e.g., from a wide gap ($\sim$2 eV) for monolayer down to a narrow gap ($\sim$0.3 eV) for bulk. On top of the DFT-GW approximation and by further including excitonic effects, it was shown by solving the Bethe-Salpeter equation (BSE) [10] that the DFT-GW-BSE band gaps of phosphorene multilayers are in good agreement with the experimental band gaps measured by photoluminescence spectroscopy [8]. The scaling of the band gap with the number of phosphorene layers provides an excellent degree of freedom for band gap tuning: A desirable band gap for a certain application can be tailored by simply varying the layer thickness. Moreover, the band gap of multilayer phosphorene remains direct at the Γ point of the Brillouin zone, irrespective of the number of phosphorene layers. This is distinctively different than multilayer TMDs which present a finite band gap at the K point of the Brillouin zone that is only direct in case of single layers. In addition, due to the puckered honeycomb lattice, phosphorene exhibits two inequivalent crystallographic directions, i.e., zigzag (parallel to atomic ridges) and the armchair (perpendicular to ridges) directions, as sketched in Fig. 1.1b. This structural anisotropy is the origin of the strong in-plane anisotropy manifested in the electronic, optical, and transport properties of phosphorene [10–13], which is not typical for graphene, monolayer hBN, and monolayer TMDs and thus perhaps makes phosphorene so special among most 2D materials.

Although phosphorene has fascinating intrinsic properties, as mentioned above, for practical applications it is always essential to tune these properties extrinsically via e.g. strain and gating. Due to the puckered lattice structure, phosphorene

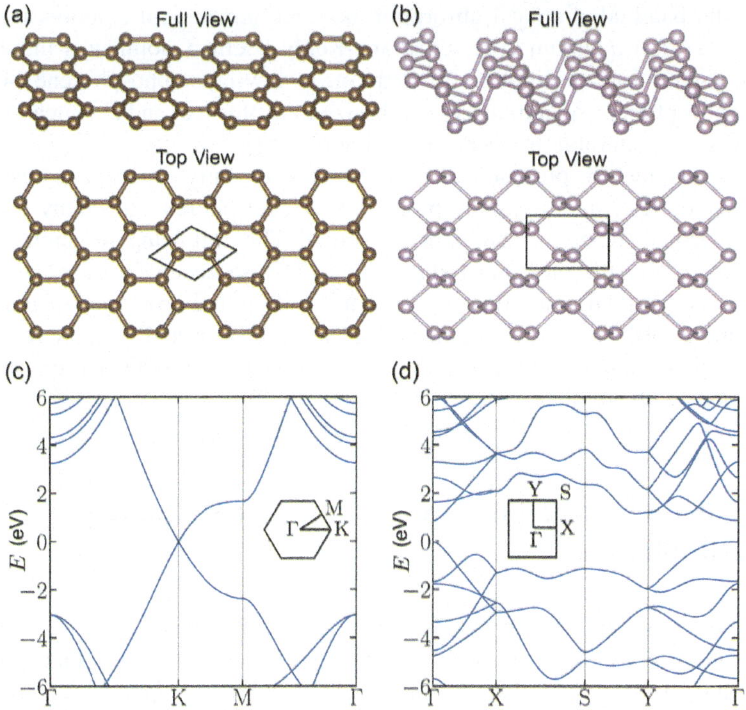

Fig. 1.1 Lattice and band structures of graphene and phosphorene: (**a**) and (**c**) for graphene; (**b**) and (**d**) for phosphorene. In (**a**) and (**b**) the black rhombohedron and rectangle represents the primitive unit cells of graphene and phosphorene, respectively

exhibits an excellent response ability to various external fields such as strain and electric field, thereby allowing for such extrinsic tuning, which have promoted a number of theoretical and experimental studies. *Theoretically*, for instance, it was shown that (i) by applying axial strain, various interesting electronic properties can be induced including, for instance, directional preference of charge transport [14], direct-indirect band-gap transition [15], semiconductor-metal phase transition [16], emergence of anisotropic Dirac-like cones [17], enhancement of room-temperature electron mobility [18], and enhancement of electron-hole interaction and excitonic effects [19]; (ii) by applying vertical electric field, the band gap of multilayer phosphorene can be closed and as a result, an electronic phase transition from normal semiconductor to Dirac semimetal can be induced [20–23], leading to the appearance of a linear energy spectrum and zeroth Landau level in gated multilayer phosphorene [22, 23]; and (iii) in addition to strain and electric field, layer stacking/twisting [24–30], quantum confinement [31–37], and lattice disorder [38–43] can also be utilized to tune the electronic, optical, and transport properties of phosphorene. *Experimentally*, for instance, it was reported that (i) strain-engineered rippling in few-layer phosphorene induces a remarkable shift of the band-edge optical absorption between tensile and compressive strain regions [44]; (ii) quantum confined Franz-Keldysh effect dominates the electro-optic modulation of mid-infrared absorption in few-layer phosphorene [45, 46]; and (iii) phosphorene quantum dots synthesized via physical and chemical methods exhibit a strong photoluminescence emission [47, 48].

In this chapter, we present a summary of our recently obtained results of the electronic, optical, and transport properties of phosphorene tuned by electrical gating, mechanical strain, and lattice disorder. The following research subjects are included: (i) self-consistent tight-binding (TB) approach for gated multilayer phosphorene [49]; (ii) gate tunable electronic and optical properties of multilayer phosphorene [50]; (iii) strain engineered linear dichroism and Faraday rotation of few-layer phosphorene [51]; and (iv) quantum transport in defective phosphorene nanoribbons with atomic vacancies [52].

1.2 Self-Consistent TB Approach for Gated Multilayer Phosphorene

It is known that applying an external electric field perpendicular to a multilayer system induces a charge redistribution over the stacked layers, which produces an internal electric field that counteracts the externally applied one (i.e., the electric-field-induced charge screening). Although the influence of a perpendicular electric field on the electronic properties of multilayer phosphorene was widely investigated [20–23], the electric-field-induced charge screening effect, which was shown to be of significant importance in multilayer graphene [53–57], remains poorly understood in multilayer phosphorene. Up to date, there are few studies exploring the screening effect on the electronic properties of multilayer phosphorene in the presence of a perpendicular electric field [58, 59]. In Ref. [58], the electronic structure of bilayer and trilayer phosphorene was obtained by first-principles calculations, where the screening effect was induced by the additional doping of charges; whereas in Ref. [59], the band gap of multilayer phosphorene was obtained by tight-binding calculations, where the screening effect was introduced by assuming a band-gap-dependent dielectric constant. However, the electric-field-induced charge screening was not included in these studies.

Here, we present a self-consistent TB approach for gated multilayer phosphorene which takes account of the electric-field-induced charge screening within an experimental set-up, where an electrical gating is applied across the phosphorene multilayer to produce a perpendicular electric field. The *self-consistency* comes from the unknown charge screening which needs to be determined by an iterative calculation of the electrostatic Hartree potential within the TB framework.

We consider multilayer phosphorene consisting of N coupled phosphorene layers that are AB stacked on top of each other, as shown in Fig. 1.2. It was shown by DFT calculations [25, 60] that this type of layer stacking is energetically the most stable for multilayer phosphorene. In the presence of a perpendicular electric field, the TB Hamiltonian of this N-layer phosphorene system is given by

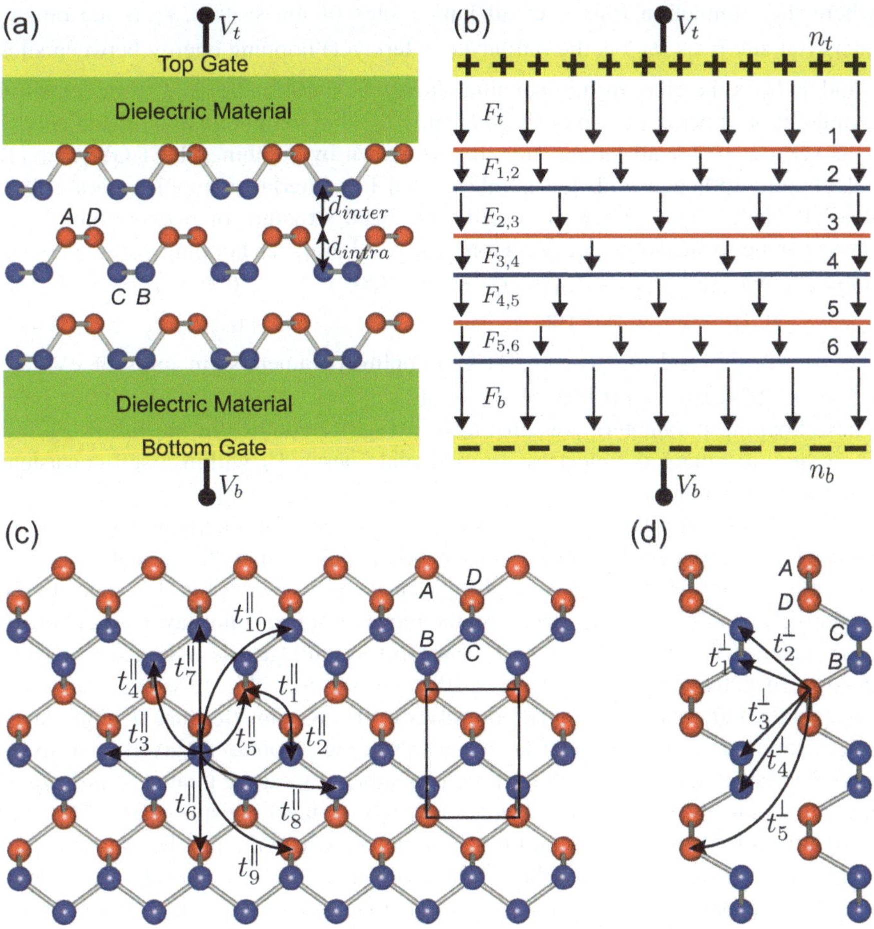

Fig 1.2 (**a**) and (**b**): Sketch of the trilayer phosphorene system, where on the top (bottom) a positively (negatively) charged gate V_t (V_b) with charge density $n_t > 0$ ($n_b < 0$) is placed. (**c**) and (**d**): Illustration of the ten intralayer hopping parameters $t_i^{\|}$ ($i = 1, 2, \ldots, 10$) and the five interlayer hopping parameters $t_i^{\perp}$ ($i = 1, 2, \ldots, 5$) used in the TB model, where the rectangle indicates the unit cell of phosphorene. Due to the puckered lattice structure, trilayer phosphorene has six atomic sublayers depicted by red (A, D) and blue (B, C) atoms, where the distance between the two adjacent sublayers is equal to either the interlayer separation (d_{inter}) or the intralayer one (d_{intra}). The top and bottom gates induce a total carrier density distributed over the different sublayers, i.e., $n = n_t + n_b = \sum_{i=1}^{6} n_i$, with n_i the carrier density on the i-th sublayer. F_t (F_b) is the electric field produced by the top (bottom) gate, and $F_{i,i+1}$ ($i = 1, 2, \ldots, 5$) is the total electric field between the two adjacent sublayers

$$H = \sum_i \varepsilon_i c_i^{\dagger} c_i + \sum_{i \neq j} t_{ij}^{\|} c_i^{\dagger} c_j + \sum_{i \neq j} t_{ij}^{\perp} c_i^{\dagger} c_j + \sum_i U_i c_i^{\dagger} c_i, \tag{1.1}$$

where the summation runs over all lattice sites of the system, ε_i is the on-site energy at site i, $t_{ij}^{\parallel}$ ($t_{ij}^{\perp}$) is the intralayer (interlayer) hopping energy between sites i and j, U_i is the electrostatic potential energy at site i, and $c_i^{\dagger}$ (c_j) is the creation (annihilation) operator of an electron at site i (j). For simplicity, the on-site energy ε_i is set to zero for all lattice sites. It was shown by combined DFT-GW and TB calculations [60] that with ten intralayer and five interlayer hopping parameters, this TB Hamiltonian can well describe the band structure of *pristine* multilayer phosphorene in the low-energy region. The ten intralayer hopping parameters (in units of eV) are $t_1^{\parallel} = -1.486$, $t_2^{\parallel} = +3.729$, $t_3^{\parallel} = -0.252$, $t_4^{\parallel} = -0.071$, $t_5^{\parallel} = +0.019$, $t_6^{\parallel} = +0.186$, $t_7^{\parallel} = -0.063$, $t_8^{\parallel} = +0.101$, $t_9^{\parallel} = -0.042$, $t_{10}^{\parallel} = +0.073$, and the five interlayer hopping parameters (in units of eV) are $t_1^{\perp} = +0.524$, $t_2^{\perp} = +0.180$, $t_3^{\perp} = -0.123$, $t_4^{\perp} = -0.168$, $t_5^{\perp} = +0.005$ [60]. These hopping parameters are illustrated in Fig. 1.2c and d. In the following, we show how to obtain the electrostatic potential energy U_i within a self-consistent Hartree approximation.

We consider multilayer phosphorene in the presence of a perpendicular electric field. This situation can be realized in an experimental setup with external top and/or bottom gates applied to multilayer phosphorene. As shown in Fig. 1.2, a positively (negatively) charged gate is placed on the top (bottom) of multilayer phosphorene of N layers ($N = 3$ in the figure). The top (bottom) gate is assumed to have a positive (negative) charge density $n_t > 0$ ($n_b < 0$) on it. These two gates are used to generate and control the carrier densities in the system. Because a single layer of phosphorene can be viewed as consisting of two atomic sublayers, due to its puckered lattice structure, both of these two sublayers can be treated as *individual* layers and thus the number of *individual* layers of N-layer phosphorene is $2N$. This is significantly different from multilayer graphene, where the number of *individual* layers is just N for N-layer graphene. Therefore, in the N-layer phosphorene system a total carrier density $n = n_t + n_b = \sum_i n_i$ is induced, where n_i is the carrier density on the i-th sublayer and the summation is over all the $2N$ sublayers.

In our model system, we assume that the top (bottom) gate produces a uniform electric field $F_t = en_t/(2\varepsilon_0\kappa)$ [$F_b = en_b/(2\varepsilon_0\kappa)$], which can be obtained from fundamental electrostatics, where e is the elementary charge, ε_0 is the permittivity of vacuum, and κ is the dielectric constant. The induced charge carriers on the phosphorene sublayers, in its turn, produce a uniform electric field $F_i = n_i e/(2\varepsilon_0\kappa)$ ($i = 1, 2, \ldots, 2N$), which counteracts the electric field produced by the external gates. The inversion asymmetry between the two adjacent sublayers i and $i+1$ is determined by a potential energy difference $\Delta_{i,i+1}$, which is given by

$$\Delta_{i,i+1} = \alpha_{i,i+1}\Big(\sum_{j=i+1}^{2N} n_j + |n_b|\Big), \tag{1.2}$$

where $\alpha_{i,i+1} = e^2 d_{i,i+1}/(\varepsilon_0 \kappa)$ with $d_{i,i+1}$ the distance between the two adjacent sublayers. In multilayer phosphorene, this inter-sublayer distance is not constant due to the puckered lattice structure of phosphorene: it is $d_{i,i+1} = d_{\text{intra}}$ if the sublayers i and $i+1$ are within the same phosphorene layer while it is $d_{i,i+1} = d_{\text{inter}}$ if these two sublayers belong to different phosphorene layers, with $d_{\text{intra}} = 0.212\,\text{nm}$ and $d_{\text{inter}} = 0.312\,\text{nm}$. The total electric field between the two adjacent sublayers is given by $F_{i,i+1} = \Delta_{i,i+1}/d_{i,i+1}$ $(i = 1, 2, \ldots, 2N-1)$. Finally, the electrostatic Hartree energies U_i, which are added to the i-th sublayer on-site elements of the N-layer TB Hamiltonian (1.10), can be obtained as

$$U_i = 0\ (i = 1);\ \ U_i = \sum_{j=1}^{i-1} \Delta_{j,j+1}\ (i > 1). \tag{1.3}$$

Here we assumed zero electrostatic potential energy on the top-most sublayer (i.e., the sublayer that is closest to the top gate).

Due to the in-plane translational invariance of the system, a Fourier transform is performed to convert the N-layer TB Hamiltonian (1.1) into momentum space, and then the converted Hamiltonian is numerically diagonalized to obtain the eigenvalues and eigenvectors of the system. All numerical calculations are performed using the recently developed TB package PYBINDING [61]. Because there are four inequivalent basis atoms (labeled A, B, C, and D) in an unit cell of phosphorene, the dimension of the Fourier-transformed TB Hamiltonian of the N-layer phosphorene system is $4N \times 4N$. Therefore, the corresponding eigenvectors are the column vectors of dimension $4N$ consisting of the coefficients of the TB wave functions,

$$c = [c_{A_1}, c_{B_1}, c_{C_1}, c_{D_1}, \ldots, c_{A_N}, c_{B_N}, c_{C_N}, c_{D_N}]^T, \tag{1.4}$$

where $c_{A_i}, c_{B_i}, c_{C_i}, c_{D_i}$ are the i-th layer coefficients for basis atoms A, B, C, D, respectively, and the symbol T denotes the transpose of a vector or matrix. Note that these TB coefficients depend on the in-plane wave vector $\mathbf{k}$. The total TB wave function of the N-layer phosphorene system is then given by

$$\Psi = \sum_{i=1}^{N} \left[c_{A_i}\psi_{A_i} + c_{B_i}\psi_{B_i} + c_{C_i}\psi_{C_i} + c_{D_i}\psi_{D_i}\right], \tag{1.5}$$

where $\psi_{A_i}, \psi_{B_i}, \psi_{C_i}, \psi_{D_i}$ are the four components of the i-th layer TB wave function. With the obtained layer-dependent coefficients $c_i = [c_{A_i}, c_{B_i}, c_{C_i}, c_{D_i}]$, the carrier densities on the sublayers of phosphorene are given by

$$\begin{aligned} n_{2i-1} &= 2\sum_{\mathbf{k}} f[E(\mathbf{k})](|c_{A_i}|^2 + |c_{D_i}|^2), \\ n_{2i} &= 2\sum_{\mathbf{k}} f[E(\mathbf{k})](|c_{B_i}|^2 + |c_{C_i}|^2), \end{aligned} \tag{1.6}$$

where $i = 1, 2, \ldots, N$, the factor 2 in front of the summations accounts for the spin degeneracy, $E(\mathbf{k})$ is the energy spectrum obtained by numerically diagonalizing the TB Hamiltonian (1.1) in momentum space, and $f[E(\mathbf{k})]$ is the Fermi-Dirac function describing the carrier distribution in the energy spectrum. In the presence of electrical gating, the carrier densities in fully occupied energy bands are changed and one has to take into account the charge density redistribution in the valence bands.

The TB Hamiltonian (1.1) depends on the gate-induced carrier densities through Eqs. (1.2) and (1.3), which in turn are calculated based on the full eigenstates of the TB Hamiltonian (1.1) in momentum space. Therefore, a self-consistent calculation following Eqs. (1.1), (1.2), (1.3) and (1.6) is required to obtain the carrier densities n_i and the Hartree energies U_i on the different sublayers of phosphorene. Since the carrier densities on the different sublayers are not known in advance, an initial guess of these densities is needed, e.g., assuming them to be equal at the beginning. Then the calculations are performed self-consistently until the carrier density per sublayer is converged. The number of self-consistent iterations depends on the total carrier density n and on the number of stacking layers N. When convergency is reached, the Fermi energy E_{F} and band structure $E_{\mathbf{k}}$ of multilayer phosphorene can be obtained, from which we can further calculate the specific electronic properties such as the fundamental band gap, the carrier effective mass, and the momentum matrix element.

Our numerical results indicate that both intralayer and interlayer charge screenings are present in gated multilayer phosphorene due to its puckered lattice structure, which is different from the result observed in gated multilayer graphene, where only interlayer charge screening is present [53–57].

1.3 Gate Tunable Electronic and Optical Properties of Multilayer Phosphorene

In this section, we present our main results of gate tunable electronic and optical properties of multilayer phosphorene. The results were obtained by using the self-consistent TB approach as mentioned above.

We consider three different gate configurations applied to multilayer phosphorene: (i) only positive top gate $V_t > 0$, $V_b = 0$ ($n_t > 0$, $n_b = 0$); (ii) only negative bottom gate $V_t = 0$, $V_b < 0$ ($n_t = 0$, $n_b < 0$); and (iii) both top and bottom gates $V_t = -V_b > 0$ ($n_t = -n_b > 0$). We calculate self-consistently the band structure of gated multilayer phosphorene and the gate-induced charge densities on the different phosphorene layers.

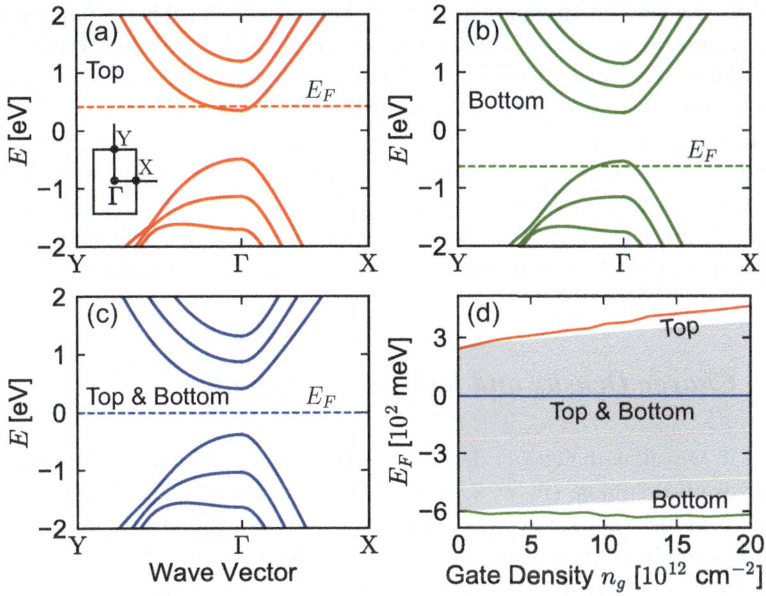

Fig. 1.3 (**a**)–(**c**) Band structure of gated multilayer phosphorene obtained by the self-consistent TB model: (**a**) in the presence of only top gate ($n_t = 1.5 \times 10^{13}\,\text{cm}^{-2}$), (**b**) in the presence of only bottom gate ($n_b = -1.5 \times 10^{13}\,\text{cm}^{-2}$), and (**c**) in the presence of both top and bottom gates ($n_t = -n_b = 1.5 \times 10^{13}\,\text{cm}^{-2}$). (**d**) Fermi energy E_F as a function of the gate charge density n_g for top gating ($n_g = n_t$), bottom gating ($n_g = n_b$), and both top and bottom gating ($n_g = n_t = -n_b$). In panels (**a**)–(**c**), the horizontal dashed lines are the self-consistently determined Fermi energies. The inset in (**a**) denotes the first Brillouin zone (BZ) of phosphorene, where Γ, X and Y are the three most important high-symmetry points. The shaded region in (**d**) represents the band-gap region

1.3.1 Band Structure and Fermi Energy

In Fig. 1.3a–c, we show the self-consistently obtained band structure and Fermi energy of three-layer phosphorene in the presence of (a) only a top gate, (b) only a bottom gate, and (c) both top and bottom gates. The gate charge densities are assumed to be $n_t = 10^{13}\,\text{cm}^{-2}$ in (a), $n_b = -10^{13}\,\text{cm}^{-2}$ in (b), and $n_t = -n_b = 10^{13}\,\text{cm}^{-2}$ in (c). As can be seen, in the presence of only a top (bottom) gate, the Fermi energy is located within the conduction (valence) band due to $n_t > 0$ ($n_b < 0$), which indicates a finite density of electrons (holes) in the system; whereas in the presence of both top and bottom gates, the Fermi energy is located within the band gap due to $n_t + n_b = 0$, which indicates no excess carriers in the system.

In Fig. 1.3d, we show the dependence of the Fermi energy (E_F) of three-layer phosphorene on the gate charge density (n_g) for the cases of only a top gate ($n_g = n_t$), only a bottom gate ($n_g = n_b$), and in the presence of both top and bottom gates ($n_g = n_t = -n_b$). The shaded region in this figure represents the band-gap region. As can be seen, applying only a top (bottom) gate enables to tune the Fermi energy of the system into the conduction (valence) band region and so the electron (hole) density of the system can be tuned with varying gate charge density. However, with both top and bottom gates applied, it is possible to tune the Fermi energy of the system into the band-gap region with no net carrier density when varying gate charge density. Notice that in the presence of only a top or bottom gate, the Fermi energy variation with the gate charge density is smaller in three-layer phosphorene than that in three-layer graphene [54, 55]. This is due to the fact that the carrier effective mass of multilayer phosphorene is larger than that of multilayer graphene.

1.3.2 Charge Density and Screening Anomaly

In Fig. 1.4, we show the carrier densities ρ_i ($i = 1, \ldots, N$) on the different layers of N-layer phosphorene ($N = 2, 3, 4, 5$) as a function of the gate charge density n_g: (a)–(d) in the presence of only a top gate ($n_g = n_t$), and (e)–(h) in the presence of both top and bottom gates ($n_g = n_t = -n_b$). The relation between the layer and sublayer densities is given by $\rho_i = n_{2i-1} + n_{2i}$. The results are very similar in the presence of only a bottom gate ($n_g = -n_b$), and can be mapped into each other by reversing the layer index and changing the carrier type.

We see that in the presence of only a top gate, as shown in Fig. 1.4a–d, the top-most layer has the largest carrier density (i.e., ρ_1) for the case of $N = 2$ layers. But this is no longer true for the cases of $N = 3, 4, 5$ layers. For instance, for $N = 4$

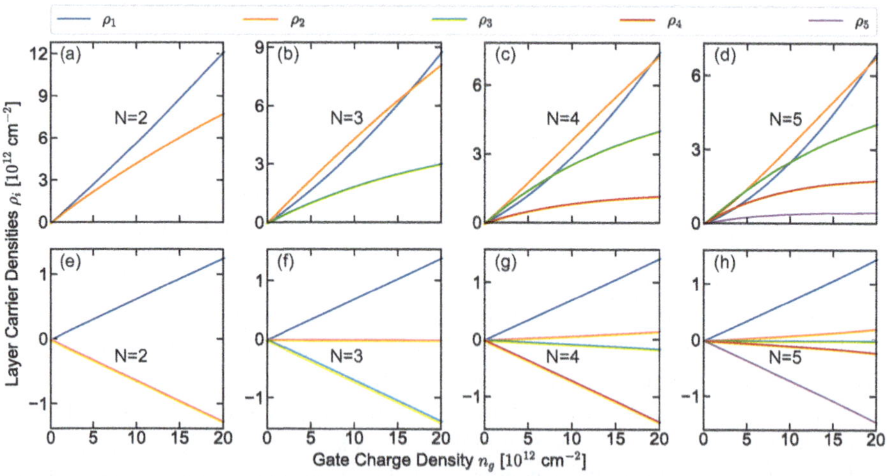

Fig. 1.4 Layer carrier densities ρ_i ($i = 1, \ldots, N$) of gated multilayer phosphorene as a function of the gate charge density n_g for $N = 2, 3, 4, 5$ layers: (**a**)–(**d**) in the presence of only a top gate ($n_g = n_t$), and (**e**)–(**h**) in the presence of both top and bottom gates ($n_g = n_t = -n_b$)

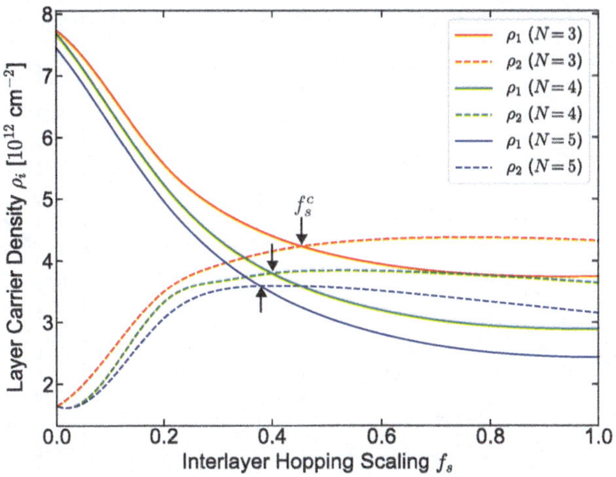

Fig. 1.5 Layer carrier densities ρ_1 and ρ_2 of top-gated multilayer phosphorene as a function of the interlayer hopping scaling factor f_s for $N = 3, 4, 5$ layers. The arrow indicates the critical value (f_s^c) of the interlayer hopping strength, below (above) which $\rho_1 > \rho_2$ ($\rho_1 < \rho_2$)

the carrier density is largest on the second-top-most layer (i.e., ρ_2). This *charge-density anomaly* in N-layer phosphorene (with $N \geq 3$) is induced by the charge transfer between the different layers due to the significant interlayer coupling. Our numerical calculations show that if the strength of the interlayer coupling is reduced below a critical value, the top-most phosphorene layer will regain the largest carrier density. In order to see this more clearly, we show in Fig. 1.5 the carrier densities on the top-most and the second-top-most phosphorene layers (ρ_1 and ρ_2) as a function of the interlayer hopping scaling factor (f_s) for the cases of $N = 3, 4, 5$ layers, where $f_s = 0\,(1)$ indicates the turning off (on) of the full interlayer hopping. As can be seen, there is indeed a critical value of the interlayer hopping strength (indicated by f_s^c in the figure), which depends on the number of layers N (for instance it is found to be $f_s^c = 0.45, 0.39, 0.37$ for $N = 3, 4, 5$).

Furthermore, we find that in the presence of both top and bottom gates, as shown in Fig. 1.4e–h, the carrier densities on the different layers exhibit an odd-even layer dependence: for the case of an even number of layers ($N = 2, 4$), the upper and lower layers with respect to the centro-symmetric plane have carrier densities opposite in sign but equal in magnitude, thereby leading to the appearance of electron-hole bilayers; whereas for the case of an odd number of layers ($N = 3, 5$), there is an additional feature that the middle layer (at the centro-symmetric plane) has zero carrier density. The layer carrier densities ρ_i increase with the gate charge density n_g in either a linear or nonlinear fashion.

In Fig. 1.6, we show the total electric fields $F_{i,i+1}$ ($i = 1, \ldots, 2N - 1$) between the two adjacent sublayers of N-layer phosphorene ($N = 2, 3, 4, 5$) as a function of the gate charge density n_g, where (a)–(d) and (e)–(h) are for the same gating configurations as in Fig. 1.4. For comparative purposes, the gate-produced

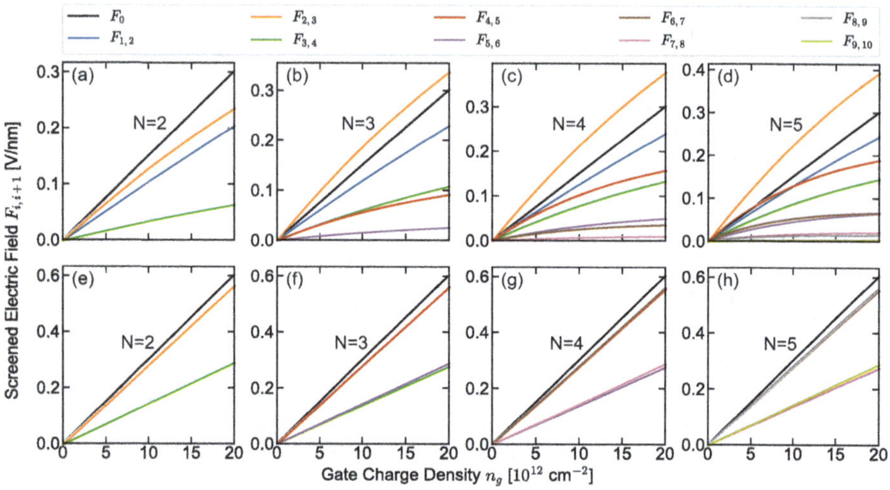

Fig. 1.6 Electric fields $F_{i,i+1}$ ($i = 1, \ldots, 2N - 1$) between the two adjacent sublayers of gated multilayer phosphorene as a function of the gate charge density n_g for $N = 2, 3, 4, 5$ layers: (**a**)–(**d**) in the presence of only a top gate ($n_g = n_t$), and (**e**)–(**h**) in the presence of both top and bottom gates ($n_g = n_t = -n_b$). In each panel, F_0 (black curve) is the electric field produced by top and/or bottom gates

(unscreened) electric field $F_0 = F_t + F_b$ is also presented in each panel. As can be seen, in all the gating configurations, most of the electric fields $F_{i,i+1}$ are significantly smaller as compared to the fields F_0 due to the screening effect, and the magnitudes of $F_{i,i+1}$ are different from each other, because the carrier densities on the different sublayers that can screen the electric field F_0 are different (see Fig. 1.7). Therefore, the screened electric field across multilayer phosphorene is not uniform, whose magnitude depends on the phosphorene sublayers.

Notice that in the presence of only a top or bottom gate and for the case $N \geq 3$, some electric fields $F_{i,i+1}$ shown in Fig. 1.6, e.g., $F_{2,3}$, $F_{4,5}$ and $F_{6,7}$, are not screened and even larger than the unscreened one F_0. This *charge-screening anomaly* is induced by the fact that there are emerging minority carriers (with opposite sign to the majority ones) on the sublayers in the presence of only a top or bottom gate. For instance, for the case of $N = 4$ layers and in the presence of only top gate, as shown in Fig. 1.7c, they are the carrier densities n_2 and n_4 on the second- and fourth-top-most sublayers, which are opposite in sign as compared to the other densities. With increasing N, the minority carriers appear on more sublayers. The emergence of minority carriers is induced by the intralayer charge transfer due to the very strong out-of-plane hopping amplitude in phosphorene. The emerging minority carriers produce an electric field, which counteracts that produced by the majority carriers and thus leads to the charge-screening anomaly.

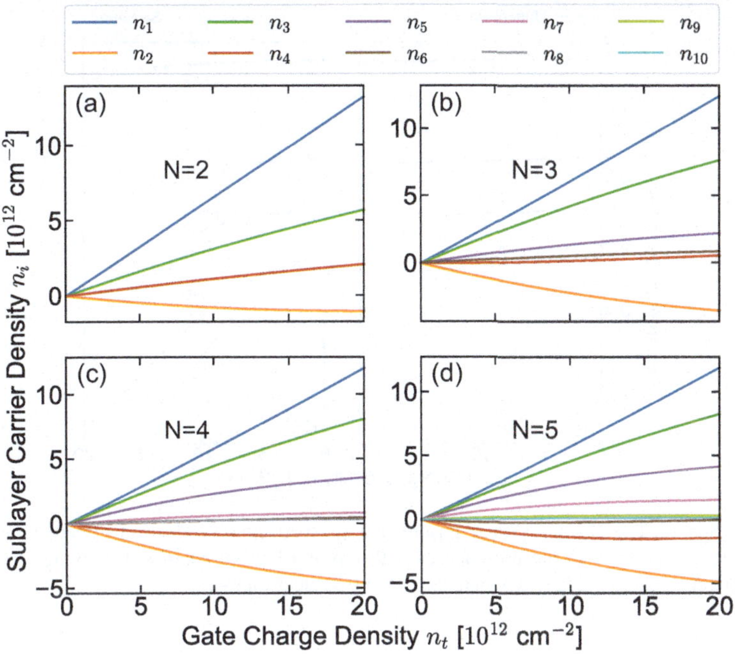

Fig. 1.7 Sublayer carrier densities n_i ($i = 1, \ldots, 2N$) of top-gated multilayer phosphorene as a function of the gate charge density n_t for $N = 2, 3, 4, 5$ layers

1.3.3 Band Gap Tuning: Theory and Experiment

In Fig. 1.8, we show both the screened and unscreened band gaps (E_g) of N-layer phosphorene ($N = 2, 3, 4, 6$) as a function of the gate charge density (n_g) in the presence of (a) only a top gate ($n_g = n_t$) and (b) both top and bottom gates ($n_g = n_t = -n_b$). As can be seen, the gate-charge-density (or gate-electric-field) tuning of the band gap is distinctively different in the absence and presence of charge screening. The unscreened band gap decreases dramatically with increasing gate charge density (or gate electric field). However, in the presence of charge screening, the magnitude of this band-gap decrease is significantly reduced and the reduction is more significant for larger number of layers. For instance, for the case of $N = 6$ layers and in the presence of both top and bottom gates, the unscreened band gap E_g becomes zero when the gate charge density n_g reaches the critical value $n_g^c \sim 1.5 \times 10^{13}$ cm^{-2}; however, due to the charge screening, the critical gate charge density n_g^c is significantly increased, which becomes larger than 2×10^{13} cm^{-2} and thus increases by more than 33%. The effect of the charge screening on the band-gap tuning with the gate electric field is qualitatively the same for all the gating configurations, as shown in Fig. 1.8a and b, i.e., it always tends to increase the band-gap value.

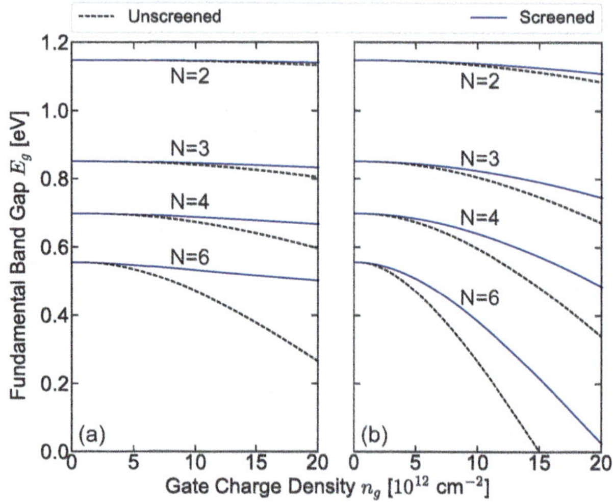

Fig. 1.8 Screened (full curves) and unscreened (dashed curves) electron effective masses of multilayer phosphorene in the presence of both top and bottom gates with varying gate charge density for $N = 2, 3, 4, 5$ layers, where m_0 is the free electron mass

We notice that in Ref. [58], similar results were obtained for this band-gap tuning in multilayer phosphorene by using first-principles calculations. However, the obtained results were restricted to the cases of bilayer and trilayer phosphorene and the screening effect was induced differently, i.e., by considering a charged system (our screening effect is induced by the gate electric field).

In order to verify the capability and accuracy of the self-consistent TB approach, we compare our theoretical band gaps of gated multilayer phosphorene with recent experimental results [59]. We consider multilayer phosphorene in the presence of dual (top and bottom) gates as in Ref. [59], and take from this experimental work all the necessary parameters for theoretical modeling and calculations, including the number of layers and the displacement field. The experiment considered the charge-neutrality condition [59], i.e., there are no excess carriers in the multilayer structure and thus the displacement field (denoted by D) is approximately uniform across the entire structure. This condition corresponds to our theoretical model where the charge densities of top and bottom gates are opposite in sign but equal in magnitude. By using the relation between the gate charge density and the gate electric field $F_{t,b} = e|n_{t,b}|/(2\varepsilon_0\kappa)$, the displacement fields of top and bottom gates can be written as $D_{t,b} = \kappa F_{t,b} = e|n_{t,b}|/(2\varepsilon_0)$. At the charge-neutrality condition ($n_t + n_b = 0$), $D = D_t = D_b$ where D is the displacement field of multilayer phosphorene, and we calculate the displacement-field tuning of the band gap of multilayer phosphorene, as was done experimentally in Ref. [59]. The magnitude of this band-gap tuning is computed as $\Delta E_g = E_g(D \neq 0) - E_g(D = 0)$.

In Fig. 1.9a, we show our theoretical results (solid lines) of ΔE_g as a function of D for N-layer phosphorene ($N = 4, 5, 6, 8$) and for comparative purposes, the

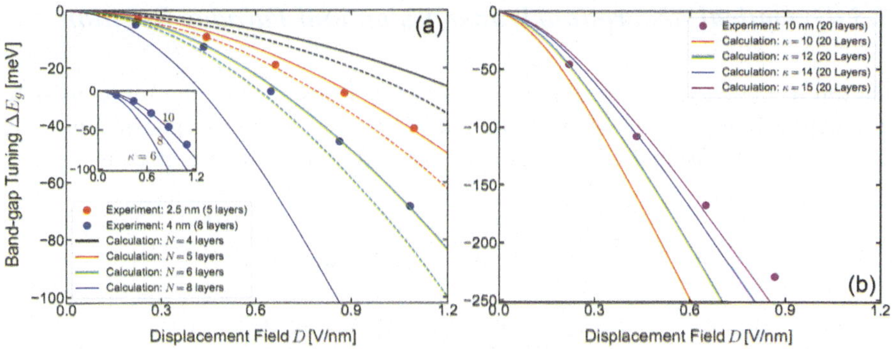

Fig. 1.9 Our theoretical results of the band-gap tuning ΔE_g of N-layer phosphorene ($N = 4, 5, 6, 8, 20$) as a function of the displacement field D (solid lines). Dashed lines and circle dots are, respectively, the theoretical and experimental results of Ref. [59] for ΔE_g as a function of D. The inset in (**a**) shows ΔE_g for the case of $N = 8$ layers for different dielectric constants κ as indicated

theoretical (dashed lines) and experimental (circle dots) ones from Ref. [59]. As seen in Fig. 1.9a, our theoretical results are in good agreement with the experimental ones for the 2.5 nm-thick sample (corresponding to $N = 5$ layers), while the simple theoretical model presented in Ref. [59] overestimates the band-gap change with displacement field. However, the 4 nm-thick sample (corresponding to $N = 8$ layers) exhibits a much smaller decrease in the band gap when compared with theory. The experimental results compare favorably with our results for $N = 6$ layers. This discrepancy could possibly be due to the uncertainty in (i) the sample thickness in the experiment and (ii) the value of the dielectric constant of the sample. In spite of this discrepancy, our theoretical results are able to capture qualitatively the main feature of the band-gap variation with displacement field (i.e., the nonlinear decrease with increasing displacement field). Notice that by increasing the dielectric constant κ from 6 to 10, we are able to reproduce the experimental results for the 4 nm-thick sample, as shown in the inset of Fig. 1.9a.

The band-gap reduction with displacement field is even more pronounced for multilayer phosphorene with $N = 20$ layers (corresponding to the 10-nm-thick sample), as shown in Fig. 1.9b. In order to fit the experimental results, we have to further increase κ from 10 to 15. This increase of the κ value is reasonable because multilayer phosphorene with larger number of layers was shown to have a larger dielectric constant [59, 62]. The physical reason is that as the band gap decreases with increasing number of phosphorene layers, the multilayer phosphorene system becomes more metallic and so the dielectric screening of the system becomes stronger, which indicates an increased dielectric constant. Therefore, with only one adjustable parameter (i.e., the dielectric constant κ) which is varied within reasonable values, our self-consistent TB calculations are able to reproduce the experimental results.

1.3.4 Optical Absorption/Transmission and Faraday Rotation

On top of the energy spectrum and wave function obtained by the self-consistent TB approach, the optical conductivity tensor of gated multilayer phosphorene can be calculated by using the linear response Kubo formula, which is given by

$$\sigma_{\alpha\beta}(\omega) = \frac{g_s e^2 \hbar}{4\pi^2 i} \sum_{m,n} \int d\mathbf{k} \frac{f[E_m(\mathbf{k})] - f[E_n(\mathbf{k})]}{E_m(\mathbf{k}) - E_n(\mathbf{k})}$$
$$\times \frac{\langle \Psi_m(\mathbf{k})|\mathbf{v}_\alpha|\Psi_n(\mathbf{k})|\langle \Psi_n(\mathbf{k})|\mathbf{v}_\beta|\Psi_m(\mathbf{k})|}{E_m(\mathbf{k}) - E_n(\mathbf{k}) + \hbar\omega + i\eta}, \tag{1.7}$$

where $\alpha, \beta = x, y$ are the tensor indices, ω is the optical frequency, $g_s = 2$ is the spin-degeneracy factor, m, n are the band indices, $\mathbf{k} = (k_x, k_y)$ is the 2D wave vector, $f(E)$ is the Fermi-Dirac function, $E_{m,n}(\mathbf{k})$ are the band energies, $\Psi_{m,n}(\mathbf{k})$ are the wave functions, $\mathbf{v}_{\alpha,\beta}$ are the α and β components of the velocity operator $\mathbf{v} = \hbar^{-1}\partial H/\partial\mathbf{k}$, and η is a finite broadening. There are four components in the optical conductivity tensor (1.11): σ_{xx} and σ_{yy} for the longitudinal optical conductivities and σ_{xy} and σ_{yx} for the transverse optical conductivities (i.e., the optical Hall conductivities). For the longitudinal components we have $\sigma_{xx} = \sigma_{yy}$ ($\sigma_{xx} \neq \sigma_{yy}$) for an isotropic (anisotropic) system. When calculating the optical conductivity tensor (1.7), we take the broadening $\eta = 10$ meV and the temperature $T = 300$ K. The real parts of the longitudinal and transverse optical conductivities are related to the optical absorption effect and optical Hall effect (e.g., Faraday rotation), respectively.

In Fig. 1.10, we show the real-part interband optical conductivities $\mathrm{Re}\,\sigma_{xx}$ and $\mathrm{Re}\,\sigma_{yy}$ of bilayer phosphorene in the absence/presence of a perpendicular electric field and without/with the field-induced charge screening. Here, we use the index pair (m, n) to denote the interband transition from the mth VB to the nth CB. As can be seen, distinct absorption peaks appear in $\mathrm{Re}\,\sigma_{xx}$ due to the strong quantum confinement along the out-of-plane direction (i.e., along the z direction). This confinement leads to the formation of quantized subbands along the z direction. The number of discrete subbands (N_S) depends on the number of stacking layers (N_L) as $N_S = 4N_L$. In the absence of the electric field, there are only two absorption peaks in $\mathrm{Re}\,\sigma_{xx}$, which are associated with the interband transitions $(1, 1)$ and $(2, 2)$; whereas in the presence of the electric field, two more absorption peaks are induced in $\mathrm{Re}\,\sigma_{xx}$, which correspond to the interband transitions $(1, 2)$ and $(2, 1)$. Based on these results, we can establish the following optical selection rules in bilayer phosphorene: (i) The interband transitions $(1, 1)$ and $(2, 2)$ are optically allowed in both the absence and presence of a perpendicular electric field; (ii) The interband transitions $(1, 2)$ and $(2, 1)$ are optically forbidden (allowed) in the absence (presence) of a perpendicular electric field. These optical selection rules are closely related to the wave-function parities (symmetries) of the conduction and valence band in the absence/presence of a perpendicular electric field. However,

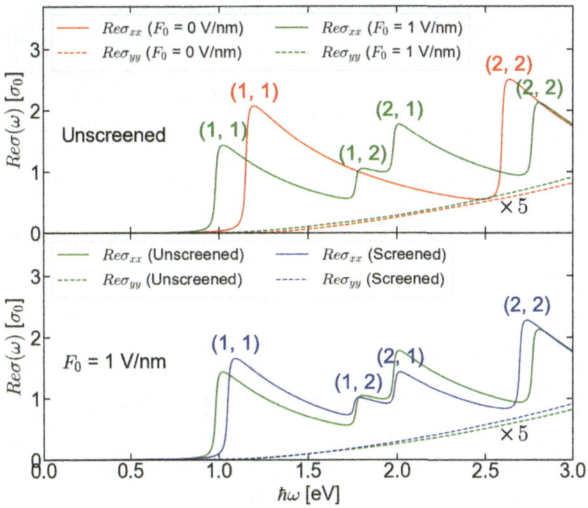

Fig. 1.10 Real parts of the interband optical conductivities, Re σ_{xx} and Re σ_{yy}, of bilayer phosphorene in the absence/presence of a perpendicular electric field and without/with the field-induced charge screening. The values of Re σ_{yy} are enlarged with a factor of 5 to make the curves more visible. Here $\sigma_0 = e^2/(4\hbar)$ is the universal optical conductivity of graphene and (m, n) indicates the interband transition from the mth VB to the nth CB

when the field-induced charge screening is taken into account, the magnitudes of these two extra absorption peaks are reduced. In addition, the cut-off absorption of Re σ_{xx} is shifted to the lower photon energy (i.e., red-shifted) by the electric field.

We also observe in Fig. 1.10 that there is a significant difference between Re σ_{xx} and Re σ_{yy}, i.e., the real part of the interband optical conductivity exhibits a significant linear dichroism. As mentioned before, Re σ_{xx} has distinct absorption peaks, which is strongly affected by the perpendicular electric field and is greatly influenced by the field-induced charge screening. However, there are no distinct absorption peaks in Re σ_{yy}, which is almost unaffected by the applied electric field and the field-induced charge screening. The absorption features of Re σ_{xx} and Re σ_{yy} in the absence/presence of a perpendicular electric field and without/with the field-induced charge screening were explained in terms of the momentum matrix elements for the corresponding interband transitions [50].

In Fig. 1.11, we show the real parts of the interband optical conductivities Re σ_{xx} and Re σ_{yy} of trilayer phosphorene in the absence/presence of a perpendicular electric field and without/with the field-induced screening. As can be seen, there are more absorption peaks in the conductivity spectrum of trilayer phosphorene when compared to that of bilayer phosphorene shown in Fig. 1.10. This is because the confinement energy is smaller and there are more discrete subbands in both the conduction and valence bands due to the increased number of stacking layers. Here we identify in the conductivity spectrum of trilayer phosphorene the following

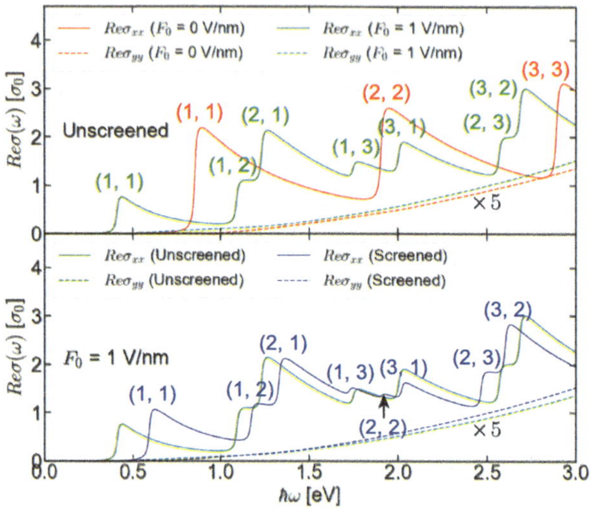

Fig. 1.11 Real parts of the interband optical conductivities, Re σ_{xx} and Re σ_{yy}, of trilayer phosphorene in the absence/presence of a perpendicular electric field and without/with the field-induced charge screening. The values of Re σ_{yy} are enlarged by a factor of 5 to make the curves more visible

interband transition channels: (1, 1), (2, 2), (3, 3) in the absence of the electric field and (1, 2), (2, 1), (1, 3), (3, 1), (2, 3), (3, 2) in the presence of the electric field.

In Fig. 1.12, we show the electric-field modulation strengths $\Delta T/T$ of the optical transmission of N-layer phosphorene ($N = 2, 3, 4$) for the armchair (x) and zigzag (y) polarizations, denoted by AC and ZZ, respectively. Here the values of $\Delta T/T$ as indicated in the colorbar are given in percentage. As seen, $\Delta T/T$ is much larger for the AC polarization than for the ZZ polarization, which indicates that the field modulation strength for the ZZ polarization is negligibly small. For both the AC and ZZ polarizations, $\Delta T/T$ oscillates as a function of the photon energy, with signs changed between negative (blue color) and positive (red color). And both of them can be divided into different regions with negative and positive values. For illustration purposes, two distinct regions are labeled by (I) and (II) in the far-top-left panel of Fig. 1.12. By analysing the field dependence of the interband optical conductivity of bilayer phosphorene, we find that the four different regions are induced by different interband transitions. For instance, a positive value of $\Delta T/T$ in region (I) is mainly induced by the interband transition (1, 1), while a negative value of $\Delta T/T$ in region (II) is mainly induced by the interband transition (2, 1). This is because the momentum matrix element of the zero-field interband transition (1, 1) [(2, 1)] can be reduced (enhanced) by applying a perpendicular field, as described before, and consequently the corresponding optical transmission can be enhanced (reduced) by the applied field, thereby leading to a positive (negative) value of the field modulation strength. Notice that the field modulation for the AC polarization becomes more efficient with increasing number of phosphorene layers,

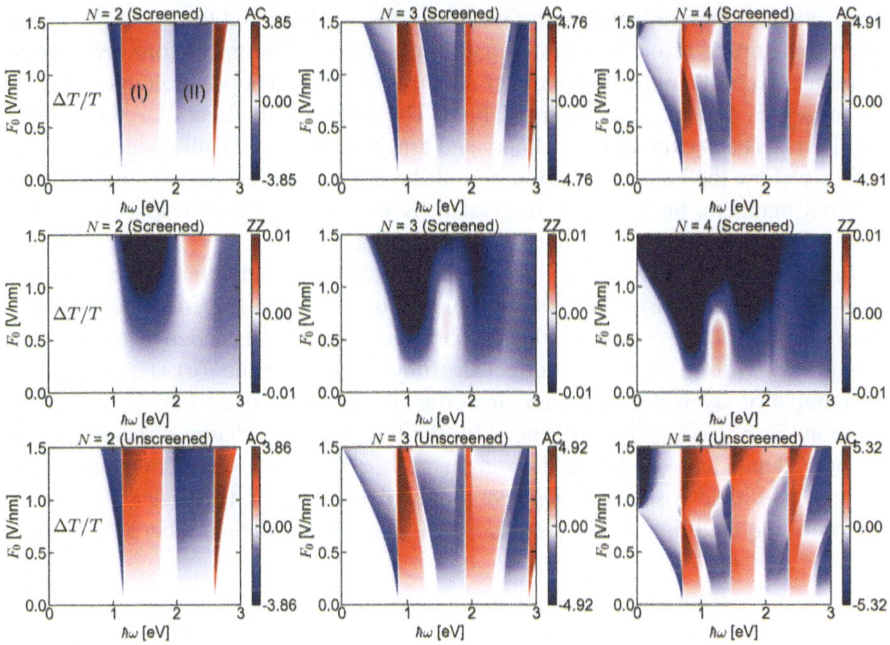

Fig. 1.12 Electric-field modulation $\Delta T/T$ of the optical transmission of N-layer phosphorene ($N = 2, 3, 4$) for the AC (armchair) and ZZ (zigzag) polarizations. Here the values of $\Delta T/T$ as indicated by the colorbar are given in percentage

as seen by comparing $\Delta T/T$ for the cases of $N = 2, 3, 4$. This is reflected by the fact that the modulation strength becomes larger and the modulation range becomes wider. For instance, the absolute value of $\Delta T/T$ increases from 3.85% in the two-layer case to 4.91% in the four-layer case, and the energy range of $\Delta T/T$ extends from 1–3 eV in the two-layer case to 0–3 eV in the four-layer case. Our numerical results indicate that with increasing field strength from 0 to 1.5 V/nm, four-layer phosphorene undergoes a semiconductor-to-semimetal transition, leading to a zero band gap in the band structure. Furthermore, we find that the field modulation strength in the presence of the field-induced charge screening is smaller when compared to that in the absence of the field-induced charge screening. This is because the strength of the applied field is reduced by the field-induced charge screening. Recent experiments reported the electric-field modulation of the linear dichroism of few-layer phosphorene [45, 46]. In agreement with our theoretical results shown in Fig. 1.12, they found that (i) the field modulation oscillates significantly with positive and negative values as a function of the photon energy for the armchair polarization, and (ii) the modulation strength is negligibly small (almost no modulation within experimental resolution) for the zigzag polarization.

When shining a beam of light on an optical medium in the presence of a magnetic field, the light polarization can be rotated in two different configurations, i.e., the

Kerr rotation when reflected at the surface of the medium and the Faraday rotation when transmitted through the interior of the medium. These rotations are a direct consequence of the optical Hall effect arising from the carrier-photon interaction in the presence of an external magnetic field. The optical Hall effect (e.g. the Faraday rotation) has been explored in ordinary and graphene quantum Hall systems [63, 64]. Here we show the Faraday rotation in multilayer phosphorene in the absence of an external magnetic field. This is motivated by the fact that the phosphorene lattice has a reduced symmetry due to its puckered structure.

In Fig. 1.13, we show the real part of the optical Hall conductivity $\mathrm{Re}\,\sigma_{xy}$ of few-layer phosphorene: (a) Effect of the field-induced charge screening for the field strength of $F_0 = 1$ V/nm in trilayer phosphorene; (b) Dependence on the field strength F_0 with the screening effect in trilayer phosphorene; (c) Dependence on the number of layers with the screening effect for the field strength of $F_0 = 1$ V/nm. As seen, $\mathrm{Re}\,\sigma_{xy}$ is greatly influenced by the field-induced charge screening and is strongly dependent on the field strength and on the layer number. Interestingly, $\mathrm{Re}\,\sigma_{xy}$ has either positive or negative values depending on the optical frequency. According to the relation $\theta_F \simeq \mathrm{Re}\sigma_{xy}/[(1+n_{\mathrm{sub}})c\epsilon_0]$ with θ_F the Faraday rotation

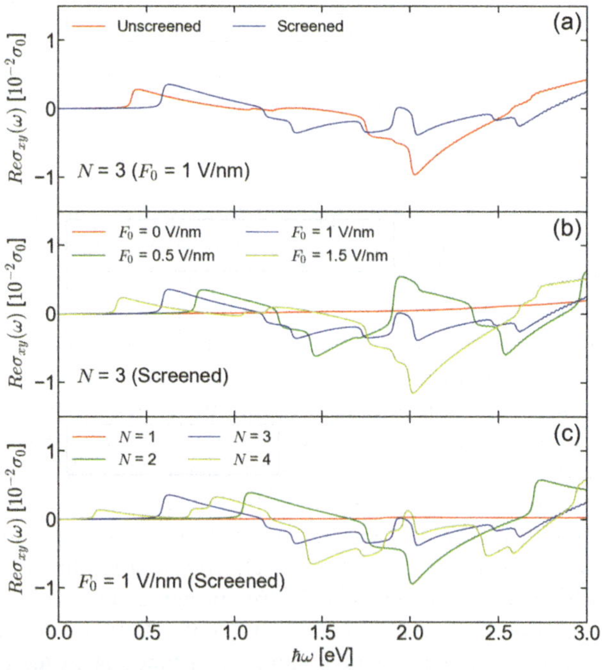

Fig. 1.13 Real parts of the interband optical Hall conductivities $\mathrm{Re}\,\sigma_{xy}$ and $\mathrm{Re}\,\sigma_{yx}$ of N-layer phosphorene: (**a**) Effect of the field-induced charge screening for the field strength of $F_0 = 1$ V/nm in trilayer phosphorene; (**b**) Dependence on the field strength F_0 with the screening effect in trilayer phosphorene; (**c**) Dependence on the layer number N with the screening effect for the field strength of $F_0 = 1$ V/nm

angle and n_{sub} the refractive index of the substrat [64], the frequency dependence of $\text{Re}\,\sigma_{xy}$ indicates that the polarization of light can be rotated in opposite directions by varying the frequency of light when transmitted through few-layer phosphorene. Notice that the optical Hall conductivity is smaller than its longitudinal counterpart by two orders of magnitude (i.e., $100\text{Re}\sigma_{xy} \sim \text{Re}\sigma_{xx}$), which agrees with the results obtained from DFT calculations for other 2D materials such as GaS and GaSe multilayers [65]. The results shown here, i.e., the strong dependence of the optical Hall conductivity on the perpendicular electric field and on the number of phosphorene layers, indicate that Faraday rotation is not only electrically tunable but also layer-thickness dependent.

1.4 Strain Engineered Linear Dichroism and Faraday Rotation of Few-Layer Phosphorene

DFT calculations [66] have shown that phosphorene can sustain tensile strain up to 27% (30%) along its zigzag (armchair) direction without breaking the lattice stability, and compared with other 2D materials, phosphorene exhibits superior mechanical flexibility with an order of magnitude smaller Young's modulus. These excellent mechanical properties of phosphorene make it promising for practical large-magnitude strain engineering. Furthermore, strain engineering of phosphorene might be even more interesting because of its intrinsic structural anisotropy.

In previous section, it was shown by self-consistent TB calculations that in few-layer phosphorene the Faraday rotation is even present in the absence of an external magnetic field due to the reduced symmetry of the phosphorene lattice (i.e., the puckered lattice), and that the linear dichroism and Faraday rotation of few-layer phosphorene can significantly be modulated by a perpendicular electric field. Here we employ the TB approach to investigate the effect of strain on the linear dichroism and Faraday rotation of few-layer phosphorene, where strain is applied uniaxially along its armchair or zigzag direction. We show how these optical properties can be engineered by strain. Our investigation is important for mechano-optoelectronic devices based on strained few-layer phosphorene.

1.4.1 *TB Hamiltonian and Strain Modification*

We consider AB-stacked few-layer phosphorene in the presence of an in-plane strain applied along the armchair or zigzag direction of the phosphorene lattice, as sketched in Fig. 1.14. It was shown experimentally [67] that such uniaxial strain can be produced uniformly. From DFT calculations [25, 60], it is known that AB stacking is energetically the most favorable for few-layer phosphorene. The TB Hamiltonian for *pristine* few-layer phosphorene is given by Eq. (1.1).

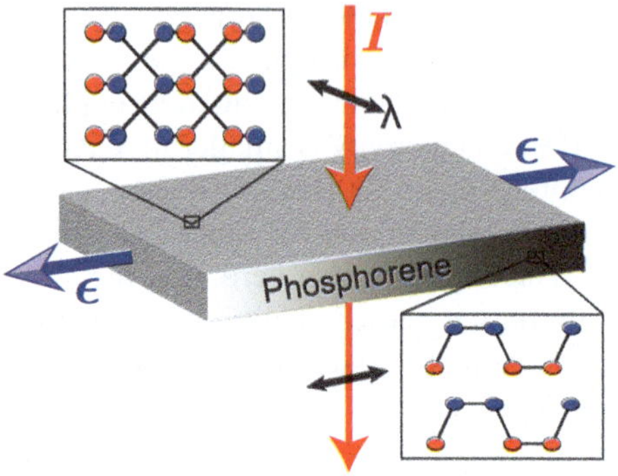

Fig. 1.14 Sketch of strained few-layer phosphorene, excited by normal incident light with intensity (I) and polarization (λ), where strain (ϵ) is applied uniaxially along the in-plane direction (e.g. the armchair or zigzag direction of the phosphorene lattice). The two black rectangles show the top and side views of the phosphorene lattice. When light transmits through the sample, its intensity is reduced due to optical absorption and its polarization is rotated due to the optical Hall effect

When applying an axial *strain* to few-layer phosphorene, the change in the relative interatomic positions can be described as

$$\mathbf{r}_{ij}^{\epsilon} = (\mathbf{I} + \boldsymbol{\epsilon}) \cdot \mathbf{r}_{ij}^{0}, \tag{1.8}$$

where $\mathbf{I}$ is the unit tensor, $\boldsymbol{\epsilon}$ is the strain tensor, $\mathbf{r}_{ij}^{\epsilon}$ is the position vector originated from site i pointing to site j, and $\mathbf{r}_{ij}^{0}$ is the unstrained position vector. Therefore, in the presence of strain, both the bond length and angle of the lattice can be changed, either of which can modify the hopping energies between the atomic orbitals. However, according to the Harrison rule [68], the hopping energies t between the p-like orbitals depend only on the bond length r as $t \propto 1/r^2$, which means that the applied strain does not change the bond angle but only change the bond length. For our case where the TB Hamiltonian of few-layer phosphorene is constructed on the basis of the p_z-like orbitals, this was confirmed by combined DFT and TB calculations [69] which showed that although the change of the bond angle is almost not noticeable in strained phosphorene, the modification of the hopping energies due to this change is much smaller than due to the change of the bond length. Therefore, the hopping energies in strained monolayer and bilayer phosphorene can reasonably be obtained as

$$t_{ij}^{\epsilon} = t_{ij}^{0} \frac{|\mathbf{r}_{ij}^{0}|^2}{|\mathbf{r}_{ij}^{\epsilon}|^2}, \tag{1.9}$$

where t_{ij}^{ϵ} and t_{ij}^{0} are the strained and unstrained hopping energies, respectively. Alternatively, one can also use an exponential formula (i.e., $t_{ij}^{\epsilon} =$

$t_{ij}^0 \exp[-\beta_{ij}(\mathbf{r}_{ij}^{\epsilon}/\mathbf{r}_{ij}^0 - 1)]$) to describe how the hopping energies are changed by the applied strain. However, this exponential formula requires the knowledge of the values of the parameters β_{ij}, which currently are not well known for strained few-layer phosphorene. In principle, these parameters can be obtained by fitting the TB band structure to the DFT one, e.g., using the maximally localized wannier functions [70].

1.4.2 Strain Tunable Longitudinal and Transverse Optical Conductivities

By diagonalizing the TB Hamiltonian in the absence/presence of strain in momentum space, the energy spectrum and the wave function of unstrained/strained few-layer phosphorene is obtained. On top of this, the optical conductivity tensor of unstrained/strained few-layer phosphorene is calculated with the linear response Kubo formula (1.7). When calculating the optical conductivity tensor of free-standing samples, we took an energy broadening of 10 meV and the lattice temperature of 300 K (unless otherwise specified). The real parts of the longitudinal and transverse optical conductivities are related to the optical absorption effect and the optical Hall effect (e.g., Faraday rotation), respectively.

Because we consider undoped few-layer phosphorene in the absence and presence of strain, there are no free charge carriers (electrons and holes) in the system and therefore the Fermi energy of the system is located within the band gap of the system. When the system is excited by normal incident light with certain polarization (as sketched in Fig. 1.14), only interband transitions from the occupied valence bands to the unoccupied conduction bands contribute to the optical conductivity of the system. In order to show the effect of strain more clearly, we consider a reasonably large strain (up to 10%) applied to few-layer phosphorene. The same order of magnitude of strain was demonstrated both theoretically [66] and experimentally [67].

In Fig. 1.15a and b, we show the longitudinal optical conductivities (σ_{xx} and σ_{yy}) of strained bilayer phosphorene for different strain magnitudes and directions, with strain applied along the armchair (ϵ_{ac}) or zigzag (ϵ_{zz}) direction. Here, we define armchair (zigzag) directions along x (y) direction and $\epsilon_{ac/zz} < 0$ ($\epsilon_{ac/zz} > 0$) as compressive (tensile) armchair/zigzag strain, and we use the band index pair (m, n) to denote the interband transition from the mth valence band (VB) to the nth conduction band (CB).

As can be seen, in the absence of strain, there is a significant difference between σ_{xx} and σ_{yy} with the former being about 10 times larger than the latter, leading to the so-called *linear dichroism* that is typical for phosphorene systems. Applying an uniaxial strain (ϵ_{ac} or ϵ_{zz}) modulates significantly σ_{xx} and σ_{yy}. The energy cutoffs of these optical conductivities are red (blue) shifted in the presence of compressive (tensile) strain along both the armchair and zigzag directions. The remarkable shift

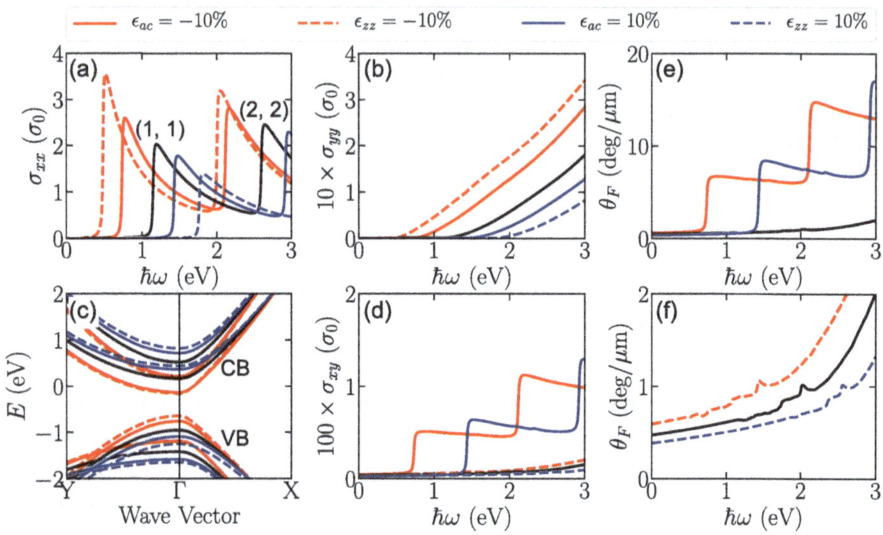

Fig. 1.15 Effect of strain on the longitudinal optical conductivities (**a**) σ_{xx} and (**b**) σ_{yy}, on the band structure (**c**), on the optical Hall conductivity σ_{xy} (**d**), and on the Faraday rotation angles θ_{F} (**e**) and (**f**) of bilayer phosphorene. Here, $\sigma_0 = e^2/(4\hbar)$ is the universal optical conductivity, the strain is applied along the armchair (ϵ_{ac}) or zigzag (ϵ_{zz}) direction, the black curves show the unstrained results, and the band index pair (m, n) denotes the interband transition from the mth valence band (VB) to the nth conduction band (CB)

of the energy cutoff between the compressive and tensile strain was also confirmed by recent experiment [44]. Intuitively we attribute this strain-induced shift in the energy cutoffs of σ_{xx} and σ_{yy} to the electronic band-gap change induced by the applied strain. In order to gain better insights, we have to look into the effect of strain on the band structure of bilayer phosphorene, as shown in Fig. 1.15c. It is clear from the band structure that the band gap does increase (decrease) in the presence of tensile (compressive) strain for both the armchair and zigzag directions. More importantly, the intensities of σ_{xx} and σ_{yy} are enlarged (reduced) by compressive (tensile) strain and optical transitions are only allowed between the CBs and VBs with the same band indices, i.e., $(m, n) = (1, 1)$ and $(m, n) = (2, 2)$.

In Fig. 1.15d, we show the optical Hall conductivity (σ_{xy}) of strained bilayer phosphorene for different strain magnitudes and directions. As seen, in the absence of strain σ_{xy} is nonzero however it is about 100 (10) times smaller than σ_{xx} (σ_{yy}), which agrees with the results obtained from DFT calculations for other 2D materials such as GaS and GaSe multilayers [65]. This small but nonzero optical Hall conductivity implies that Faraday rotation is present in bilayer phosphorene *even at zero magnetic field*. This is due to the reduced symmetry of the phosphorene lattice (i.e. the puckered structure). According to the relation $\theta_{\mathrm{F}} \simeq \sigma_{xy}/[(1+n_{\mathrm{sub}})c\epsilon_0]$ with θ_{F} the Faraday rotation angle and n_{sub} the refractive index of the substrate [64], the frequency dependence of σ_{xy} indicates that the polarization of light can be rotated by simply varying the frequency of light transmitted through bilayer phosphorene. Now we look at the effect of strain on the optical Hall conductivity. As seen, the intensity of σ_{xy} is almost unaffected by compressive or tensile *zigzag* strain. However, it is enlarged by an order of magnitude by compressive or tensile *armchair* strain.

In Fig. 1.15e and f, we show the Faraday rotation angles in units of degree/μm, corresponding to the optical Hall conductivities for the armchair and zigzag strain, as shown in Fig. 1.15d. When calculating the Faraday rotation angle θ_F, we took $n_{sub} = 1$ throughout the paper, which models a free-standing sample. Recently, by using DFT calculations, magneto-optical effects (e.g. Faraday rotation) were predicted for the ferromagnetic monolayer $Cr_2Ge_2Te_6$ [71]. We notice that the rotation angle θ_F is smaller in both armchair- and zigzag-strained phosphorene ($\sim$10 and $\sim$1 degree/μm) than in ferromagnetic monolayer of $Cr_2Ge_2Te_6$ ($\sim$120 degree/μm). The difference is mainly due to (i) different band properties of monolayer phosphorene and monolayer $Cr_2Ge_2Te_6$ and (ii) different driving mechanisms: uniaxial strain (in phosphorene) versus local magnetic moment (in monolayer $Cr_2Ge_2Te_6$).

In addition, we see from Fig. 1.15a–f that the band structure, the optical conductivity, and the Faraday rotation exhibit distinctively different responses to different strain directions (i.e. armchair and zigzag directions). This is a consequence of the structural anisotropy of the phosphorene lattice. Therefore, strain engineering in phosphorene might be even more interesting than in other 2D materials. Our numerical results indicate that the tuning of the Faraday rotation becomes less significant with decreasing magnitude of the applied strain, e.g., the Faraday rotation angle with 1% strain is smaller than that with 10% strain.

1.5 Quantum Transport in Defective Phosphorene Nanoribbons with Atomic Vacancies

Recently phosphorene nanostructures (such as nanodots, nanorings and nanoribbons) have drawn a lot of attention. Among them, phosphorene nanoribbons (PNRs) are of particular interest. It was shown by DFT and TB calculations that these one-dimensional (1D) nanoribbons exhibit not only the usual confinement effect but also peculiar edge effects, both of which play an important role in their electronic, optical, magnetic and transport properties [31, 72–76].

In spite of the fast growing knowledge on perfect PNRs, as mentioned above, less attention has been paid to their defective counterparts. Recently, point defects (both intrinsic and extrinsic) in phosphorene and bulk BP were theoretically investigated using DFT calculations [38–40, 42]. More recently, atomic vacancies were experimentally observed on the surface of bulk BP [77, 78], and it was found that such vacancies exhibit strongly anisotropic and highly delocalized resonant states [77, 78]. However, compared to defective phosphorene and bulk BP, far less work [41, 43] has been done on defective PNRs.

From a practical point of view, when PNRs are fabricated, they will inevitably have some defects (e.g., edge vacancies produced during lithography or plasma etching process), which certainly affect the electronic and transport properties of PNRs and thus the performance of PNR-based electronic devices. From a fundamental point of view, it is essential to have a better understanding of the electronic and transport properties of defective PNRs and of how to control these properties. More importantly, point defects like atomic vacancies were already experimentally observed on the surface of bulk BP [77, 78]. All these motivate us to investigate the electronic transport properties of defective PNRs with atomic vacancies in a theoretical perspective. As is known, such point defects are one of the most important lattice imperfections in realistic materials [79]. Moreover, for 2D materials their atomically thin nature is expected to be more beneficial for introducing atomic vacancies. For instance, a high concentration of such vacancies (up to 2%) was experimentally observed in monolayer MoS_2 [80, 81]. This concentration could be even higher in phosphorene, due to the much lower formation energy of its atomic vacancies [42].

Here we investigated theoretically the electronic transport properties of defective PNRs containing atomic vacancies. Our theoretical study was carried out by means of atomistic quantum transport simulations based on the TB approach combined with the scattering matrix (S-matrix) formalism. Both cases of a single atomic vacancy and a random distribution of atomic vacancies are considered in the present work. In case of a single vacancy, we study the effects of vacancy type and its spatial location on the electronic and transport properties of defective PNRs. In case of a random distribution of vacancies, we study the influence of vacancy concentration on the electronic and transport properties of defective PNRs. Due to the randomness nature of such vacancies, the electronic and transport properties were calculated and averaged over an ensemble of many different vacancy configurations. Moreover, the effects of ribbon length and width were also studied for defective PNRs.

1.5.1 Model System and TB Hamiltonian

The model system under consideration is composed of two leads and a scattering region, as shown in Fig. 1.16. The scattering region is a finite-size PNR of length L and width W, which contains lattice defects induced by atomic vacancies that are located at the edges or in the bulk of the PNR (see Fig. 1.16). The TB Hamiltonian of the considered system is given by [82]

$$H = \sum_i \varepsilon_i c_i^\dagger c_i + \sum_{i \neq j} t_{ij} c_i^\dagger c_j, \tag{1.10}$$

where the summation runs over all lattice sites, ε_i is the on-site energy at site i, t_{ij} is the hopping energy between sites i and j, and $c_i^\dagger$ (c_j) is the creation (annihilation) operator of an electron at site i (j).

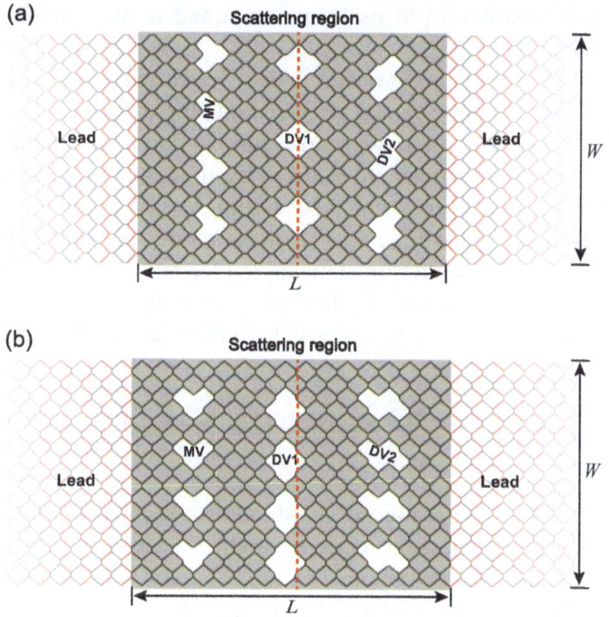

Fig. 1.16 Model system under consideration: (**a**) armchair and (**b**) zigzag PNRs. The system is composed of two semi-infinite leads of width W (light-red shaded region) and a finite scattering region of length L and width W (dark-gray shaded region). Here, three types of atomic vacancies are taken into account, i.e., mono-vacancy (MV), di-vacancy of type I (DV1), and di-vacancy of type II (DV2), which can be located at the edges or in the bulk of the scattering region. Two types of di-vacancies (DV1 and DV2) are distinguished where DV1 (DV2) keeps (breaks) the sublattice symmetry of the PNR

In the absence of atomic vacancies, the on-site energies ε are set to zero for all lattice sites and the hopping energies between two different sites are included up to the fifth nearest neighbors, of which the values are $t_1 = -1.220\,\text{eV}$, $t_2 = 3.665\,\text{eV}$, $t_3 = -0.205\,\text{eV}$, $t_4 = -0.105\,\text{eV}$, and $t_5 = -0.055\,\text{eV}$ [82]. In the presence of atomic vacancies, both the on-site and hopping energies are modified. Since a single vacancy can be modeled by removing a single atom from the perfect lattice (as shown in Fig. 1.16), the on-site energy at this atomic site as well as the hopping energies from this atomic site to its nearest neighbors will vanish.

1.5.2 S-Matrix Formalism and Quantum Conductance

In what follows, we show briefly how to calculate the conductance of the PNR system via the S-matrix formalism. The detailed description of this formalism can be found in Ref. [83]. Mathematically, the S-matrix formalism is equivalent to the non-equilibrium Green's function formalism due to the Fisher–Lee relation [84], but it has a simpler structure in its formulation and is more stable in numerical computations [85].

In a two-terminal system composed of two leads and a scattering region, the transport modes (or channels) can be classified into three types: incoming, outgoing and evanescent ones. The first two carry current corresponding to propagating modes while the last one correspond to decaying modes. We denote the number of incoming, outgoing, and evanescent modes with N_{in}, N_{out} and N_{ev}, respectively, and the corresponding wave functions of these modes with ψ^{in}, ψ^{out}, and ψ^{ev}. Moreover, we denote the scattering states in the leads and in the scattering region with Ψ^L and Ψ^S, respectively. With these notations and taking the transport direction along the x-axis, the scattering state in the leads can be written as [85]

$$\Psi_n^L(x) = \psi_n^{\text{in}}(x) + \sum_{m=1}^{N_{\text{out}}} S_{mn}\psi_m^{\text{out}}(x) + \sum_{k=1}^{N_{\text{ev}}} \mathcal{S}_{kn}\psi_k^{\text{ev}}(x), \tag{1.11}$$

where S_{mn} ($\mathcal{S}_{kn}$) gives the scattering amplitude from an incoming mode n to an outgoing mode m (an evanescent mode k), both of which are the elements of the S-matrix. The S-matrix can be obtained by matching the wave function in the leads with the one in the scattering region [85], i.e., $\Psi_n^L(x = x_I) = \Psi_n^S(x = x_I)$, where $x = x_I$ denotes the intersection of the lead and the scattering region.

Once the S-matrix is obtained, the conductance of the system at zero temperature can be calculated by using the Landauer formula [83]

$$G_{ab} = \frac{e^2}{h} T_{ab} = \frac{e^2}{h} \sum_{n\in a;m\in b} |S_{mn}|^2, \tag{1.12}$$

where a and b label the two leads of the system and T_{ab} is the transmission coefficient from lead a to lead b. It is clear that the conductance value depends on the number of available transport modes. In the present work, transport calculations were performed using KWANT [85], while other properties (e.g. density of states) were calculated using PYBINDING [61].

1.5.3 Effects of Single Vacancy

We first consider defective PNRs with single vacancy. We study how such point defect affects the electronic and transport properties of PNRs. This is not only realistic (because single vacancy has been experimentally observed on the surface of bulk BP [77]), but also can provide basic insights into the understanding of the effects of many vacancies (e.g. randomly distributed vacancies).

We first study the effect of the type of vacancy on the DOS and conductance of defective APNR and ZPNR. Three types of single vacancies are considered: (i) mono-vacancy (MV), (ii) di-vacancy of type I (DV1), and (iii) di-vacancy of type

II (DV2). For simplicity, we assume MV, DV1 and DV2 to be located in the center of the scattering region. Notice that two types of di-vacancies (DV1 and DV2) are distinguished where DV1 (DV2) preserves (breaks) the sublattice symmetry of PNR. Here, the sublattice symmetry (asymmetry) is defined in terms of the number of different sublattice atoms in the PNR. Although the unit cell of phosphorene has four sublattice atoms, it was shown [73] that in the tight-binding model they can be reduced to two inequivalent ones (labeled A and B) in the unit cell due to the D_{2h} point group invariance. For convenience, we use N_A and N_B to denote the number of atoms in sublattices A and B of PNR. In the absence of vacancies, we have $N_A = N_B$ and the sublattice symmetry of PNR is preserved. In the presence of a MV, created by removing one sublattice atom (A or B), we have $N_A \neq N_B$ and thus the sublattice symmetry of PNR is broken. Since a DV can be created by removing two sublattice atoms (A, B or A, A or B, B), in the presence of a DV2, created by removing two A (or two B) sublattice atoms, we have $N_A \neq N_B$ which breaks the sublattice symmetry of PNR; whereas in the presence of a DV1, created by removing both A and B sublattice atoms, we have $N_A = N_B$ which preserves the sublattice symmetry of PNR.

In Fig. 1.17, we show the results for the cases of no-vacancy (NV), MV, DV1, and DV2 as indicated: (a), (c) are for APNR and (b), (d) are for ZPNR. Here, the ribbon length and width are set to $L = 10$ nm and $W = 4$ nm for all the simulated PNRs. In order to see clearly the effect of a single vacancy on the DOS, the local DOS (LDOS) calculated at neighboring sites around the vacancy is superimposed on the DOS. It is clear that, for both APNR and ZPNR, introducing a single vacancy decreases the conductance value and smoothes the conductance step. As expected, we find that: (i) the DV2 has a larger effect on the DOS and conductance than the MV, because a DV can be viewed as consisting of two MVs that are very close to each other; and (ii) both the MV and DV2 have a larger impact on the DOS and conductance than the DV1, because the former two break the sublattice symmetry, while the latter one preserves such a symmetry. These results are consistent with those previously found for graphene [86, 87]. Note that all types of single vacancies (i.e., MV, DV1, and DV2) have little influence on the DOS of PNR outside the band gap (i.e., the bulk DOS of PNR), because the lattice region removed by a single vacancy is negligibly small as compared to the whole part of PNR. Nevertheless, they can affect the bulk and edge states of PNR via the vacancy scattering, and can thus affect the bulk and edge conductance of PNR.

For defective APNR, as shown in Fig. 1.17a and c, both the MV and DV2 result in resonant peaks in the DOS within the band gap, which correspond to quasi-localized vacancy states. However, the DV1 does not give rise to such peaks because this two-atom vacancy preserves the sublattice symmetry of PNR (while such a symmetry is broken by both the MV and DV2). The results are very different for defective ZPNR, as shown in Fig. 1.17b and d. We see that both the MV and DV2 induce resonant peaks in the edge DOS within the band gap, and simultaneously cause anti-resonant dips in the edge conductance. The anti-resonant dips in the edge conductance are mainly due to the vacancy scattering on the edge propagating modes. Intuitively, the vacancy located in the bulk center of ZPNR should not have a considerable

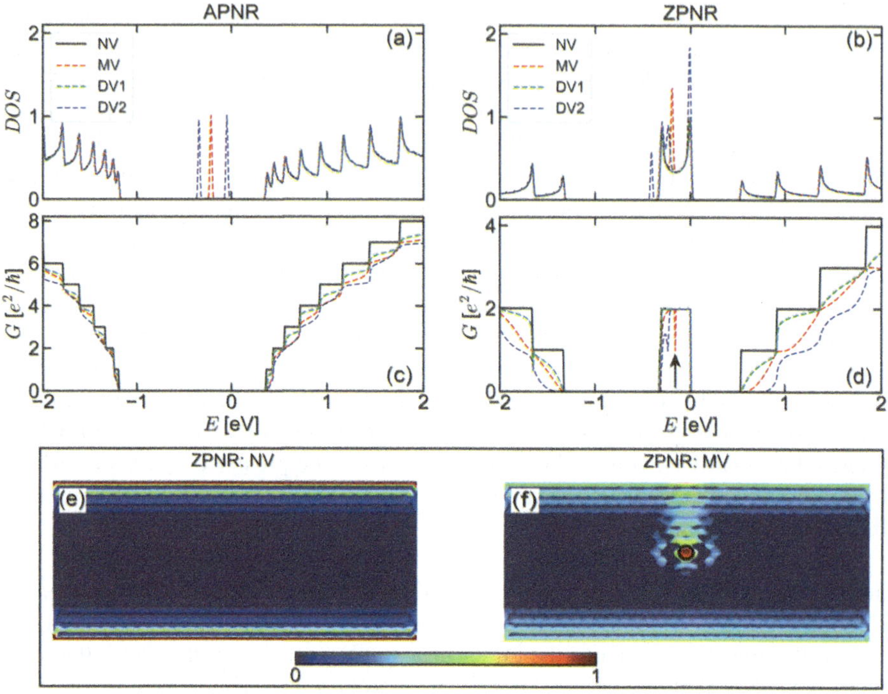

Fig. 1.17 (**a**)–(**d**) DOS and conductance of defective APNR and ZPNR for different types of a single vacancy placed in the center of the PNR, together with the results when no vacancy (NV) is present; (**e**) and (**f**) Spatial LDOS of edge propagating modes in the ZPNR without and with a single vacancy. Here, NV, MV, DV1, and DV2 denote the cases of no vacancy, mono-vacancy, di-vacancy of type I, and di-vacancy of type II, respectively. In panel (**d**), the black arrow indicates the energy of those edge propagating modes at which their spatial LDOS is calculated. In panel (**f**), the black circle indicates the single vacancy and its position

influence on its edge conductance, because the vacancy position is far from the edges of ZPNR. In order to understand this counter-intuitive result, we have to calculate the spatial LDOS of edge propagating modes of ZPNR and to examine how it is affected by a bulk central vacancy. The results are plotted in Fig. 1.17e and f, corresponding to the cases of NV and MV, respectively. From these two panels, we see that the quasi-localized vacancy state is considerably coupled to the edge state at the upper boundary of ZPNR, which may lead to a considerable edge-vacancy scattering and thus an anti-resonant dip in the edge conductance. Note that there is an asymmetry in the coupling of the vacancy state with the edge state. This is because the quasi-localized vacancy state is spatially highly anisotropic, due to the inequivalent sublattices at which the vacancy is created. For instance, the spatial LDOS of the vacancy state shown in Fig. 1.17f is mostly distributed between the center and the upper edge of ZPNR. This anisotropic feature of the quasi-localized vacancy state was experimentally observed on the surface of BP [77]. We further find that with increasing ribbon width W, such a vacancy-edge coupling decreases (and even disappears) and consequently, the edge conductance becomes less affected (and even unaffected) by the bulk central vacancy.

Moreover, for defective ZPNR shown in Fig. 1.17b and d, there is a one-to-one correspondence between the resonant peak(s) in the edge DOS and the anti-resonant dip(s) in the edge conductance for the MV case, and those peak(s) and dip(s) have the same energy positions. But there is no such correspondence for the DV2 case, e.g., two resonant peaks in the edge DOS with only one of them giving rise to an anti-resonant dip in the edge conductance. This is because the other peak is not located within the energy range of the edge DOS, although it is located within the band gap, and thus this peak has no influence on the edge conductance. Again, the DV1 does not give rise to such resonant peaks in the edge DOS, and thus the edge conductance is almost unaffected by this two-atom vacancy. Although MV and DV have distinctive influences on the edge conductance, they have almost the same effect on the bulk conductance, i.e., both of them cause a significant decrease of the conductance value.

Next we study the effect of the vacancy position on the DOS and conductance of defective APNR and ZPNR. Without loss of generality, we show the results for the MV case, because a DV can be viewed as two closely positioned MVs. In Fig. 1.18, we plot the DOS and conductance of defective APNR and ZPNR for different MV positions, together with the results of their perfect counterparts with no vacancy (NV). Here, the ribbon length and width are set to $L = 10\,\text{nm}$ and $W = 4\,\text{nm}$ for all simulated PNRs. The vacancy (MV) is assumed to move along the vertical central line of the scattering region (indicated by the red dashed line in Fig. 1.16) when changing its position, characterized by a vertical distance d from the center of PNR. It can be clearly seen from this figure that, the DOS and conductance of the defective PNR depend sensitively on the MV position. For defective ZPNR, as shown in Fig. 1.18b and d, by moving such vacancy from its center to its edge (i.e., changing d from 0.2 to 1.6 nm), the resonant peak(s) in the edge DOS as well as the anti-resonant dip(s) in the edge conductance are both shifted in energy. This is not surprising for defective ZPNR because of the one-to-one correspondence between resonant peak(s) in the edge DOS and anti-resonant dip(s) in the edge conductance. However, there is no such correspondence for defective APNR due to the absence of the in-gap (edge) conductance, as shown in Fig. 1.18a and c, where the MV induces only a resonant peak in the DOS within the band gap, whose energy position is very slightly shifted by varying its position.

Apart from on the DOS and conductance of edge propagating modes, it is also interesting to look into the effects of vacancy position on the DOS and conductance of bulk propagating modes. As shown in Fig. 1.18c and d, for the different conductance plateaus, changing the MV position gives rise to different consequences. For instance, the value of the first conductance plateau decreases continuously by moving the MV from the edge of the ZPNR ($d = 1.6\,\text{nm}$) to the center of the ZPNR ($d = 0.2\,\text{nm}$). This result can be readily understood in an intuitive way. As the MV is moved closer to the center of the ZPNR, it certainly affects the bulk conductance more significantly. However, for the second or higher conductance plateaus, the situation becomes somewhat complicated. The

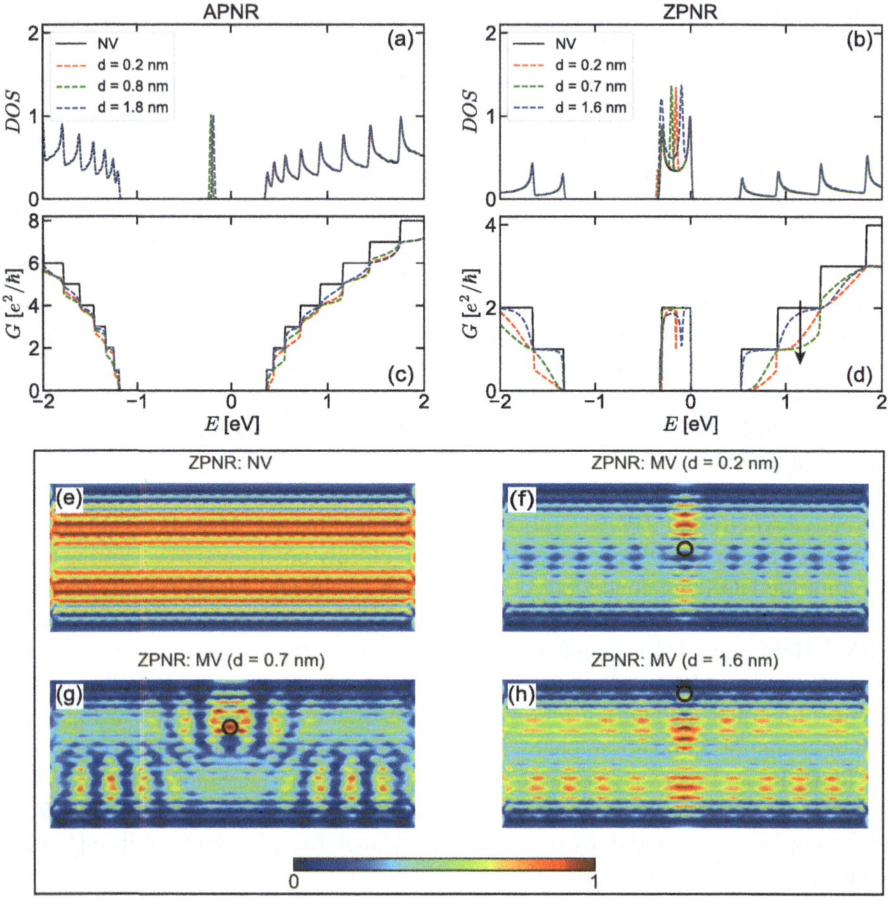

Fig. 1.18 (**a**)–(**d**) DOS and conductance of defective APNR and ZPNR for different positions (d) of the mono-vacancy (MV) from the center of the PNR, together with the results of their perfect counterparts with no vacancy (NV). (**e**)–(**h**) Spatial LDOS of bulk propagating modes in the ZPNR for the cases of NV and MV. In panel (**d**), the black arrow indicates the energy of those bulk propagating modes at which their spatial LDOS is calculated. In panels (**f**)–(**h**), the black circles indicate the different vacancy positions

conductance value does not decrease continuously by moving the MV from the edge of the ZPNR to the center of the ZPNR, e.g., the central MV ($d = 0.2$ nm) does not induce the largest decrease of the value of the second conductance plateau. Intuitively, this result is not easy to be understood. Again, one has to examine how the MV position affects the spatial LDOS of bulk propagating modes in the second conduction or valence subbands. The results are plotted in Fig. 1.18e–h, corresponding to the cases of NV, MV ($d = 0.2$ nm), MV ($d = 0.7$ nm), and MV ($d = 1.6$ nm), respectively. As seen, the MV renders the bulk propagating modes to be less extended due to the vacancy scattering, which depends sensitively on the

MV position. This may explain our counter-intuitive result that why the central MV ($d = 0.2$ nm) does not have the largest impact on the second conductance plateau. Similar results obtained for the defective APNR can be understood in the same way.

1.5.4 Random Distribution of Vacancies

Now we turn to the effects of a random distribution of vacancies on the DOS and conductance of PNRs. Such vacancies can be used to simulate fabrication-induced and intrinsic (native) lattice defects in realistic PNR samples. Due to the randomness nature of such vacancies, we calculate the averaged properties by simulating many different defective PNRs for a given ribbon size and a given vacancy concentration. The number of simulated samples is chosen sufficiently large to decrease sample-to-sample fluctuations and to obtain converged results. It was shown previously [43] that converged results can be obtained for an ensemble of 50 to 200 random distributions, and in the present work we took an ensemble of 100 random distributions.

Random vacancies in the simulated PNRs are implemented by randomly removing single atoms from both the edge and bulk regions of the PNRs. In Fig. 1.19, we show the effects on the DOS and conductance of APNR and ZPNR for different vacancy concentrations P as indicated. Here, the ribbon length and width are set to be $L = 10$ nm and $W = 4$ nm for all the simulated samples, and the vacancy concentration P is defined as the ratio of the number of removed atoms to the number of total atoms. As seen from this figure, with increasing vacancy concentration, the conductance of both APNR and ZPNR is strongly suppressed, and no quantized plateaus are observed. For the APNR case, due to the strong decrease of the conductance, the transport gap increases from 1.52 to 1.61 eV and to 2.05 eV with increasing P from 0% to 1% and to 5%. In particular, for the case of ZPNR, the edge conductance is fully suppressed as the vacancy concentration is increased to $P = 5\%$, leading to a *global* transport gap that is not present in perfect ZPNR. This indicates that with increasing vacancy concentration ZPNR undergoes a metal-to-semiconductor transition. The physical reason behind this is that random vacancies act as randomly distributed short-range scatterers and induce very strong back scattering, leading to the so-called Anderson localization. Therefore, in the presence of random atomic vacancies and with increasing vacancy concentration, three different transport regimes, i.e., ballistic, diffusive and Anderson localization, can clearly be observed in APNRs and ZPNRs, as described above.

We note that in the presence of a random distribution of vacancies, the DOS of both APNR and ZPNR is almost fully extended over the whole band gap, as shown in Fig. 1.19a and b. This is due to the presence of many quasi-localized vacancy states that are overlapping among each other. For defective APNR, the presence of such vacancy states makes it difficult to extract the band gap from the DOS spectrum, as shown in Fig. 1.19a. However, from the conductance spectrum it is possible to extract a distinct transport gap, as shown in Fig. 1.19c.

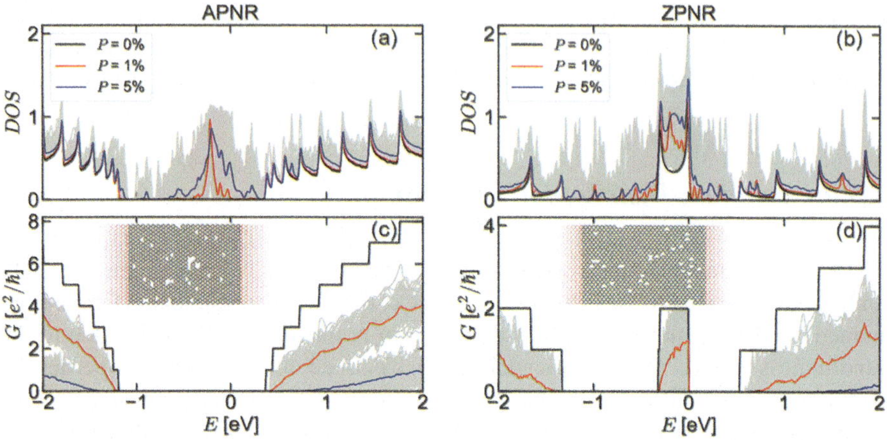

Fig. 1.19 DOS and conductance of defective PNR with random vacancies for different vacancy concentrations P as indicated, together with the results of their perfect counterparts with no vacancy ($P = 0$): (**a**), (**c**) for the APNR case and (**b**), (**d**) for the ZPNR case. Here, each panel contains the individual results for all the simulated PNRs with vacancies (gray lines), the averaged results of them (color lines), and the result for perfect PNR (black line). The inset of panels (**c**) and (**d**) shows schematically the APNR and ZPNR containing a random distribution of vacancies

As is known, ribbon length and width are important parameters in tuning the electronic, optical and transport properties of nanoribbons. Here, in order to investigate the effects of ribbon length and width on the conductance of defective PNR with random vacancies, we introduce a physical quantity which characterizes the ratio of the change in the conductance of defective PNR relative to that of perfect one. Mathematically, this quantity is defined as $\Delta G/G = (G - G')/G$, with G and G' being the conductance of perfect and defective PNRs, respectively. Thus, a larger value of $\Delta G/G$ means a higher sensitivity of PNR to vacancy disorder.

In Fig. 1.20, we show the relative change of the conductance of defective PNRs, $\Delta G/G$, for different ribbon widths W and vacancy concentrations P as indicated: (a), (c), (e) for the APNR case and (b), (d), (f) for the ZPNR case. Here, the ribbon length is fixed at $L = 10\,\text{nm}$ for all the simulated PNRs. From this figure, we find that (i) for fixed vacancy concentration, both narrower APNRs and ZPNRs are more sensitive to vacancy disorder than their wider counterparts, because vacancies are distributed more sparsely in the wider PNRs; (ii) for fixed ribbon width and vacancy concentration, ZPNRs are more sensitive to vacancy disorder than APNRs, because the carrier (electron or hole) effective mass of APNR is smaller than that of ZPNR, and therefore, the relaxation time (due to vacancy scattering) is longer for APNR than for ZPNR; and (iii) with increasing vacancy concentration, the difference between the sensitivities of APNR and ZPNR to vacancy disorder becomes smaller and smaller, and eventually disappears (i.e., $\Delta G/G = 1$) for all the ribbon widths due to strong Anderson localization.

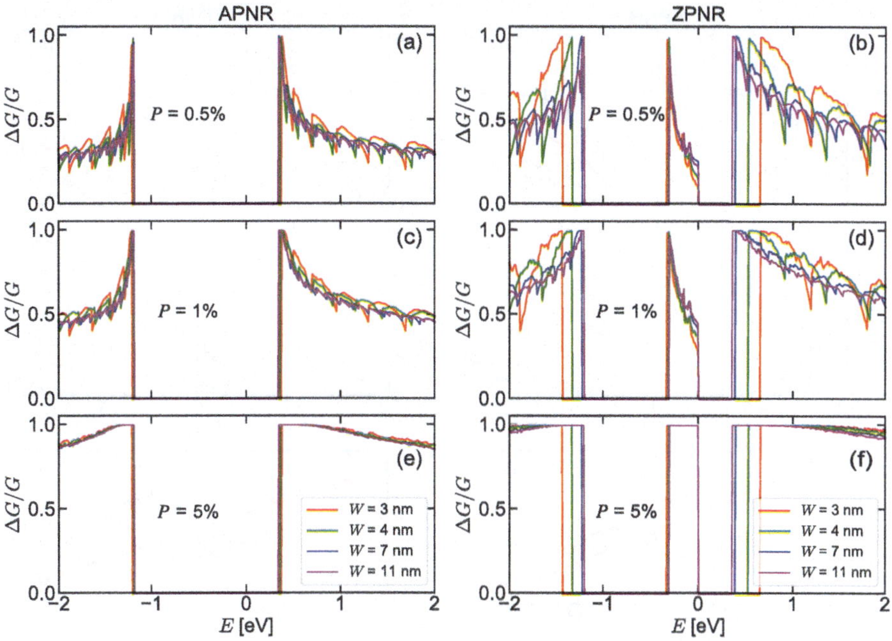

Fig. 1.20 Ratio of the change in the conductance of defective PNR with random vacancies relative to that of perfect PNR, for different ribbon widths W and vacancy concentrations P as indicated: (**a**), (**c**), (**e**) for the APNR case and (**b**), (**d**), (**f**) for the ZPNR case. Here, the ribbon length is set to be $L = 10\,\text{nm}$ for all the simulated PNRs

In Fig. 1.21, we show the conductance of defective PNRs with random vacancies for different ribbon lengths L and vacancy concentrations P as indicated: (a), (c), (e) for the APNR case and (b), (d), (f) for the ZPNR case. Here, the ribbon width is fixed at $W = 4\,\text{nm}$ for all the simulated PNRs. From this figure, we see that for fixed vacancy concentration, both longer APNRs and ZPNRs are more sensitive to vacancy disorder than their shorter counterparts. This result can be understood as follows: The conductance of defective PNR not only depends on the ribbon width W, which determines the number of available transport modes, but also depends on the ribbon length L, over which the vacancies are distributed. It is known that in a 1D system, the conductance decreases exponentially with the length of the system as $G = G_0 \exp(-L/L_0)$, with L_0 being the localization length [88]. As shown in the panels (e) and (f), our numerical simulations (indicated by color dots) agree with this prediction (indicated by color lines). However, the agreement is better for the edge conductance than the bulk conductance. This is because edge propagating modes are 1D transport channels as they are localized at the boundaries of PNR, while bulk propagating modes are mainly distributed in the bulk region of PNR and thus they are more 2D-like.

In addition, using the relation $G = G_0 \exp(-L/L_0)$, we can fit the localization length L_0 of defective PNR in the presence of vacancy disorder. For instance, from

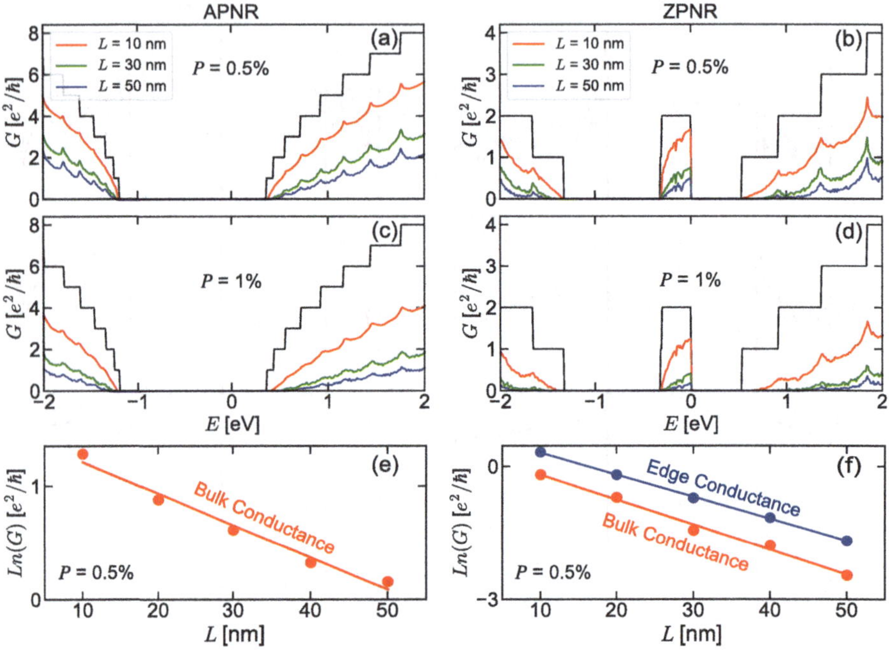

Fig. 1.21 Conductance of defective PNR with random vacancies for different ribbon length L and vacancy concentrations P as indicated: (**a**), (**c**), (**e**) for the APNR case and (**b**), (**d**), (**f**) for the ZPNR case. Here, the ribbon width is set to $W = 4$ nm for all the simulated PNRs. In panels (**a**)–(**d**), black lines are the results of perfect PNRs with no vacancy ($P = 0$), which are presented for reference purposes. In panels (**e**) and (**f**), red (blue) dots are the numerical results of bulk (edge) conductance, and corresponding lines are the fitted ones

the panels (e) and (f), the fitted localization lengths are known to be $L_0 = 35.5$ nm for the bulk conductance of APNR, $L_0 = 20.1$ nm for the edge conductance of ZPNR, and $L_0 = 15.6$ nm for the bulk conductance of ZPNR. Here, the values of L_0 are extracted at energy $E = 1.2$ eV for the bulk conductance and at energy $E = -0.1$ eV for the edge conductance. Generally, this length is energy dependent and in transport experiments it can be measured around the Fermi energy. Moreover, our numerical calculations indicate that the localization length L_0 decreases for both the bulk and edge conductance with increasing vacancy concentration P, as expected.

References

1. Novoselov KS, Jiang D, Schedin F, Booth TJ, Khotkevich VV, Morozov SV, Geim AK (2005) Two-dimensional atomic crystals. PNAS 102:10451
2. Novoselov KS, Mishchenko A, Carvalho A, Neto AHC (2016) 2D materials and van der Waals heterostructures. Science 353(6298):aac9439

3. Castellanos-Gomez A (2015) Black phosphorus: narrow gap, wide applications. J Phys Chem Lett 6:4280
4. Carvalho A, Wang M, Zhu X, Rodin AS, Su H, Neto AHC (2016) Phosphorene: from theory to applications. Nat Rev Mater 1:1
5. Cartz L, Srinivasa SR, Riedner RJ, Jorgensen JD, Worlton TG (1979) Effect of pressure on bonding in black phosphorus. J Chem Phys 71:1718
6. Morita A (1986) Semiconducting black phosphorus. Appl Phys A 39:227
7. Li L, Yu Y, Ye GJ, Ge Q, Ou X, Wu H, Feng D, Chen XH, Zhang Y (2014) Nat Nanotechnol. Black Phosphorus Field-Effect Transistors 9:372
8. Liu H, Neal AT, Zhu Z, Luo Z, Xu X, Tománek D, Ye PD (2014) Phosphorene: an unexplored 2D semiconductor with a high hole mobility. ACS Nano 8:4033
9. Castellanos-Gomez A, Vicarelli L, Prada E, Island JO, Narasimha-Acharya KL, Blanter SI, Groenendijk DJ, Buscema M, Steele GA, Alvarez JV, Zandbergen HW, Palacios JJ, van der Zant HSJ (2014) Isolation and characterization of few-layer black phosphorus. 2D Mater 1:025001
10. Tran V, Soklaski R, Liang Y, Yang L (2014) Layer-controlled band gap and anisotropic excitons in few-layer black phosphorus. Phys Rev B 89:235319
11. Qiao J, Kong X, Hu ZX, Yang F, Ji W (2014) High-mobility transport anisotropy and linear dichroism in few-layer black phosphorus. Nat Commun 5:4475
12. Rodin AS, Carvalho A, Castro Neto AH (2014) Strain-induced gap modification in black phosphorus. Phys Rev Lett 112:176801
13. Yuan H, Liu X, Afshinmanesh F, Li W, Xu G, Sun J, Lian B, Curto AG, Ye G, Hikita Y, Shen Z, Zhang SC, Chen X, Brongersma M, Hwang HY, Cui Y (2015) Polarization-Sensitive broadband photodetector using a black phosphorus vertical p–n junction. Nat Nanotechnol 10:707
14. Fei R, Yang L (2014) Strain-engineering the anisotropic electrical conductance of few-layer black phosphorus. Nano Lett 14:2884
15. Peng X, Wei Q, Copple A (2014) Strain-engineered direct-indirect band gap transition and its mechanism in two-dimensional phosphorene. Phys Rev B 90:085402
16. Manjanath A, Samanta A, Pandey T, Singh AK (2015) Semiconductor to metal transition in bilayer phosphorene under normal compressive strain. Nanotechnology 26:075701
17. Wang C, Xia Q, Nie Y, Guo G (2015) Strain-induced gap transition and anisotropic dirac-like cones in monolayer and bilayer phosphorene. J Appl Phys 117:124302
18. Morgan Stewart H, Shevlin SA, Catlow CRA, Guo ZX (2015) Compressive straining of bilayer phosphorene leads to extraordinary electron mobility at a new conduction band edge. Nano Lett 15:2006
19. Çakır D, Sahin H, Peeters FM (2014) Tuning of the electronic and optical properties of single-layer black phosphorus by strain. Phys Rev B 90:205421
20. Liu Q, Zhang X, Abdalla LB, Fazzio A, Zunger A (2015) Switching a normal insulator into a topological insulator via electric field with application to phosphorene. Nano Lett 15:1222
21. Pereira JM, Katsnelson MI (2015) Landau levels of single-layer and bilayer phosphorene. Phys Rev B 92:075437
22. Yuan S, van Veen E, Katsnelson MI, Roldán R (2016) Quantum Hall effect and semiconductor-to-semimetal transition in biased black phosphorus. Phys Rev B 93:245433
23. Wu JY, Chen SC, Gumbs G, Lin MF (2017) Field-Induced diverse quantizations in monolayer and bilayer black phosphorus. Phys Rev B 95:115411
24. Dai J, Zeng XC (2014) Bilayer phosphorene: effect of stacking order on bandgap and its potential applications in thin-film solar cells. J Phys Chem Lett 5:1289
25. Çakır D, Sevik C, Peeters FM (2015) Significant effect of stacking on the electronic and optical properties of few-layer black phosphorus. Phys Rev B 92:165406
26. Shu H, Li Y, Niu X, Wang J (2016) The stacking dependent electronic structure and optical properties of bilayer black phosphorus. Phys Chem Chem Phys 18:6085
27. Cao T, Li Z, Qiu DY, Louie SG (2016) Switchable transport and optical anisotropy in 90° twisted bilayer black phosphorus. Nano Lett 16:5542

28. Kang P, Zhang WT, Michaud-Rioux V, Kong XH, Hu C, Yu GH, Guo H (2017) Moiré impurities in twisted bilayer black phosphorus: effects on the carrier mobility. Phys Rev B 96:195406
29. Pan D, Wang TC, Xiao W, Hu D, Yao Y, Simulations of twisted bilayer orthorhombic black phosphorus, Phys. Rev. B 96, 041411 (2017).
30. Sevik C, Wallbank JR, Gülseren O, Peeters FM, Çakır D (2017) Gate induced monolayer behavior in twisted bilayer black phosphorus. 2D Mater 4:035025
31. Tran V, Yang L (2014) Scaling laws for the band gap and optical response of phosphorene nanoribbons. Phys Rev B 89:245407
32. Zhang R, Zhou XY, Zhang D, Lou WK, Zhai F, Chang K (2015) Electronic and magneto-optical properties of monolayer phosphorene quantum dots. 2D Mater 2:045012
33. Niu X, Li Y, Shu H, Wang J (2016) Anomalous size dependence of optical properties in black phosphorus quantum dots. J Phys Chem Lett 7:370
34. Li LL, Moldovan D, Xu W, Peeters FM (2017) Electric- and magnetic-field dependence of the electronic and optical properties of phosphorene quantum dots. Nanotechnology 28:085702
35. Li LL, Moldovan D, Xu W, Peeters FM (2017) Electronic properties of bilayer phosphorene quantum dots in the presence of perpendicular electric and magnetic fields. Phys Rev B 96:155425
36. Li LL, Moldovan D, Vasilopoulos P, Peeters FM (2017) Aharonov–Bohm oscillations in phosphorene quantum rings. Phys Rev B 95:205426
37. Abdelsalam H, Saroka VA, Lukyanchuk I, Portnoi ME (2018) Multilayer phosphorene quantum dots in an electric field: energy levels and optical absorption. J Appl Phys 124:124303
38. Hu W, Yang J (2015) Defects in phosphorene. J Phys Chem C 119:20474
39. Wang G, Pandey R, Karna SP (2015) Effects of extrinsic point defects in phosphorene: B, C, N, O, and F adatoms. Appl Phys Lett 106:173104
40. Wang V, Kawazoe Y, Geng WT (2015) Native point defects in few-layer phosphorene. Phys Rev B 91:045433
41. Wu Q, Shen L, Yang M, Cai Y, Huang Z, Feng YP (2015) Electronic and transport properties of phosphorene nanoribbons. Phys Rev B 92:035436
42. Cai Y, Ke Q, Zhang G, Yakobson BI, Zhang YW (2016) Itinerant atomic vacancies in phosphorene. J Am Chem Soc 138:10199
43. Poljak M, Suligoj T (2016) Immunity of electronic and transport properties of phosphorene nanoribbons to edge defects. Nano Res 9:1723
44. Quereda J, San-Jose P, Parente V, Vaquero-Garzon L, Molina-Mendoza AJ, Agraït N, Rubio-Bollinger G, Guinea F, Roldán R, Castellanos-Gomez A (2016) Strong modulation of optical properties in black phosphorus through strain-engineered rippling. Nano Lett 16:2931
45. Peng R, Khaliji K, Youngblood N, Grassi R, Low T, Li M (2017) Midinfrared electro-optic modulation in few-layer black phosphorus. Nano Lett 17:6315
46. Sherrott MC, Whitney WS, Jariwala D, Biswas S, Went CM, Wong J, Rossman GR, Atwater HA (2019) Anisotropic quantum well electro-optics in few-layer black phosphoru. Nano Lett 19:269
47. Vishnoi P, Mazumder M, Barua M, Pati SK, Rao CNR (2018) Phosphorene quantum dots. Chem Phys Lett 699:223
48. Ge S, Zhang L, Fang Y (2019) Photoluminescence mechanism of phosphorene quantum dots (PQDs) produced by pulsed laser ablation in liquids. Appl Phys Lett 115:092107
49. Li LL, Partoens B, Peeters FM (2018) Tuning the electronic properties of gated multilayer phosphorene: a self-consistent tight-binding study. Phys Rev B 97:155424
50. Li LL, Partoens B, Xu W, Peeters FM (2018) Electric-field modulation of linear dichroism and Faraday rotation in few-layer phosphorene. 2D Mater 6:015032
51. Li LL, Peeters FM (2019) Strain engineered linear dichroism and Faraday rotation in few-layer phosphorene. Appl Phys Lett 114:243102
52. Li LL, Peeters FM (2018) Quantum transport in defective phosphorene nanoribbons: effects of atomic vacancies. Phys Rev B 97:075414

53. McCann E (2006) Asymmetry gap in the electronic band structure of bilayer graphene. Phys Rev B 74:161403
54. Avetisyan AA, Partoens B, Peeters FM (2009) Electric field tuning of the band gap in graphene multilayers. Phys Rev B 79:035421
55. Avetisyan AA, Partoens B, Peeters FM (2009) Electric-field control of the band gap and Fermi energy in graphene multilayers by top and back gates. Phys Rev B 80:195401
56. Koshino M, McCann E (2009) Gate-induced interlayer asymmetry in ABA-stacked trilayer graphene. Phys Rev B 79:125443
57. Zhang F, Sahu B, Min H, MacDonald AH (2010) Band structure of ABC-stacked graphene trilayers. Phys Rev B 82:035409
58. Jhun B, Park CH (2017) Electronic structure of charged bilayer and trilayer phosphorene. Phys Rev B 96:085412
59. Deng B, Tran V, Xie Y, Jiang H, Li C, Guo Q, Wang X, Tian H, Koester SJ, Wang H, Cha JJ, Xia Q, Yang L, Xia F (2017) Efficient electrical control of thin-film black phosphorus bandgap. Nat Commun 8:14474
60. Rudenko AN, Yuan S, Katsnelson MI (2015) Toward a realistic description of multilayer black phosphorus: from GW approximation to large-scale tight-binding simulations. Phys Rev B 92:085419
61. Moldovan D, Peeters FM (2016) Pybinding: a Python package for tight-binding calculations. https://doi.org/10.5281/zenodo.56818
62. Kumar P, Bhadoria BS, Kumar S, Bhowmick S, Chauhan YS, Agarwal A (2016) Thickness and electric-field-dependent polarizability and dielectric constant in phosphorene. Phys Rev B 93:195428
63. Morimoto T, Hatsugai Y, Aoki H (2009) Optical Hall conductivity in ordinary and graphene quantum hall systems. Phys Rev Lett 103:116803
64. Ikebe Y, Morimoto T, Masutomi R, Okamoto T, Aoki H, Shimano R (2010) Optical Hall effect in the integer quantum hall regime. Phys Rev Lett 104:256802
65. Li F, Zhou X, Feng W, Fu B, Yao Y (2018) Thickness-dependent magneto-optical effects in hole-doped GaS and GaSe multilayers: a first-principles study. New J Phys 20(4):043048
66. Wei Q, Peng X (2014) Superior mechanical flexibility of phosphorene and few-layer black phosphorus. Appl Phys Lett 104:251915
67. Roldán R, Castellanos-Gomez A, Cappelluti E, Guinea F (2015) Strain engineering in semiconducting two-dimensional crystals. J Phys Condens Matter 27:313201
68. Harrison WA (2004) Elementary electronic structure. World Scientific, Singapore
69. Taghizadeh Sisakht E, Fazileh F, Zare MH, Zarenia M, Peeters FM (2016) Strain-induced topological phase transition in phosphorene and in phosphorene nanoribbons. Phys Rev B 94:085417
70. Marzari N, Mostofi AA, Yates JR, Souza I, Vanderbilt D (2012) Maximally localized Wannier functions: theory and applications. Rev Mod Phys 84:1419
71. Fang Y, Wu S, Zhu ZZ, Guo GY (2018) Large magneto-optical effects and magnetic anisotropy energy in two-dimensional $Cr_2Ge_2Te_6$. Phys Rev B 98:125416
72. Carvalho A, Rodin AS, Neto AHC (2014) Phosphorene nanoribbons. Europhys Lett 108:47005
73. Ezawa M (2014) Topological origin of quasi-flat edge band in phosphorene. New J Phys 16:115004
74. Zhu Z, Li C, Yu W, Chang D, Sun Q, Jia Y (2014) Magnetism of zigzag edge phosphorene nanoribbons. Appl Phys Lett 105:113105
75. Zhang J, Liu HJ, Cheng L, Wei J, Liang JH, Fan DD, Shi J, Tang XF, Zhang QJ (2014) Phosphorene nanoribbon as a promising candidate for thermoelectric applications. Sci Rep 4:1
76. Grujić MM, Ezawa M, Tadić MŽ, Peeters FM (2016) Tunable skewed edges in puckered structures. Phys Rev B 93:245413
77. Kiraly B, Hauptmann N, Rudenko AN, Katsnelson MI, Khajetoorians AA (2017) Probing single vacancies in black phosphorus at the atomic level. Nano Lett 17:3607

78. Riffle JV, Flynn C, St. Laurent B, Ayotte CA, Caputo CA, Hollen SM (2018) Impact of vacancies on electronic properties of black phosphorus probed by STM. J Appl Phys 123:044301
79. Freysoldt C, Grabowski B, Hickel T, Neugebauer J, Kresse G, Janotti A, Van de Walle CG (2014) First-principles calculations for point defects in solids. Rev Mod Phys 86:253
80. Hong J, Hu Z, Probert M, Li K, Lv D, Yang X, Gu L, Mao N, Feng Q, Xie L, Zhang J, Wu D, Zhang Z, Jin C, Ji W, Zhang X, Yuan J, Zhang Z (2015) Exploring atomic defects in molybdenum disulphide monolayers. Nat Commun 6:1
81. Vancsó P, Magda GZ, Pet Jõ, Noh JY, Kim YS, Hwang C, Biró LP, Tapasztó L (2016) The intrinsic defect structure of exfoliated MoS 2 single layers revealed by scanning tunneling microscopy. Sci Rep 6:Article number: 29726
82. Rudenko AN, Katsnelson MI (2014) Quasiparticle band structure and tight-binding model for single- and bilayer black phosphorus. Phys Rev B 89:201408
83. Datta S (1995) Electronic transport in mesoscopic systems. Cambridge University Press, Cambridge
84. Wimmer M (2009) Quantum transport in nanostructures: from computational concepts to spintronics in graphene and magnetic tunnel junctions. Ph.D. thesis, University of Regensburg
85. Groth CW, Wimmer M, Akhmerov AR, Waintal X (2014) Kwant: a software package for quantum transport. New J Phys 16:063065
86. Li TC, Lu SP (2008) Quantum conductance of graphene nanoribbons with edge defects. Phys Rev B 77:085408
87. Petrović MD, Peeters FM (2016) Quantum transport in graphene Hall bars: effects of vacancy disorder. Phys Rev B 94:235413
88. Lee PA, Ramakrishnan TV (1985) Disordered electronic systems. Rev Mod Phys 57:287

Chapter 2
Detection of CBRN Agents Through Nanocomposite Based Photonic Crystal Sensors

S. Bellucci

Abstract The goal of our recent project is to develop cheap but effective photonic crystal structures formed by a periodic distribution of nanoparticles in polymer matrix for highly sensitive detection of chemical and biological agents. The volume photonic crystal structures are fabricated using holographic method in original nanocomposites developed by authors. Project main steps are: (i) theoretical analysis and design; (ii) fabrication and characterization of label-free sensors; (iii) functionalization of photonic crystal structures with graphene nanoflakes, and (iv) testing of enhancement effects in Raman spectroscopy. The project realization will promote emerging nanotechnologies for early detection of environmental contamination.

During the first part of the investigation, the following results were obtained. The improved method for determining parameters of the waveguide-grating (period) and the incident plane wave (angle of incidence and wavelength) providing resonant conditions, under which the reflection coefficient from the grating is close to unity, was developed. The pressing of the initially low-viscous composite between two glass substrates was found as the simplest and optimal method for the fabrication of thin (0.7–2 μm) photosensitive layers with high thickness uniformity and surface quality. VIS and UV Holographic exposure of thin layers demonstrated the formation of volume diffraction gratings of sufficiently high refractive index contrast.

The main task of the second part of the investigation consists in the design, fabrication and research of photonic crystal structures with improved resonant properties; characterization of label-free photonic crystal sensors.

Keywords Photonic crystal sensors · Nanocomposite

S. Bellucci (✉)
INFN-Laboratori Nazionali di Frascati, Frascati, Italy
e-mail: bellucci@lnf.infn.it; stefano.bellucci@lnf.infn.it

J. Bonča, S. Kruchinin (eds.), *Advanced Nanomaterials for Detection of CBRN*, NATO Science for Peace and Security Series A: Chemistry and Biology,
https://doi.org/10.1007/978-94-024-2030-2_2

2.1 Introduction

INFN Frascati National Laboratories (LNF), in collaboration with the National Academy of Science of Kiev and the Lviv Polytechnic in Ukraine and with the Fraunhofer Institute of Potsdam-Golm in Germany, proposed developing effective photonic crystal structures, made by a distribution of nanoparticles in polymer matrix for photonic crystal powered sensor applications and fluorescent and Raman spectroscopy.

In the last years the expertise in nanotechnology of the LNF NEXT group – that has been committed in the field of carbon nanostructured materials for electronic applications [1] since decades – made it possible to achieve recognitions for DNA bio-sensors design and for the application of high-tech devices based on innovative materials, such as carbon nanotubes and the first stable two-dimensional material with exceptional properties, i.e. graphene, in biological, medical and environmental fields.

The new project Nanocomposite based photonic crystal sensors of biological and chemical agents, directed by Stefano Bellucci – scientific responsible for INFN in the Graphene Flagship project of the European Commission, and founder and leader of the NEXT group – is financed by NATO in the frame of the Science for Peace programme. It aims at building prototypes of photonic crystal structures, designed for fast, non-destructive detection of very small amounts (of the order of 100 femtograms, down to even single molecules) of biological and chemical toxins and to monitor with high-sensitivity environmental pollution caused by biological and chemical agents.

The project suggests a new approach based on the deposition of special recognition layers, such as graphene nano-flakes and rare earth oxides, on the photonic crystal surface to detect the various agents chemically related to the recognition layer. Unlike resonant textured structures which are widely used nowadays – we propose using photonic crystal resonant structures for waveguides with periodic modulation of the permittivity. The photonic crystal structures in photosensitive nanocomposites will be realized by holographic lithography. We will use original photopolymer nanocomposites, developed by German and Ukrainian research teams participating in the Project and that include nanoparticles of different nature. Photonic crystal structures are formed by diffusion redistribution of the nanoparticles, such as graphene nano-flakes, in the polymer matrix during exposure to an interference pattern. The periodic structure and its thickness, as well as the refractive index of the material can be varied in such a way that the structure can support guided-mode resonances at designated wavelength region.

2.2 Implementation of the Investigation

To implement the project is necessary at this stage to perform the following tasks:

- Computer design of the photonic crystal sensors based on the resonance phenomena in the waveguide structure with grating using real parameters of photonic crystal waveguide structures in theoretical calculations; prediction of their characteristics; development of the recommendations for manufacturing of photonic crystal sensor.
- Improvement of polymer nanocomposites for fabrication of photonic crystal structures.
- Development of refined one-step all-optical technology for formation of photonic crystal structures in nanocomposite waveguides
- Manufacturing of the photonic crystal waveguide structure prototypes with optimal grating-waveguide parameters (thickness, period, amplitude of refractive index modulation) theoretically calculated for sensing applications.
- Research of the resonant properties of photonic crystal structures.
- Investigation of the characteristics of photonic crystal sensor based on nanocomposite.

2.3 Summary of Accomplishments

Approximate methods for determining the parameters of the created grating as sensitive sensor element for measuring the refractive index of the research medium has been developed. It is possible to calculate the resonance wavelengths at the non-zero angle of incidence using obtained approximate equations.

- The exact values of the resonant wavelengths at different angles of the incidence on the grating were found by the RCWA method. They are very close to the obtained values of wavelengths using developed approximate method.
- It was established that the sensitivity on both the wavelength and angle increases when refractive index of the substrate decreases. The sensitivity increases by 2.8 times when refractive index of the substrate decrease from 1.515 to 1.45.
- Suitable organic-inorganic nanocomposites were prepared and optimized. Optimization of the exposure conditions allows obtaining the photonic crystal structures with amplitude modulation of refractive index 0.01÷0.017.
- Gratings-waveguides are characterized by a high homogeneity of the nanoparticles distribution over the depth. Observed surface corrugation with a height of about 5 nm does not influence their resonant properties.
- Waveguide photonic crystal structures based on volume gratings recorded in a nanocomposite are characterized by the presence of resonance peaks in the reflection spectra. Spectral position of the resonances depends on the angle of incidence of radiation on the structure. The reflection coefficient reaches 23%.

The spectral half-width of the resonance peak does not exceed 0.012 nm and is limited by the resolution of spectral equipment.

- The Q-factor of photonic crystal structures determined from experimental measurements exceeds 50,000.
- The characteristics of the sensor based on nanocomposite photonic crystal waveguide structure are examined. A modified measurement method was proposed. The modified method provides the sensor sensitivity 122 nm/RIU and minimal detectable change in the refractive index Δnmin of about 1×10^{-4} RIU.

2.4 Characterization of Polymeric Films by Contact Angle Method

The sessile drop method concerned with the application of a drop, using a high-precision syringe, on the surface to be examined. The side view of the drop on the surface was taken via a camera. The contact angle is the balance between the actions of the three interfacial tensions in the three-phase points of contact. In the system considered, three separation surfaces can be distinguished (Fig. 2.1), each of which:

- Liquid-vapor surface and relative liquid-vapor surface tension γlv.
- Solid-vapor surface and the relative solid-vapor surface tension γsv.
- Solid-liquid surface and the relative surface tension γsl.

The possibility of estimating the surface tension of a solid across the contact angle is based on Young's proposed relationship (Fig. 2.2).

The DataPhysics OCA 15 PRO tool was used. The contact angle for each sample was obtained using the instrument software. The measurement range of the instrument is between 0° and 180°, with an accuracy of 0.01°, a surface energy between 1×10/m and 2×10^3/m and a resolution of ± 0.01 mN/m. For each sample 10 drops of ultrapure water degassed with volume 3 µl were delivered and with a dosing rate of 1 µl/ate using a 1× magnification.

In the particular case of water, we can define:

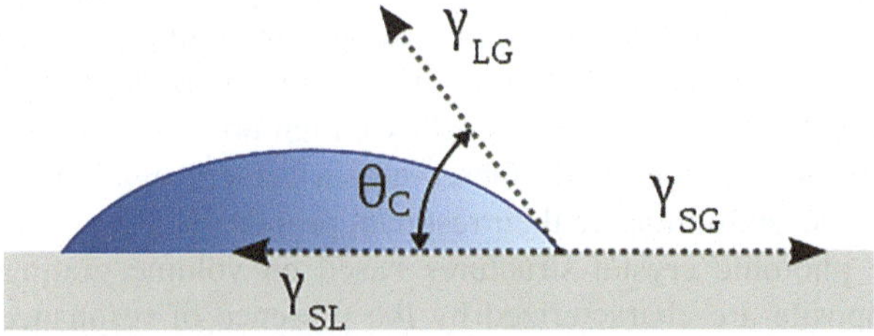

Fig. 2.1 The three interfacial tensions (γ) and contact angle (θc)

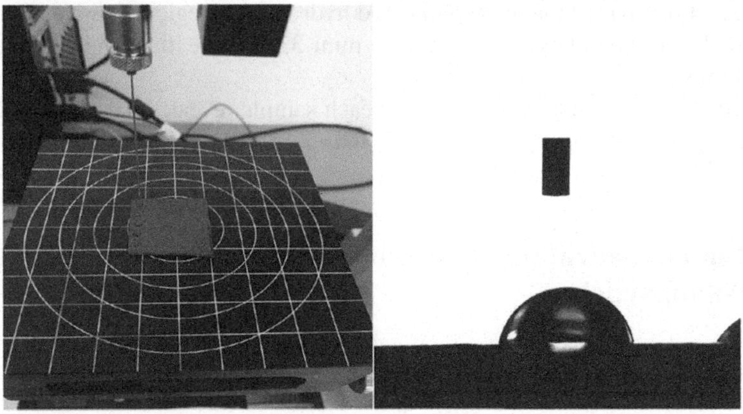

Fig. 2.2 Deposition of drop (left) in DataPhysics Optical Contact Angle (OCA) 15 PRO setup and an optical image (right) of a drop on a substrate

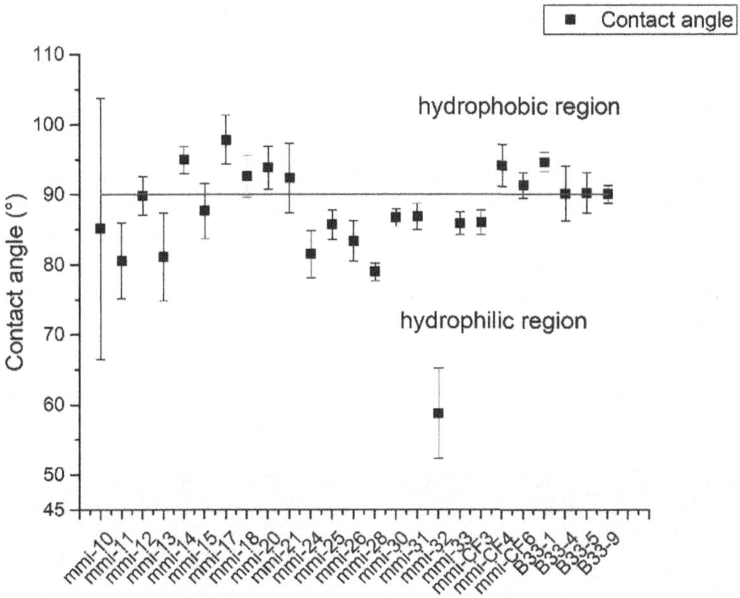

Fig. 2.3 Results of Contact angle

- Hydrophilic surface 0° < θ < 90°.
- 90° hydrophobic surface < θ < 150°.
- Superhydrophobic surface θ > 150°.

Furthermore, the sessile drop method is suitable to check the homogeneity of a surface analyzing the standard deviation.

The results of the analysis are reported in the graph (Fig. 2.3).

The red line divided the hydrophilic and hydrophobic region: samples are placed all around this boundary, excepted for mmi-32 sample that show a moderate hydrophilicity.

The analysis of standard deviation for each sample revealed a high homogeneity of samples, excepted for mmi-10,-11,-13 and -32.

2.5 Functionalization of Graphene Nanoplatelets with Metal Nanoparticles

In order to functionalize graphene nanoplatelets (GNP), as obtained by the method developed at INFN by the NEXT group [2–19], with rare earth, we have carried out a preliminary study of electrochemical deposition of a typical metal, copper (Cu), to demonstrate the capability of GNP to reduce and bind metal nanoparticles.

A Palmsens potentiostat/galvanostat was used for the electrodeposition experiment. The GNP were used as negative electrode to guarantee the efficacy of electrochemical deposition.

After deposition, the copper is present as single nanostructure, with an average diameter 100 nm, covering the most part of the surface of GNP as visible from SEM images Fig. 2.4

We have explored the possibility to change the covered area and nanoparticle dimension adding other component, as a surfactant (CTAB) and a polymer (PVP).

Using CTAB, the coverage appeared inhomogeneous and nanoparticles coalesced in crystalline structure, as shown in SEM images Fig. 2.5

Nevertheless, some interesting formation of Cu nanoparticles appeared on the edge of GNP (Fig. 2.6).

The presence of PVP during the deposition has affected the deposition of Cu, too. With respect to CTAB, the GNP surface coverage appeared most homogeneous

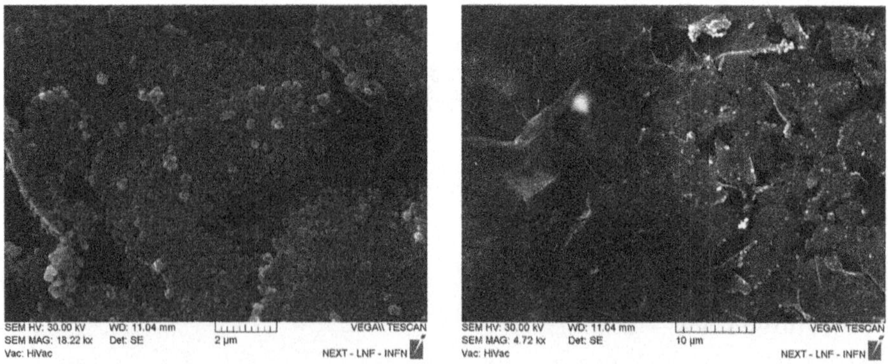

Fig. 2.4 SEM images of the pristine GNP film after electrodeposition

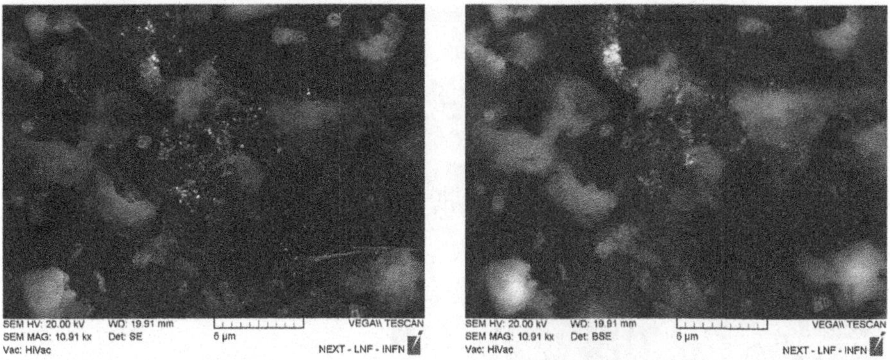

Fig. 2.5 SEM images of the GNP:CTAB film after electrodeposition

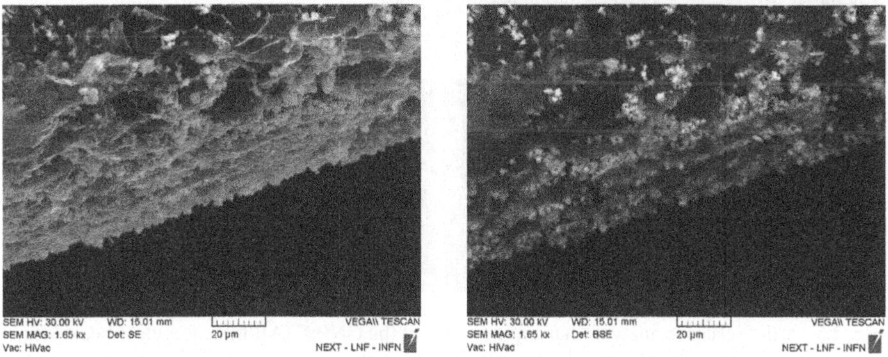

Fig. 2.6 SEM images of the edge area of the GNP:CTAB film

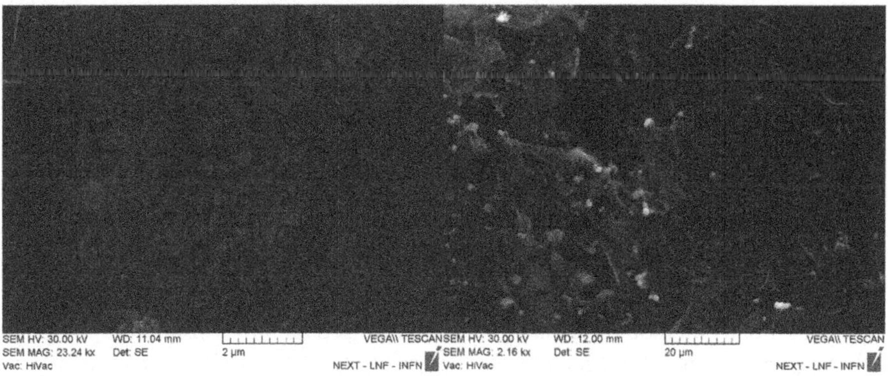

Fig. 2.7 SEM images of the GNP:PVP film after electrodeposition

than CTAB sample and less than pristine ones. The structure appeared less define than pristine one and coated from PVP, see Fig. 2.7.

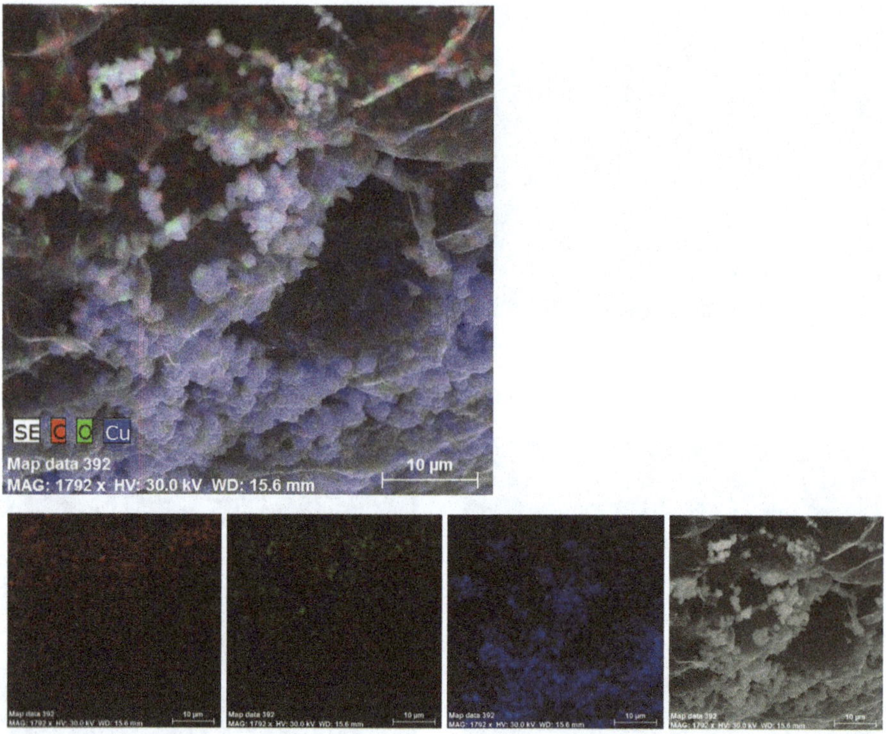

Fig. 2.8 X fluorescence analysis in mapping mode of the GNP:CTAB film after electrodeposition; the corresponding SEM image analyzed (rightmost panel)

Finally, we have carried out a microanalysis on Cu nanoparticles to control the oxidation degree. The results, shown in Fig. 2.8, revealed a partial oxidation of metal, green color: this result is comfortable for next step, in which we will decorate GNP with rare earth to obtain rare earth oxides.

2.6 Optimization of GNP Deposition

In order to deposit functionalized and pristine GNP, we have preliminarily studied spray coating as method of deposition. We have used an airbrush in order to obtain a homogeneous and thin coating of the substrates and GNP, both pristine and stabilized with Arabic gum (AG), as "paint" to modulate the wettability of the photonic crystal using hydrophobic (GNP) or hydrophilic (AG) nanoparticles. To the aim of tuning the wettability of the substrate, an important role can be played by the distribution of the nanoparticles, hence by their pattern as formed on the substrate.

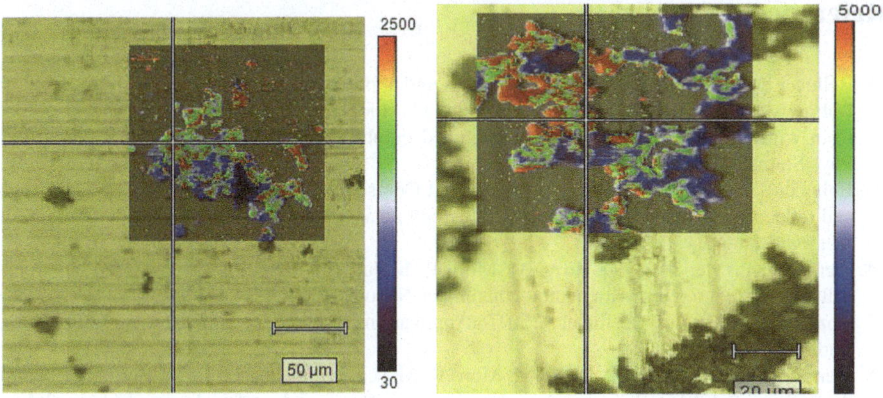

Fig. 2.9 Raman mapping of pristine (left) and AG (right) stabilized GNP

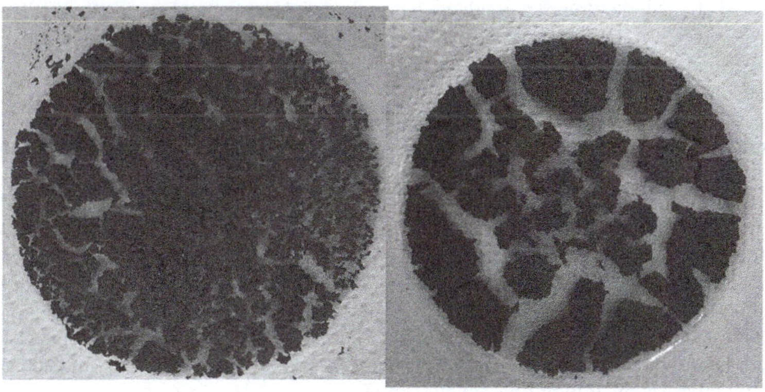

Fig. 2.10 Preliminary results of deposition by de-wetting, using isopropylic alcool as medium, and accelerating de-wetting in oven

We have checked the coating using a Raman mapping. The following maps (Fig. 2.9) show a spread-out droplet made of hydrophobic (GNP) or hydrophilic (AG) nanoparticles, coating the photonic crystal substrate: anyway, the thickness of film seems to be thin.

Recently, we have focused our attention on the nanoparticles self-assembling mediated by de-wetting process, that leads to the formation of a peculiar pattern [20]. Following the procedure outlined in works cited in the cited review article [20], we have obtained very preliminary results, using isopropylic alcool as medium, and accelerating de-wetting in oven, see Fig. 2.10.

Acknowledgments I acknowledge partial funding from NATO SPS-G5351. I wish to thank A. Cataldo, T. Smirnova, O. Sakhno, V. Fito, for their collaboration to the present investigation.

References

1. Bellucci S (2005) Carbon nanotubes: physics and applications. Phys Status Solidi (C) 2(1):34
2. Dabrowska A, Bellucci S, Cataldo A, Micciulla F, Huczko A (2014) Nanocomposites of epoxy resin with graphene nanoplates and exfoliated graphite: Synthesis and electrical properties. Phys Status Solidi (B) 251(12):2599
3. Bellucci S, Bovesecchi G, Cataldo A, Coppa P, Corasaniti S, Potenza M (2019) Transmittance and reflectance effects during thermal diffusivity measurements of gnp samples with the flash method. Materials 12(5):696
4. Pierantoni L, Mencarelli D, Bozzi M, Moro R, Bellucci S (2014) Graphene-based electronically tuneable microstrip attenuator. Nanomater Nanotechnol 4:1
5. Bozzi M, Pierantoni L, Bellucci S (2015) Applications of graphene at microwave frequencies. Radioengineering 24:661
6. Pierantoni L, Mencarelli D, Bozzi M, Moro R, Moscato S, Perregrini L, Micciulla F, Cataldo A, Bellucci S (2015) Broadband microwave attenuator based on few layer graphene flakes. IEEE Trans Microw Theory Tech 63(8):2491
7. Maffucci A, Micciulla F, Cataldo A, Miano G, Bellucci S (2016) Bottom-up realization and electrical characterization of a graphene-based device. Nanotechnology 27(9):095204
8. Yasir M, Bozzi M, Perregrini L, Bistarelli S, Cataldo A, Bellucci S (2016) Innovative tunable microstrip attenuators based on few-layer graphene flakes. In: 16th Mediterranean microwave symposium (MMS), pp 1–4
9. Yasir M, Bistarelli S, Cataldo A, Bozzi M, Perregrini L, Bellucci S (2017) Enhanced tunable. Microstrip attenuator based on few layer graphene flakes. IEEE Microw Wirel Components Lett 27(4):332
10. Yasir M, Savi P, Bistarelli S, Cataldo A, Bozzi M, Perregrini L, Bellucci S (2017) A planar antenna with voltage-controlled frequency tuning based on few-layer graphene. IEEE Antennas Wirel Propag Lett 16:2380
11. Yasir M, Bistarelli S, Cataldo A, Bozzi M, Perregrini L, Bellucci S (2018) Tunable phase shifter based on few-layer graphene flakes. IEEE Microw Wirel Components Lett 29(1):47
12. Levin V, Morokov E, Petronyuk Y, Cataldo A, Bistarelli S, Micciulla F, Bellucci S (2017) Cluster microstructure and local elasticity of carbon-epoxy nanocomposites studied by impulse acoustic microscopy. Polym Eng Sci 57(7):697
13. Levin V, Petronyuk Y, Morokov E, Chernozatonskii L, Kuzhir P, Fierro V, Celzard A, Mastrucci M, Tabacchioni I, Bistarelli S, Bellucci S (2016) The cluster architecture of carbon in polymer nanocomposites observed by impulse acoustic microscopy. Phys Status Solidi (B) 253(10):1952
14. Repetsky SP, Vyshyvana IG, Kuznetsova EY, Kruchinin SP (2018) Energy spectrum of graphene with adsorbed potassium atoms. Int J Mod Phys B 32:1840030
15. Repetsky SP, Vyshyvana IG, Kruchinin SP, Bellucci S (2018) Influence of the ordering of impurities on the appearance of an energy gap and on the electrical conductance of graphene, Sci Rep 8:9123
16. Rodionov VE, Shnidko IN, Zolotovsky A, Kruchinin SP (2013) Electroluminescence of Y_2O_3:Eu and Y_2O_3:Sm films. Mater Sci 31:232
17. Levin V, Petronyuk Y, Morokov E, Chernozatonskii L, Kuzhir P, Fierro V, Celzard A, Bellucci S, Bistarelli S, Mastrucci M, Tabacchioni I (2016) Bulk microstructure and local elastic properties of carbon nanocomposites studied by impulse acoustic microscopy technique. AIP Conf Proc 1736(1):020056
18. Levin VM, Petronyuk YS, Morokov ES, Celzard A, Bellucci S, Kuzhir PP (2015) What does see the impulse acoustic microscopy inside nanocomposites? Phys Proc 70:703
19. Bellucci S, Maffucci A, Maksimenko S, Micciulla F, Migliore M et al (2018) Electrical permittivity and conductivity of a graphene nanoplatelet contact in the microwave range. Materials 11(12):2519
20. Stannard A (2011) Dewetting-mediated pattern formation in nanoparticle assemblies. J Phys Condens Matter 23:083001

Chapter 3
Impurity Ordering Effects on Graphene Electron Properties

S. P. Repetsky, I. G. Vyshyvana, S. P. Kruchinin, R. M. Melnyk, and A. P. Polishchuk

Abstract The impurity ordering effects on graphene electron properties within a one-band model of strong coupling are studied. We carried out analytical and numerical calculations of the energy spectrum of electrons in the cases of low and high orderings. The limiting case of weak scattering for the varying degrees of scattering potential is analyzed. The ordering of impurity atoms causes the appearance of a gap in the energy spectrum of electrons. The gap width is proportional to the ordering parameter and the scattering potential for moderate magnitudes of the latter, but, as the scattering potential increases, its more complex behavior is observed. It is established that the regions of localization of electronic states are at the edges of the gap and the edges of the energy spectrum. The case of weak scattering potential allows the analytic investigation at the gap edges. Within the Lifshitz one-electron tight-binding model, the electrical conductivity of graphene is investigated. When the Fermi level enters the gap region of the energy spectrum, the electrical conductivity becomes zero, and the metal-dielectric transition occurs. If the Fermi level is in the region of the energy band, the electron relaxation time and electrical conductivity go to infinity, when the order parameter reaches its maximum value.

Keywords Graphene · Energy gap · Density of states · Ordering parameter · Green's function · Metal-insulator transition

---

S. P. Repetsky · I. G. Vyshyvana
Institute of High Technologies, Taras Shevchenko National University of Kyiv, Kyiv, Ukraine

S. P. Kruchinin (✉)
Bogolyubov Institute for Theoretical Physics, Kyiv, Ukraine

R. M. Melnyk
National University of Kyiv-Mohyla Academy, Kyiv, Ukraine

A. P. Polishchuk
Aerospace Faculty, National Aviation University, Kyiv, Ukraine

J. Bonča, S. Kruchinin (eds.), *Advanced Nanomaterials for Detection of CBRN*,
NATO Science for Peace and Security Series A: Chemistry and Biology,
https://doi.org/10.1007/978-94-024-2030-2_3

3.1 Introduction

Last time, the possibility of a targeted modification of graphene with the help of purposely-introduced impurities, created defects, and atoms or chemical functional groups deposited on a surface attracts a particular attention. In this case, broad potentialities for a change in the physical properties of graphene are opened due to the controlled introduction of impurities by the method of ionic implantation. Thus, graphene becomes a basis generating a new class of functional materials. Such materials find sometimes the unexpected applications to nanoelectromechanical systems, systems of accumulation of hydrogen, etc. It is expected that graphene has all possibilities to become a successor of silicon in electronic devices in the near future. This would allow one to significantly increase the level of their miniaturization and working frequencies. The quasirelativistic spectrum of charge carriers underlies the uniqueness of graphene, but hampers, at the same time, the use of graphene in field transistors due to the absence of a gap in its spectrum. It is known that the impurities can induce the appearance of a gap in the energy spectrum. The width of the gap depends on the type of impurities and their concentration.

Most studies of the energy spectrum of graphene are based on the density functional theory. The most significant achievements are related to the self-consistent meta-gradient approximation and the method of projection of adjoint waves [1] realized in the VASP and GAUSSIAN softwares [1]. The numerical calculations executed by this method showed the opening of a gap in the energy spectrum of graphene caused by the presence of an impurity. However, the clarification of the nature of this effect requires, in addition to the mentioned numerical calculations, the analytic studies of the influence of impurities on the energy spectrum and properties of graphene.

The electronic structures of the isolated monolayer of graphene, two- and three-layer graphenes, and graphene grown on ultrathin layers of hexagonal boron nitride (h-BN) were calculated in [2] in the frame of the density functional theory with the use of the method of pseudopotential. It was shown that the energy gap 57 meV in width appears in graphene grown on a monolayer of h-BN.

Graphenes with impurities of aluminium, silicon, phosphorus, and sulfur were studied within the analogous method in work [3], where it was shown, in particular, that graphene with a 3% impurity of phosphorus has a gap 0.67 eV in width. In work [4] with the use of the QUANTUM-ESPRESSO software, the possibilities of the opening of a gap in the energy spectrum of graphene at the introduction of the impurities of boron and nitrogen (the gap width is 0.49 eV), as well as the impurities of atoms of boron and atoms of lithium adsorbed on the surface (the gap width is 0.166 eV), were demonstrated.

It is obvious that it is insufficient to restrict ourselves by numerical calculations in order to understand the nature of the influence of impurities on the energy spectrum and properties of graphene. They should be also described within a simple, but adequate model presenting the exact analytic solutions.

In the Lifshitz tight-binding one-electron model, the theory of reconstruction of the spectrum of graphene caused by an increase in the concentration of point impurities was developed in work [5]. Moreover, the possibility of the metal-dielectric transition in such a system was predicted. The results of the analytic consideration of a reconstruction of the spectrum were confirmed with the help of a numerical experiment. This made it possible to verify the existence of a quasigap filled with localized states and showed its dominant role in the localization of the scattering by pairs and triples of impurity centers.

In [6, 7], the splitting in the energy spectrum of graphene with a zigzag boundary was studied. This spectrum describes the electron waves that propagate along the boundary and decay with an increasing distance from it. It was shown that the electronic spectra of graphene with isolated vacancies exhibit a similar behavior. The split electronic energy spectrum is accompanied by the formation of a sharp resonance state on the local density curve. It was shown that a similar resonance also arises in the phonon spectrum near the intersection point of the acoustic and optical branches of elastic waves polarized perpendicularly to the plane of the graphene monolayer. In the frequency range under consideration, these phonons practically do not interact with differently polarized phonons, have high group velocities, and make a dominant contribution to the electron-phonon interaction. The presented results demonstrate the possibility of increasing the critical temperature of the superconducting transition in graphene by a controlled creation of defects such as a vacancy or a zigzag border.

The numerical calculations within the Kubo–Greenwood quantum-mechanical formalism in the Lifshitz tight-binding one-electron model were performed in [8–12] to study the influence of impurity atoms or atoms adsorbed on the surface on the electronic structure and electrical conductance of graphene. In those works, the method of reduction of the Hamiltonian to the three-diagonal form was developed to study the influence of completely ordered impurity atoms on the energy spectrum and electrical conductance of graphene in the ballistic and diffusive modes of conductance. In work [10], it was found that the gap 0.45 eV in width appears in the energy spectrum of electrons of graphene deposited on a potassium substrate. There, it was assumed that the appearance of this gap is associated with a change in the symmetry of the crystal. This assumption was corroborated in work [13], where the influence of the atomic ordering on the energy spectrum and electrical conductance of an alloy was analytically studied in the Lifshitz tight-binding one-electron model. It was also established [13] that, for a long-range ordering of the alloy, the gap arises in the energy spectrum of electrons. The gap width is equal to the difference of the scattering potentials of components of the alloy. It was also found that the metal-dielectric transition appears in the alloy in the case where the Fermi level falls in the domain of the gap at a long-range atomic ordering.

When as the gap appears in the energy spectrum of graphene in the case where the Fermi level falls in the domain of the gap, the velocity of an electron on the Fermi level can decrease. This leads to a decrease in the electrical conductance, which can worsen the functional characteristics of graphene as a material for field transistors.

Within the Lifshitz tight-binding one-electron model, the influence of the ordering of impurities on the energy spectrum and electrical conductance of graphene was considered in work [14]. It was established that the ordering of substitutional atoms on nodes of the crystal lattice causes the appearance of a gap $\eta\,|\delta|$ in width in the energy spectrum of graphene centered at the point $y\,\delta$, where η is the ordering parameter, δ is the difference of the scattering potentials of impurity atoms and carbon, and y is the impurity concentration. If the Fermi level falls in the domain of the gap, then the electrical conductance $\sigma_{\alpha\alpha} \to \infty$ at the ordering of graphene, i.e., the metal-dielectric transition arises. If the Fermi level is located outside the gap, then the electrical conductance increases with the order parameter η by the rule $\sigma_{\alpha\alpha} \sim \left(y^2 - \frac{1}{4}\eta^2\right)^{-1}$. As the ordering of impurity atoms $\eta \to 1$ at the concentration $y = 1/2$, the electrical conductance of graphene $\sigma_{\alpha\alpha} \to \infty$, i.e., graphene transits in the state of ideal conductance. We note that the conclusions in work [14] were based on the results of analytic studies of the energy spectrum and electrical conductance of graphene performed in the approximation of coherent potential. However, the convergence region of the decomposition cluster used in [14] for the Green function and the applicability of the coherence potential approximation were not analyzed. Both themes will be studied in the present work.The features of the energy spectrum of electrons in the region of the gap arising upon the ordering of impurity atoms will be investigated.

3.2 Theoretical Model

The Hamiltonian in a Lifshitz one-electron strong bond model describing single-electron states of graphene with substitutional impurity can be represented as [14]

$$H = \sum_{ni} |ni\rangle v_{ni} \langle ni| + \sum_{ni,n'i' \neq ni} |ni\rangle h_{ni,n'i'} \langle n'i'|, \tag{3.1}$$

where $h_{ni,n'i'}$ is a non-diagonal matrix element of the Hamiltonian (jump integral) in the Vane representation, which is independent of the random distribution of atoms in the assumed approximation of the diagonal disorder, the diagonal matrix element v_{ni} is v^A or v^B depending on which atom A or B is at the node ni, n is the number of elementary cell, and i is the number of the sublattice node in the unit cell.

In expression (3.1), we now add and subtract the translational invariant operator $\sum_{ni} |ni\rangle \sigma_i \langle ni|$, where σ_i is the diagonal matrix element of the Hamiltonian of some effective ordered medium (coherent potential), which depends on the sublattice number. As a result, the graphene Hamiltonian can be represented as

$$
\begin{aligned}
H &= \tilde{H} + \tilde{V}, \\
\tilde{H} &= \sum_{ni} |ni\rangle \sigma_i \langle ni| + \sum_{ni,n'i' \neq ni} |ni\rangle h_{ni,n'i'} \langle n'i'|, \\
\tilde{V} &= \sum_{ni} \tilde{v}_{ni}, \quad \tilde{v}_{ni} = |ni\rangle (v_{ni} - \sigma_i) \langle ni|.
\end{aligned} \tag{3.2}
$$

Green's function is an analytic function in the upper half-plane of the complex energy z. The function is defined by the expression

$$G(z) = (z - H)^{-1} \tag{3.3}$$

and satisfies the equation

$$G = \tilde{G} + \tilde{G} T \tilde{G}, \tag{3.4}$$

where $\tilde{G}$ is Green's function of the effective medium corresponding to the Hamiltonian $\tilde{H}$ in (3.2). The scattering matrix T can be represented as an infinite series [13]

$$T = \sum_{(n_1 i_1)} t^{n_1 i_1} + \sum_{(n_1 i_1) \neq (n_2 i_2)} T^{(2)\, n_1 i_1, n_2 i_2} + \ldots \tag{3.5}$$

Here,

$$T^{(2)\, n_1 i_1, n_2 i_2} = \left[I - t^{n_1 i_1} \tilde{G} t^{n_2 i_2} \tilde{G} \right]^{-1} t^{n_1 i_1} \tilde{G} t^{n_2 i_2} \left[I + \tilde{G} t^{n_1 i_1} \right], \tag{3.6}$$

the scattering operator on one node

$$t^{n_1 i_1} = \left[I - \tilde{v}_{in} \tilde{G} \right]^{-1} \tilde{v}_{in}, \tag{3.7}$$

and I is identity matrix. The members of series (3.5) describe the processes of multiple scattering of electrons on clusters of one, two, three, etc. scattering centers.

As was shown in work [13], the contributions of the processes of electron scattering on clusters to the density of states and the electrical conductance decrease, as the number of atoms in a cluster increases. These contributions are guided by some small parameter $\gamma_i(\varepsilon)$. The parameter $\gamma_i(\varepsilon)$ is small in wide regions of the crystal characteristics, except for narrow energy intervals on the edges of the spectrum and on the edges of the energy gap. The expression for the specified $\gamma_i(\varepsilon)$ parameter will br given below.

Neglecting the contribution of scattering processes on clusters of three or more atoms that are small by the specified parameter $\gamma_i(\varepsilon)$, the density of one-electron states of graphene can be represented as [14]

$$g(\varepsilon) = \frac{1}{v} \sum_{i,\lambda} P^{\lambda 0i} g^{\lambda 0i}(\varepsilon),$$

$$g^{\lambda 0i}(\varepsilon) = -\frac{2}{\pi} \operatorname{Im} \left\{ \tilde{G} + \tilde{G} t^{\lambda 0i} \tilde{G} + \right.$$

$$\left. + \sum_{(lj) \neq (0i),\, \lambda'} P^{\lambda' lj / \lambda 0i} \tilde{G} \left[t^{\lambda' lj} + T^{(2)\lambda 0i,\, \lambda' lj} \right] \tilde{G} \right\}_{0i,0i}, \tag{3.8}$$

where $v = 2$ is the number of graphene sublattices.

Using the Cuban–Greenwood formula and neglecting the contribution of scattering processes on clusters of three or more atoms, we present the static electrical conductivity of graphene ($T = 0$) [14] as

$$\sigma_{\alpha\beta} = -\frac{e^2 \hbar}{2\pi \Omega_1} \sum_{s,s'=+,-} (2\delta_{ss'} - 1) \sum_i \left\{ [v_\beta \tilde{K}(\varepsilon^s, v_\alpha, \varepsilon^{s'})] + \right.$$

$$+ \sum_\lambda P^{\lambda 0i} \tilde{K}(\varepsilon^{s'}, v_\beta, \varepsilon^s) t^{\lambda 0i}(\varepsilon^s) \tilde{K}(\varepsilon^s, v_\alpha, \varepsilon^{s'}) t^{\lambda 0i}(\varepsilon^{s'}) +$$

$$+ \sum_\lambda P^{\lambda 0i} \sum_{lj \neq 0i,\, \lambda'} P^{\lambda' lj / \lambda 0i} \left[[\tilde{K}(\varepsilon^{s'}, v_\beta, \varepsilon^s) v_\alpha \tilde{G}(\varepsilon^{s'})] T^{(2)\lambda 0i, \lambda' lj}(\varepsilon^{s'}) + \right.$$

$$+ [\tilde{K}(\varepsilon^s, v_\alpha, \varepsilon^{s'}) v_\beta \tilde{G}(\varepsilon^s)] T^{(2)\lambda 0i, \lambda' lj}(\varepsilon^s) +$$

$$+ \tilde{K}(\varepsilon^{s'}, v_\beta, \varepsilon^s) \left[t^{\lambda' lj}(\varepsilon^s) \tilde{K}(\varepsilon^s, v_\alpha, \varepsilon^{s'}) t^{\lambda 0i}(\varepsilon^{s'}) + \right.$$

$$+ (t^{\lambda}_{0i}(\varepsilon^s) + t^{\lambda'}_{lj}(\varepsilon^s)) \tilde{K}(\varepsilon^s, v_\alpha, \varepsilon^{s'}) T^{(2)\lambda 0i, \lambda' lj}(\varepsilon^{s'}) +$$

$$+ T^{(2)\lambda' lj, \lambda 0i}(\varepsilon^s) \tilde{K}(\varepsilon^s, v_\alpha, \varepsilon^{s'}) t^{\lambda 0i}(\varepsilon^{s'}) +$$

$$+ T^{(2)\lambda' lj, \lambda 0i}(\varepsilon^s) \tilde{K}(\varepsilon^s, v_\alpha, \varepsilon^{s'}) T^{(2)\lambda 0i, \lambda' lj}(\varepsilon^{s'}) +$$

$$\left. \left. \left. + T^{(2)\lambda' lj, \lambda 0i}(\varepsilon^s) \tilde{K}(\varepsilon^s, v_\alpha, \varepsilon^{s'}) T^{(2)\lambda' lj, \lambda 0i}(\varepsilon^{s'}) \right] \right] \right\}_{0i,0i} \Bigg|_{\varepsilon=\mu} \tag{3.9}$$

$\tilde{K}(\varepsilon^s, v_\alpha, \varepsilon^{s'}) = \tilde{G}(\varepsilon^s) v_\alpha \tilde{G}(\varepsilon^{s'})$, $\tilde{G}(\varepsilon^+) = \tilde{G}_r(\varepsilon)$, $\tilde{G}(\varepsilon_1^-) = \tilde{G}_a(\varepsilon) = \tilde{G}_r^*(\varepsilon)$, $\tilde{G}_r(\varepsilon)$ and $\tilde{G}_a(\varepsilon)$ are the retarded and advanced Green's functions, $\Omega_1 = 2\Omega_0$ is the graphene unit cell volume, Ω_0 is the one atom volume,

$$\tilde{G}_{njn'j'}(\varepsilon) = \frac{1}{N} \sum_{\mathbf{k}} \tilde{G}_{jj'}(\mathbf{k}, \varepsilon) \exp(i\mathbf{k}(r_{n'j'} - r_{nj})), \tag{3.10}$$

$\tilde{G}_{jj'}(\mathbf{k}, \varepsilon)$ is the Fourier transform of Green's function, r_{nj} is the position vector of the node nj. The wave vector $\mathbf{k}$ changes within the Brillouin zone.

The operator of electron velocity α-projection is given by the expression

$$\upsilon_{\alpha ii'}(\mathbf{k}) = \frac{1}{\hbar}\frac{\partial h_{ii'}(\mathbf{k})}{\partial k_\alpha}, \quad (3.11)$$

the Fourier transform of the jump integral $h_{jj'}(\mathbf{k})$ is calculated for the nearest atom neighbors

$$h_{jj'}(\mathbf{k}) = \gamma_1 \sum_{n'\neq n} \exp(i\mathbf{k}(r_{n'j'} - r_{nj})), \quad (3.12)$$

$\gamma_1 = (pp\pi)$ is the jump integral [15], and r_{nj} is the position vector of the node nj.

The Fermi level μ is determined from the equation

$$\langle Z \rangle = \int_{-\infty}^{\mu} g(\varepsilon)\, d\varepsilon. \quad (3.13)$$

Here, $\langle Z \rangle$ is the average number of electrons per atom whose energy values belong to the energy band.

In expressions (3.8) and (3.9), $P^{\lambda 0i}$ is the probability of filling the node $0i$ of the crystal lattice, $i = 1, 2$, by atoms of sort $\lambda = A, B$,

$$P^{B01} = y_1 = y + \frac{1}{2}\eta, \quad P^{B02} = y_2 = y - \frac{1}{2}\eta, \quad P^{A01} = 1 - P^{B01}, \quad (3.14)$$

y is the concentration of impurity atoms, and η is the far order parameter.

In expressions (3.8) and (3.9), $P^{\lambda' lj/\lambda 0i}$ is the probability of filling the node lj by an atom of sort λ' provided that the atom of the sort λ fills the node $0i$, and $P^{\lambda' lj/\lambda 0i}$ is the parameter of paired interatomic correlations in atoms of crystalline lattice nodes filled with atoms.

The probabilities are determined by the interatomic pair correlations $\varepsilon^{BB}_{lj,0i}$ via [16]

$$P^{\lambda'/\lambda}_{lj,0i} = P^{\lambda'}_{lj} + \frac{\varepsilon^{BB}_{lj,0i}}{P^{\lambda}_{0i}} (\delta_{\lambda' B} - \delta_{\lambda' A})(\delta_{\lambda B} - \delta_{\lambda A}), \quad (3.15)$$

where δ is the Kronecker delta-function. Note that the interatomic pair correlations also satisfy

$$\varepsilon^{BB}_{lj,0i} = \langle (c^B_{lj} - c^B_j)(c^B_{0i} - c^B_i) \rangle. \quad (3.16)$$

Here, c_{lj}^B is a bit randomize number which takes the value of one, if an atom of sort B is at the node, or zero otherwise, $A_j^B = \left\langle A_{0j}^B \right\rangle = P^{B0j}$. Brackets mean the averaging over the distribution of impurity atoms at the nodes of the crystalline lattice.

The coherent potential is determined by the condition $\left\langle t^{n_1 i_1} \right\rangle = 0$. Hence, the equation for the coherent potential reads [14]

$$\sigma_i = \langle \upsilon_i \rangle - (\upsilon_A - \sigma_i)\, \tilde{G}_{0i,0i}(\varepsilon)\, (\upsilon_B - \sigma_i); \quad \langle \upsilon_i \rangle = (1 - y_i)\, \upsilon_A + y_i\, \upsilon_B. \tag{3.17}$$

Putting $\upsilon_A = 0$ in expression (3.17), we obtain

$$\langle \upsilon_i \rangle = y_i\, \delta, \tag{3.18}$$

where

$$\delta = \upsilon_B - \upsilon_A \tag{3.19}$$

is the difference of the scattering potentials of graphene components.

For the analytic description of the energy spectrum and the electrical conductivity of graphene, we consider only the first constituents in expressions (3.8), (3.9), which make the major contribution to the density of states and the electrical conductivity. Thus,

$$g(\varepsilon) = -\frac{2}{\pi \nu} \mathrm{Im} \sum_i \tilde{G}_{0i,0i}(\varepsilon) = -\frac{2}{\pi \nu N} \mathrm{Im} \sum_{i,\mathbf{k}} \tilde{G}_{ii}(\mathbf{k}, \varepsilon), \tag{3.20}$$

$$\begin{aligned} \sigma_{\alpha\alpha} &= -\frac{e^2 \hbar}{2\pi V_1} \sum_i \left[\upsilon_\alpha (\tilde{G}(\varepsilon) - \tilde{G}^*(\varepsilon))\, \upsilon_\alpha (\tilde{G}(\varepsilon) - \tilde{G}^*(\varepsilon)) \right]_{0i,0i} = \\ &= -\frac{e^2 \hbar}{2\pi V_1 N} \sum_{i,\mathbf{k}} \left[\upsilon_\alpha(\mathbf{k}) (\tilde{G}(k, \varepsilon) - \tilde{G}^*(\mathbf{k}, \varepsilon))\, \upsilon_\alpha(\mathbf{k}) (\tilde{G}(\mathbf{k}, \varepsilon) - \right. \\ &\left. - \tilde{G}^*(\mathbf{k}, \varepsilon)) \right]_{0i,0i}. \end{aligned} \tag{3.21}$$

The wave vector in formulas (3.20), (3.21) varies within the Brillouin zone. The operator of the α-projection of the velocity

$$\upsilon_{\alpha i i'}(\mathbf{k}) = \frac{1}{\hbar} \frac{\partial h_{ii'}(\mathbf{k})}{\partial k_\alpha}, \tag{3.22}$$

The Fourier transform of the jump integral $h_{ii'}(\mathbf{k})$ is calculated for the nearest atom neighbors.

The Fourier transform of Green's function is as follows:

$$\tilde{G}_{11}(\mathbf{k},\varepsilon)=\frac{\varepsilon-\sigma_2}{D(\mathbf{k},\varepsilon)},\quad \tilde{G}_{12}(\mathbf{k},\varepsilon)=\frac{h_{21}(\mathbf{k})}{D(\mathbf{k},\varepsilon)},$$
$$\tilde{G}_{21}(\mathbf{k},\varepsilon)=\frac{h_{12}(\mathbf{k})}{D(\mathbf{k},\varepsilon)},\quad \tilde{G}_{22}(\mathbf{k},\varepsilon)=\frac{\varepsilon-\sigma_1}{\varepsilon-\sigma_2}\tilde{G}_{11}(\mathbf{k},\varepsilon), \tag{3.23}$$
$$D(\mathbf{k},\varepsilon)=(\varepsilon-\sigma_1)(\varepsilon-\sigma_2)-h_{12}(\mathbf{k})h_{21}(\mathbf{k}).$$

In this model, the value of the wave vector lying in a region around the Dirac point is mainly due to the energy spectrum of electrons in the middle of the zone. The Brillouin zone has two such areas. For these areas,

$$h_{12}(\mathbf{k})=h_{21}(\mathbf{k})=\hbar\upsilon_{\mathrm{F}}k, \tag{3.24}$$

$\upsilon_{\mathrm{F}}=\dfrac{3|\gamma_1|a_0}{2\hbar}$ is the electron velocity at the Fermi level, $\gamma_1=(pp\pi)$ is the jump integral [15], and a_0 is the distance between the nearest neighbors.

Substituting (3.23), (3.24) in (3.20) and replacing the sum over the wave vector by an integral [14], we have

$$\tilde{G}_{01,01}(\varepsilon)=-\frac{S_1(\varepsilon-\sigma_2)}{\pi\hbar^2\upsilon_{\mathrm{F}}^2}\ln\sqrt{1-\frac{w^2}{(\varepsilon-\sigma_1)(\varepsilon-\sigma_2)}},$$
$$\tilde{G}_{02,02}(\varepsilon)=-\frac{S_1(\varepsilon-\sigma_1)}{\pi\hbar^2\upsilon_{\mathrm{F}}^2}\ln\sqrt{1-\frac{w^2}{(\varepsilon-\sigma_1)(\varepsilon-\sigma_2)}}, \tag{3.25}$$

$w=3|\gamma_1|$ is the half-width of the energy band of pure graphene, $S_1=\frac{3\sqrt{3}a_0^2}{2}$ is the area of the unit cell of graphene.

Consider the effect of ordering atoms on the energy spectrum of graphene electrons with a substitutional admixture in the limiting case of weak scattering $|\delta/w|\ll 1$. In this case, the solution of the system of equations (3.17), (3.25) is [14]

$$\tilde{G}_{01,01}(\varepsilon)=-\frac{S_1(\varepsilon-\sigma_2')}{\pi\hbar^2\upsilon_{\mathrm{F}}^2}\ln\sqrt{1-\frac{w^2}{(\varepsilon-\sigma_1')(\varepsilon-\sigma_2')}},$$
$$\tilde{G}_{02,02}(\varepsilon)=-\frac{S_1(\varepsilon-\sigma_1')}{\pi\hbar^2\upsilon_{\mathrm{F}}^2}\ln\sqrt{1-\frac{w^2}{(\varepsilon-\sigma_1')(\varepsilon-\sigma_2')}},$$
$$\sigma_1'=y_1\delta-y_1(1-y_1)\delta^2\frac{S_1(\varepsilon-y_2\delta)}{\pi\hbar^2\upsilon_{\mathrm{F}}^2}\ln\sqrt{1-\frac{w^2}{(\varepsilon-y_1\delta)(\varepsilon-y_2\delta)}}, \tag{3.26}$$
$$\sigma_2'=y_2\delta-y_2(1-y_2)\delta^2\frac{S_1(\varepsilon-y_1\delta)}{\pi\hbar^2\upsilon_{\mathrm{F}}^2}\ln\sqrt{1-\frac{w^2}{(\varepsilon-y_1\delta)(\varepsilon-y_2\delta)}},$$
$$\mathrm{sign}(\varepsilon-\sigma_1')=-\mathrm{sign}(\varepsilon-\sigma_2');$$

and

$$
\begin{aligned}
\tilde{G}_{01,01}(\varepsilon) &= -\frac{S_1(\varepsilon-\sigma_2'-i\sigma_2'')}{\pi\hbar^2 \upsilon_F^2}\ln\sqrt{\frac{w^2}{|(\varepsilon-\sigma_1')(\varepsilon-\sigma_2')|}-1}-i\frac{S_1|\varepsilon-\sigma_2'|}{2\hbar^2\upsilon_F^2},\\
\tilde{G}_{02,02}(\varepsilon) &= -\frac{S_1(\varepsilon-\sigma_1'-i\sigma_1'')}{\pi\hbar^2 \upsilon_F^2}\ln\sqrt{\frac{w^2}{|(\varepsilon-\sigma_1')(\varepsilon-\sigma_2')|}-1}-i\frac{S_1|\varepsilon-\sigma_1'|}{2\hbar^2\upsilon_F^2},\\
\sigma_1' &= y_1\delta - y_1(1-y_1)\delta^2\frac{S_1(\varepsilon-y_2\delta)}{\pi\hbar^2\upsilon_F^2}\ln\sqrt{\frac{w^2}{|(\varepsilon-y_1\delta)(\varepsilon-y_2\delta)|}-1},\\
\sigma_1'' &= -y_1(1-y_1)\delta^2\frac{S_1|\varepsilon-y_2\delta|}{2\hbar^2\upsilon_F^2},\\
\sigma_2' &= y_2\delta - y_2(1-y_2)\delta^2\frac{S_1(\varepsilon-y_1\delta)}{\pi\hbar^2\upsilon_F^2}\ln\sqrt{\frac{w^2}{|(\varepsilon-y_1\delta)(\varepsilon-y_2\delta)|}-1},\\
\sigma_2'' &= -y_2(1-y_2)\delta^2\frac{S_1|\varepsilon-y_1\delta|}{2\hbar^2\upsilon_F^2},\\
&\mathrm{sign}(\varepsilon-\sigma_1') = \mathrm{sign}(\varepsilon-\sigma_2').
\end{aligned}
\tag{3.27}
$$

In (3.26)–(3.27), σ_i' and σ_i'' are the real and imaginary parts of the coherent potentials σ_i, $i = 1, 2$.

The analysis of formulae (3.26)–(3.27) shows that if the impurity atoms are ordered in the crystal lattice of graphene, then there is a gap of width $\eta|\delta|$ in the energy spectrum, and the center of the gap is at $y\delta$. The energy values of ε corresponding to the edges of the energy gap are determined from the equations $\varepsilon - \sigma_1' = 0$, $\varepsilon - \sigma_2' = 0$. It follows from Eqs. (3.14) that the maximum value of the ordering parameter is equal $\eta_{\max} = 2y$, $y < 1/2$. At the complete ordering of the impurity atoms, the width of the gap is equal to $2y|\delta|$, being proportional to the concentration of the impurity y and the modulus of the scattering potential of the graphene components δ. For $y = 1/2$, the width of the gap takes the maximum value. For $\delta > 0$ and $\delta < 0$, the gap is located, respectively, to the right and left of the Dirac point on the energy scale. As can be seen from formulas (3.20) and (3.26), the density of electron states $g(\varepsilon) = 0$ in the approximation of the coherent potential for this region of energy values.

As follows from expressions (3.20) and (3.27), the density of states goes to infinity in a vicinity of the edge of the gap. This is due to the presence of the first component in the expression for Green's function $\tilde{G}_{0i,0i}$ and the second component in the coherent potential σ_i' (3.27). The width of this energy region is [14]

$$
\left|\frac{\Delta\varepsilon(\eta)}{w}\right| = \frac{w}{\eta|\delta|}\exp\left(-\frac{2y\pi w^2}{3\sqrt{3}\eta\delta^2(1-y+\eta/2)(y-\eta/2)}\right);\quad 0<\eta\le 2y.
\tag{3.28}
$$

The peak width estimation (3.28) is made under the condition that the density of states on the slope of the peak is twice more than its value at the point of the adjacent minimum.

The distance from the peak to the edge of the energy zone increases with the probability of finding the impurity atoms on the carbon lattice y_2 and the impurity scattering parameter δ. The electron state density curve at the edge of the energy band also has a characteristic peak width (3.33). With increasing the probability of finding the impurity atoms on the carbon lattice y_2 (with decreasing the order parameter η), the impurity peak on the state energy density curve splits, which is caused by the interaction of the impurity atoms in the case of their high concentration. This is illustrated by the results of numerical calculations of the electron density in the case of a large value of the impurity atom concentration and the impurity scattering parameter in Fig. 3.1.

Beyond the specified peak, the density of states increases linearly with the distance to the gap edge [14]:

$$g(\varepsilon) = \frac{S_1(\varepsilon - y\delta)}{\pi \hbar^2 \upsilon_F^2}, \qquad \left|\frac{\Delta\varepsilon(\eta)}{w}\right| < \left|\frac{\varepsilon - \sigma_i'}{w}\right| \le \left|\frac{\delta}{w}\right|. \tag{3.29}$$

If the Fermi level falls into the gap region, then the number of free charge carriers tends to zero. In this case, when the impurity is ordered, the electrical conductivity $\sigma_{\alpha\alpha} \to 0$, as it follows from formulas (3.21) and (3.26), i.e., a metal-dielectric transition occurs.

Let us evaluate the electrical conductivity of graphene in the case where the Fermi level is out of the gap. Substituting (3.23), (3.24) in (3.20) and replacing the sum over the wave vector by an integral [14], we obtain

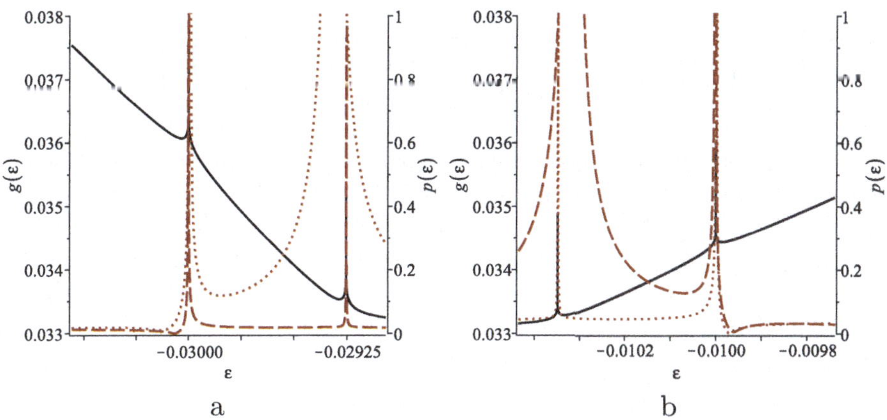

Fig. 3.1 Density of states $g(\varepsilon)$ (black line), parameter $p_i(\varepsilon)$ (red dotted line $i = 1$, red dashed line $i = 2$) as a function of the energy ε on the left edge of the gap (**a**) and on the right edge of the gap (**b**). The scattering parameter $\delta/w = -0.1$, the substitutional impurity concentration $y = 0.2$, the order parameter $\eta = 0.2$

$$\sigma_{\alpha\alpha} = \frac{2e^2\hbar\upsilon_F^2}{\pi^2 a_0^2 d\left(y^2 - \frac{1}{4}\eta^2\right)\delta^2}, \tag{3.30}$$

where d is the thickness of graphene. The factor d in the denominator of the right-hand side of formula (3.30) can be omitted.

In [13], it was shown that the contribution of electron scattering processes on clusters to the density of states and the electrical conductance are guided by some small parameter $p_i(\varepsilon)$, except for narrow energy intervals on the edges of the spectrum and on the edges of the energy gap. In [14], the analytic study of the effect of impurity ordering on the energy spectrum and the electrical conductance of graphene in the approximation of coherent potential was executed. In order to estimate the corrections of this approximation caused by the contribution of electron scattering processes on clusters of two, three, etc. atoms, the following parameter was introduced in [17]:

$$\begin{aligned} p_i(\varepsilon) &= \left|\left\langle\left(t^{0i}(\varepsilon)\right)^2\right\rangle \sum_{lj\neq 0i} \tilde{G}_{0i,lj}(\varepsilon)\,\tilde{G}_{lj,0i}(\varepsilon)\right|; \\ \left\langle\left(t^{0i}(\varepsilon)\right)^2\right\rangle &= (1-y_i)\left(t^{A0i}(\varepsilon)\right)^2 + y_i\left(t^{B0i}(\varepsilon)\right)^2. \end{aligned} \tag{3.31}$$

This parameter was analyzed in [13], where its following representation was given:

$$\begin{aligned} p_i(\varepsilon) &= \left|Q_i(\varepsilon)/(1+Q_i(\varepsilon))\right|; \\ Q_i(\varepsilon) &= -\frac{\left\langle\left(t^{0i}(\varepsilon)\right)^2\right\rangle}{1+\left\langle\left(t^{0i}(\varepsilon)\right)^2\right\rangle\left(\tilde{G}_{0i,0i}(\varepsilon)\right)^2}\times \\ &\times\left(\frac{1}{1+\left\langle\left(t^{0i}(\varepsilon)\right)^2\right\rangle\left(\tilde{G}_{0i,0i}(\varepsilon)\right)^2}\frac{d}{d\varepsilon}\tilde{G}_{0i,0i}(\varepsilon) + \left(\tilde{G}_{0i,0i}(\varepsilon)\right)^2\right). \end{aligned} \tag{3.32}$$

The parameter p_i is small, with the exception of narrow energy intervals at the edges of the gap. It follows from formula (3.27) that, as the energy value tends to the edge of the gap, $d\tilde{G}_{0i,0i}(\varepsilon)/d\varepsilon \to \infty$, and the parameter $p_i \to 1$ (3.32). The parameter $p_i(\varepsilon)$ may be greater than 1, if $1 + Q_i(\varepsilon)$ is close to zero (see Fig. 3.1).

The parameter $p_i(\varepsilon)$ takes values $p_i(\varepsilon) \geq 1/2$ in a narrow range at the edge of the energy gap:

$$\begin{aligned} \left|\frac{\Delta\varepsilon_1'(\eta)}{w}\right| &= \exp\left(-\frac{\pi w^4}{3^3\delta^4(1-y)^2(-\eta^2/4)(y^2-\eta^2/4)}\right); \\ \left|\frac{\Delta\varepsilon_2'(\eta)}{w}\right| &= \exp\left(-\frac{\pi w^2}{3\sqrt{3}\delta^2(1-y-\eta/2)(y+\eta/2)}\right). \end{aligned} \tag{3.33}$$

Expressions (3.33) are obtained from (3.7), (3.27), and (3.32). The characteristic $\Delta\varepsilon'(\eta)$ describes two regions of the localization of electron states at each edges of the gap (see Fig. 3.1). Thus, the processes of scattering on multisite clusters give a significant contribution to the density of states at the energies of electrons lying in the intervals (3.33).

We note that formulas (3.29), (3.30) for the density of states and electrical conductance of graphene cannot be used, if the Fermi level falls in the intervals of energies (3.33) at the gap edges.

The above expressions (3.29), (3.30) were obtained in the case of a small value of the scattering potential $|\delta/w| \ll 1$. The influence of the ordering of impurity atoms on the energy spectrum and the electrical conductivity of graphene for an arbitrary value of the scattering potential is more complex.

3.3 Calculation Results for a Particular Magnitude of the Scattering Potential

Figure 3.2 shows the results of numerical calculations of the density of state $g(\varepsilon)$ of graphene performed according to (3.8) for the following values: the substitutional impurity concentration $y = 0.2$, the scattering potential $\delta/w = -0.2$ and -0.6, the ordering parameter $\eta = 0$ and different values of paired interatomic correlations in the first coordination sphere $\varepsilon_{lj\ 0i}^{BB} = \varepsilon^{BB}$. Energy values are given in units of half-width of the energy band w. The electron density of states $g(\varepsilon)$ of graphene (Fig. 3.2, solid curve) is calculated in the approximation of the coherent potential. In this case, only the first component of the sum in formula (3.8) was taken into

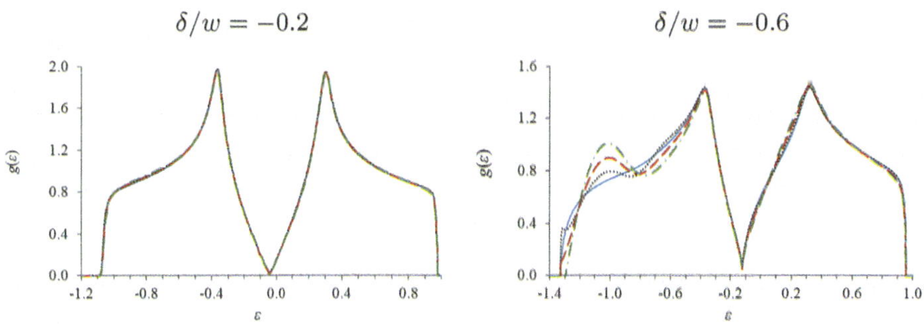

Fig. 3.2 Dependence of the density of electronic states $g(\varepsilon)$ on the energy ε. The value of the substitutional impurity concentration $y = 0.2$, the scattering potential $\delta/w = -0.2$ on the left-hand side (lhs) and $\delta/w = -0.6$ on the right-hand side (rhs), the ordering parameter $\eta = 0$, different values of the parameter of pair interatomic correlations $\varepsilon_{lj,0i}^{BB} = \varepsilon^{BB}$. The density of states calculated in the approximation of coherent potential (solid line) with account for the scattering on pairs of atoms within the first coordination sphere: completely disordered arrangement of impurity atoms $\varepsilon^{BB} = 0$ (dotted line), with the interatomic pair correlations $\varepsilon^{BB} = -0.05$ (dashed line) and with $\varepsilon^{BB} = -0.1$ (dash-dotted line)

account. The dotted curve shows the density of states $g(\varepsilon)$ calculated with regard for the scattering on pairs of atoms located within the first coordination sphere in the case of completely disordered arrangement of impurity atoms on the graphene lattice, $\varepsilon^{BB} = 0$, $\eta = 0$. The dotted curve and dash-dotted curve show the density of states $g(\varepsilon)$ calculated with regard for the scattering on pairs of atoms, when the paired interatomic correlation $\varepsilon^{BB} = -0.05$ and $\varepsilon^{BB} = -0.1$, the order parameter $\eta = 0$. The curve describing the density of electron states in the approximation of the coherent potential and curves accounting for the scattering on pairs of atoms coincide in the case of small values of the scattering parameter (Fig. 3.2, $\delta/w = -0.2$). For $\delta/w = -0.6$ (on the right-hand side in Fig. 3.2), a characteristic dip of the density of state occurs. Its value increases with the correlation parameter ε^{BB}.

The density of states $g(\varepsilon)$ was calculated also with regard for the scattering on pairs of atoms, which are located within three coordination spheres and within ten coordination spheres. The results practically coincide with those of calculations that involve the scattering on pairs within the first coordination sphere (Fig. 3.2). From whence, we can conclude that the area of influence of impurity electronic states of graphene in the specified model is limited by the first coordination sphere, if the substitutional impurity concentration is not more than $y = 0.2$.

Figure 3.3 shows the plot for the density of states $g(\varepsilon)$ vs. the energy ε (Fig. 3.3a) and plot for the conductivity $\sigma_{xx}(\mu) \cdot d$ vs. the Fermi level μ, d is the thickness of the graphene layer (Fig. 3.3b, c). The energy and the Fermi level are given in units of the half-width band energy. The calculations of $g(\varepsilon)$ and $\sigma_{xx}(\mu)$ are executed in accordance with (3.8), (3.9). The substitutional impurity concentration $y = 0.2$, the order parameter $\eta = 0.3$, the parameter of pair interatomic correlations $\varepsilon^{BB} = 0$, the scattering potential $\delta/w = -0.2$ (Fig. 3.3, lhs) and $\delta/w = -0.6$ (Fig. 3.3, rhs), and $\sigma_{xx}(\mu) \cdot d$ is given in units of $e^2 \cdot \hbar^{-1}$. The electrical resistance of a graphen layer

$$R = \frac{1}{\sigma_{xx} d} \frac{l}{L}, \tag{3.34}$$

where l is the length of the graphene layer along the x-axis, and L is the layer width. The x-axis is directed from the carbon atom to its nearest neighbor.

As the substitutional impurity atoms become ordered, a gap appears in the density of states (Fig. 3.3), the order parameter $\eta = 0.3$. In the region of the gap, the density of states $g = 0$. The electrical conductivity of graphene $\sigma_{xx}(\mu)$ for the Fermi level lying in the gap region is zero. For the Fermi level outside the gap region, the electrical conductivity of graphene differs from zero and increases with the density of states on the Fermi level. In contrast to the case of weak scattering $|\delta/w| \ll 1$ described above, for which the gap width increases linearly with the scattering potential $|\delta/w|$, the behavior of the energy gap width in the case of strong scattering is more complex. The width of the slit decreases, as the absolute value of the scattering potential increases (Fig. 3.3). The dependence of the electrical conductivity on the scattering potential and the order parameter is

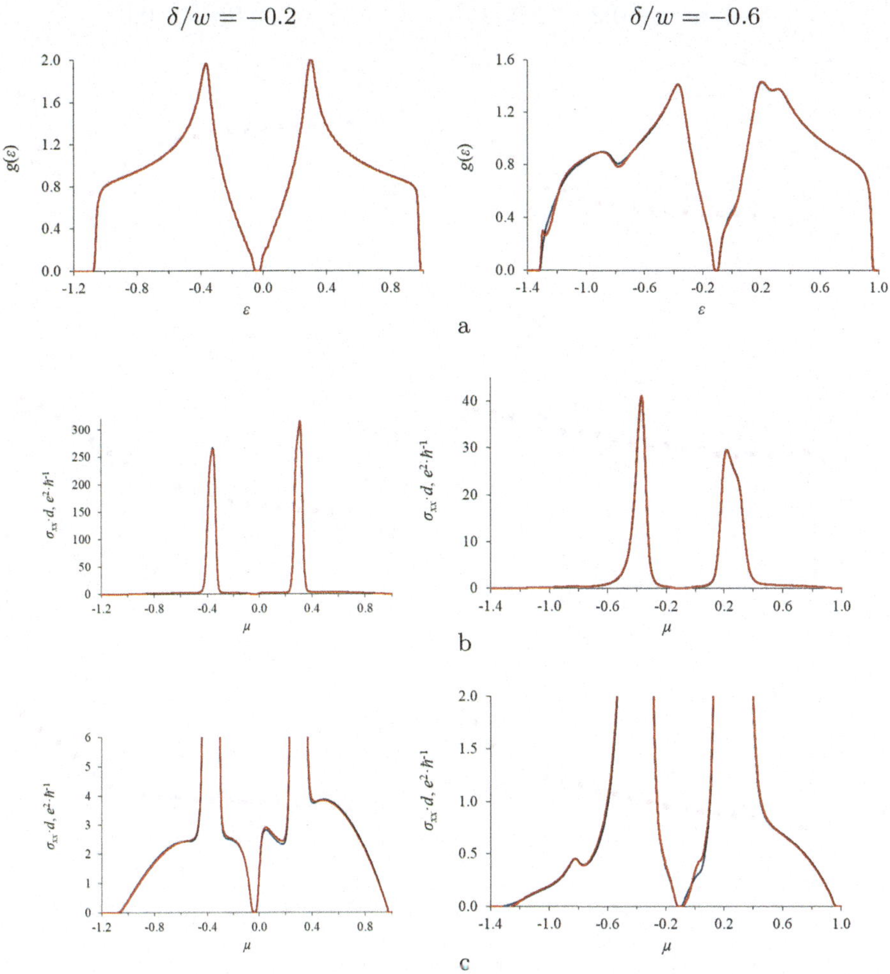

Fig. 3.3 (**a**) density of states $g(\varepsilon)$ as a function of the energy ε, (**b**) conductivity $\sigma_{xx}(\mu) \cdot d$ as a function of the Fermi level, d is the thickness of a graphene layer; (**c**) plot (**b**) is enlarged along the vertical axis in the region of values close to the origin. The scattering parameter $\delta/w = -0.2$ (lhs) and $\delta/w = -0.6$ (rhs). The scattering parameter $\delta/w = -0.2$ (lhs) and $\delta/w = -0.6$ (rhs). The substitutional impurity concentration $y = 0.2$, the order parameter $\eta = 0.3$. Blue line presents calculations in the approximation of coherent potential. Red line: calculations with regard for the electron scattering processes on the pairs of atoms of the first coordination sphere

also more complex than in the case of weak scattering, for which the electrical conductivity is described by formula (3.30).

In order to clarify the nature of the dependence of the electrical conductivity on the scattering potential δ and the order parameter η, we give the electrical conductivity of graphene $\sigma_{xx}(\mu)$ vs. the ordering parameter of impurity atoms η in Fig. 3.4 for different magnitudes δ of the scattering potential. The number of

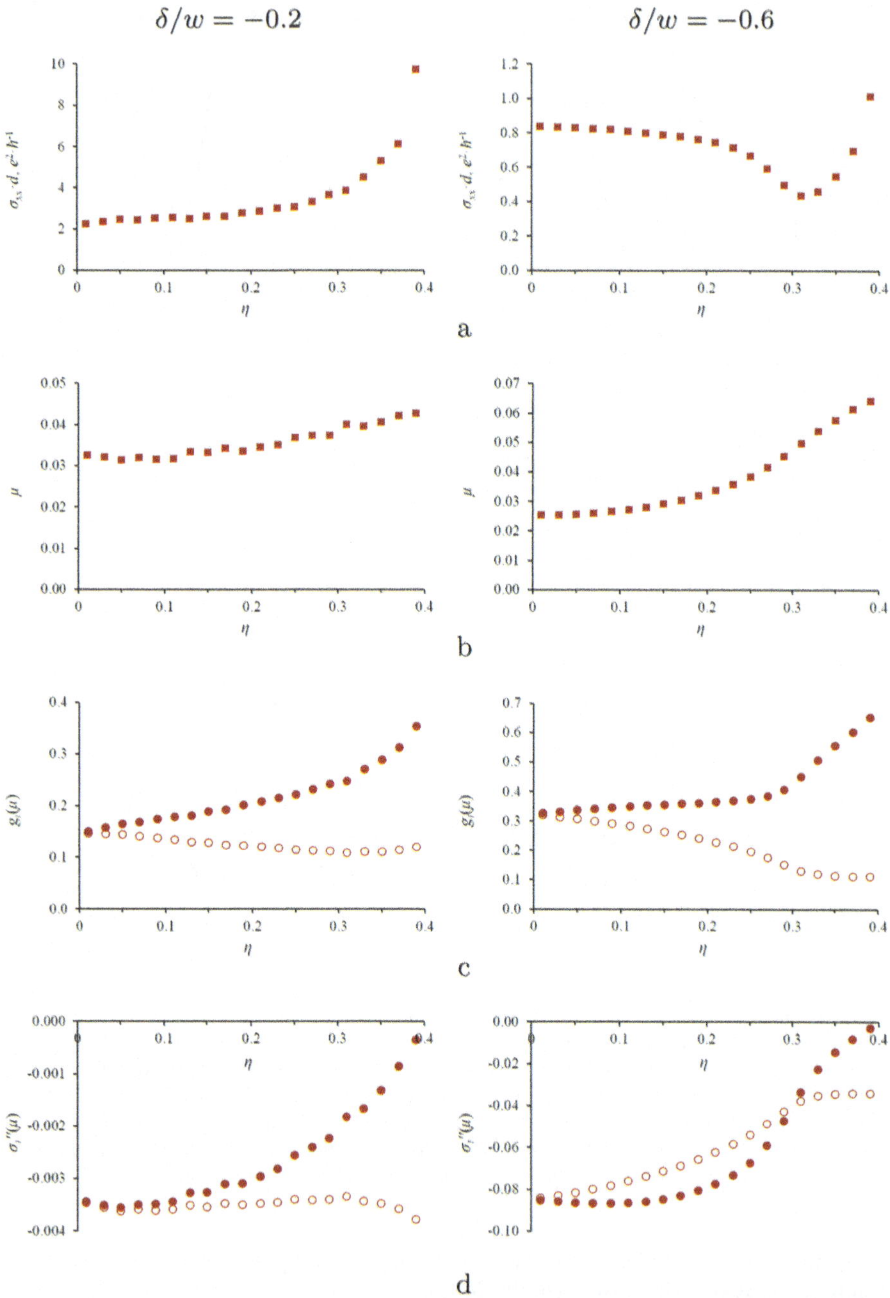

Fig. 3.4 (**a**) conductivity $\sigma_{xx}(\mu) \cdot d$, (**b**) Fermi level μ, (**c**) partial density of states $g_i(\mu)$ (circle – $g_1(\mu)$, disc – $g_2(\mu)$) and (**d**) imaginary part of the partial coherent potential $\sigma_i''(\mu)$ (circle – $\sigma_1''(\mu)$, disc – $\sigma_2''(\mu)$) as functions of order parameter η. The sublattice $i = 1$ contains impurity atoms, and the sublattice $i = 2$ contains only carbon atoms in the case of complete ordering. The substitutional impurity concentration $y = 0.2$, the scattering parameter $\delta/w = -0.2$ (on the left-hand side) and $\delta/w = -0.6$ (on the right-hand side)

electrons per atom, whose energy values are in the region of the energy band, is equal $\langle Z \rangle = 1.01$. At this value of $\langle Z \rangle$, the Fermi level μ lies to the right from the energy gap. Figure 3.4b shows the Fermi level $\mu(\eta)$ depending on the ordering parameter of the impurity η. The value of the Fermi level $\mu(\eta)$ is calculated by formula (3.13). Figure 3.4c shows the dependence of the partial density of states $g_i(\mu)$ at the Fermi level on the impurity ordering parameter η, and i is a sublattice number. Figure 3.4d shows the dependence of the imaginary part of the coherent potential $\sigma_i''(\mu)$ at the Fermi level on the ordering parameter η.

As can be seen from Fig. 3.4, the electrical conductivity of graphene increases with the ordering of the impurity η, which is caused mainly by the increase in the density of states at the Fermi level. To find out the nature of the change in the electrical conductivity of graphene $\sigma_{xx}(\mu)$ with a change in the impurity ordering parameter η, we turn to the limiting case of weak scattering $|\delta/w| \ll 1$.

In the coherent potential approximation in the case of weak scattering $|\delta/w| \ll 1$, relation (3.9) yields

$$\sigma_{\alpha\alpha} = \frac{e^2\hbar}{3\Omega_0} \sum_i \frac{g_i(\mu)\, |\upsilon_{\alpha 12}(\mu)|^2}{|\sigma_i''(\mu)|}. \tag{3.35}$$

For three-dimensional crystals with a simple lattice in the approximation of effective mass, we substitute the expressions for $g(\mu)$ and $\upsilon_\alpha(\mu)$ in (3.35) and get the well-known formula

$$\sigma_{\alpha\alpha} = e^2 n\tau(\mu)/m^*, \tag{3.36}$$

where n is the number of electrons per unit volume, whose energy is less the Fermi level, m^* is the electron effective mass, and $\tau(\mu)$ is the relaxation time of the electronic states, which is determined by the ratio

$$|\sigma''(\mu)|\tau(\mu) = \hbar. \tag{3.37}$$

In the case of weak scattering $|\delta/w| \ll 1$, the numerical calculations of $\sigma_{xx}(\mu)$ qualitatively are in good agreement with formulas (3.31), (3.35), and (3.30). When the order parameter η tends to its maximum value, the electrical conductivity $\sigma_{xx}(\mu)$ goes to infinity. As can be seen from formula (3.35), this is caused by an increase in the state density at the Fermi level $g_2(\mu)$ and an increase in the relaxation time, as the order parameter η increases (with $\eta \to \eta_{\max} = 2y$, the imaginary part of the coherent potential $\sigma_2''(\mu) \to 0$).

We note that formulas (3.8), (3.9) for the density of states and electrical conductance of graphene cannot be used, if the Fermi level falls in interval (3.33) of energies at the gap edges. The densities of states that are calculated with regard for the scattering on the pairs of atoms, which are located within the three coordination spheres and within the ten coordination spheres, coincide with the results of calculations which involve the scattering on the pairs within the first

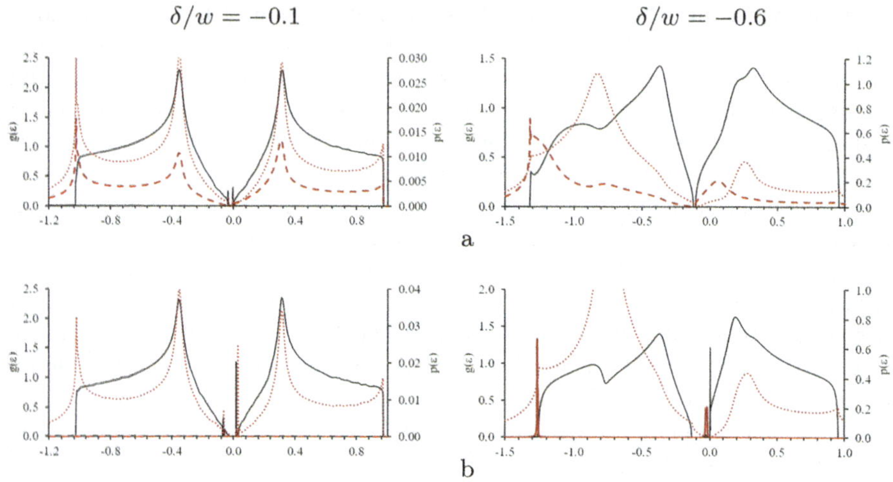

Fig. 3.5 Density of states $g(\varepsilon)$ (black line), the parameter $p_i(\varepsilon)$ (dotted red line $i = 1$, dashed red line $i = 2$) as a function of the energy ε. The scattering parameter $\delta/w = -0.1$ (lhs) and $\delta/w = -0.6$ (rhs), the substitutional impurity concentration $y = 0.2$, the order parameter $\eta = 0.2$ (**a**) and $\eta = 0.399$ (**b**)

coordination sphere. The region of impurity electronic states of graphene in the specified model is limited by the space of the first coordination sphere.

The results of numerical calculations of the density of state $g(\varepsilon)$ and parameter $p(\varepsilon)$ are shown in Fig. 3.5. The substitutional impurity concentration $y = 0.2$, the scattering potential $\delta/w = -0.1$ (the week scattering) on the left-hand side of Fig. 3.5 and $\delta/w = -0.6$ on the right-hand side. Here, we consider two cases: the low ordering of impurity atoms in graphene $\eta = 0.2$ and the higher ordering $\eta = 0.399$ ($\eta_{max} = 2y$). The region of impurity electronic states of graphene is limited by the space of the first coordination sphere.

Figure 3.5 shows that the regions of localization of electronic states are at the edges of the gap and the edges of the energy spectrum according to the analytic results presented in Fig. 3.1. For $\eta = 0.399$, only 0.25% of impurity atoms enter the carbon sublattice, and y_1/y_2 (3.14) is very small. Thus, the curve $p_2(\varepsilon)$ almost merges with the abscissa. But here, the localization of electronic states on the edges of bands is visible.

In the special case of $\eta = 0.399$ and $\delta/w = -0.6$, the width of the gap is less than $\eta|\delta|$. In the case of strong scattering, the dependence of the gap width on the order parameter and the scattering potential can work approximately. The localization of electronic states is poorly recognized on the edges of the gap and becomes noticeable only with the use of a small energy step in numerical calculations.

In order to describe the energy spectrum and the electrical conductivity of graphene with impurities, we used the method of the theory of disordered systems. This method is based on the cluster decomposition for the one-part Green function (state density) and the two-part Green function (electrical conductivity). For the zero one-node approximation, a coherent potential approximation describing the state of the electron in some efficient ordered environment is chosen. The corrections to the approximation of coherent potential are due to the contribution of the electron scattering on clusters of two, three, etc. atoms. In [13, 17], it was shown that the contribution of the scattering on a cluster decreases with an increase in the number of atoms in the cluster by some small parameter. This parameter is small in the wide range of variation in system's characteristics except for narrow energy intervals at the edges of the spectrum and the edges of the gap (3.33). This occurs when the impurity is ordered. As shown in [13, 17], these areas (3.33) are the regions of localization of electronic states. In the Van Hove regions of the energy spectrum of a pure crystal, the peaks on the state density expand due to the splitting of energy levels, while reducing the symmetry of the crystal with the introduction of disordered impurity atoms. However, the Van Hove regions do not always coincide with the regions of localization of electronic states, as in the case for binary alloys with the bcc lattice [13] or graphene with a substitutional impurity (Figs. 3.3 and 3.5).

3.4 Conclusions

In the Lifshitz tight-binding one-electron model, the influence of substitutional impurity atoms on the energy spectrum and electrical conductance of graphene is studied. It is established that the ordering of substitutional impurity atoms on nodes of the crystal lattice causes the appearance of the gap in the energy spectrum of graphene $\eta|\delta|$ in width centered at the point $y\delta$, where η is the parameter of ordering, δ is the difference of the scattering potentials of impurity atoms and carbon atoms, and y is the impurity concentration.

It is shown that if the ordering parameter η is close to $\eta_{\max} = 2y,\ \ y < 1/2$ then the density of electron states has peaks on the edges of the energy gap. Those peaks correspond to impurity levels.

If the Fermi level falls in the region of the gap, then the electrical conductance $\sigma_{\alpha\alpha} \to 0$ at the ordering of graphene, i.e., the metal-dielectric transition arises.

If the Fermi level is located outside the gap, then the electrical conductance increases with the parameter of order η. At the complete ordering of a substitutional impurity $\eta \to \eta_{\max} = 2y$, the electrical conductivity $\sigma_{\alpha\alpha} \to \infty$, imaginary part of the coherent potential $\sigma_2''(\mu) \to 0$ (sublattice $i = 2$ contain only carbon atoms in the case of complete ordering), and the relaxation time $\tau \to \infty$.

The analytic expressions for the density of electron states and the electrical conductivity of graphene obtained in the case of weak scattering $|\delta/w| \ll 1$ are compared with the results of numerical calculations for different scattering potentials $|\delta/w|$, w is the half-width of the energy band. The numerical calculation shows that the area of influence of impurity electronic states of graphene in a specified model is spatially limited by the first coordination sphere.

In the case where the ordering parameter η differs from its maximum value $\eta_{\max} = 2y$, some fraction of the impurity atoms is at the sublattice sites on which the carbon atoms are located. This leads to a peak on the energy density curve for the $g(\varepsilon)$ state, which is associated with the occurrence of localized impurity states. The distance from the peak to the energy band edge increases with the probability of finding the impurity atoms on the carbon lattice y_2 and the impurity scattering parameter δ. The curve of the density of electronic states at the edge of the energy zone has also a second characteristic peak located farther from the center of the gap. As the probability of finding the impurity atoms on the carbon lattice y_2 (decrease in the order parameter η) increases, the mentioned peak on the energy density curve splits which is caused by the mutual interaction of impurity atoms in the case of their large concentration. It is established that the regions of localization of electronic states are at the edges of the gap and the edges of the energy spectrum.

References

1. Sun J, Marsman M, Csonka GI, Ruzsinszky A, Hao P, Kim Y-S, Kresse G, Perdew JP (2011) Self-consistent meta-generalized gradient approximation within the projector-augmented-wave method. Phys Rev B 84:035117
2. Yelgel C, Srivastava GP (2012) Ab initio studies of electronic and optical properties of graphene and graphene-BN interface. Appl Surf Sci 258:8338–8342
3. Denis PA (2010) Band gap opening of monolayer and bilayer graphene doped with aluminium, silicon, phosphorus, and sulfur. Chem Phys Lett 492:251
4. Repetsky S, Vyshyvana I, Nakazawa Y, Kruchinin S, Bellucci S (2019) Electron transport in carbon nanotubes with adsorbed chromium impurities. Materials 12(3):524
5. Xiaohui D, Yanqun W, Jiayu D, Dongdong K, Dengyu Z (2011) Electronic structure tuning and band gap opening of graphene by hole/electron codoping. Phys Lett A 365:3890–3894
6. Gospodarev IA, Grishaev VI, Manzhelii EV, Sirenko VA, Syrkin ES, Feodosyev SB (2020) Effect of size quantization upon electron spectra of graphene nanoribbons. Low Temp Phys 46(2):231–240
7. Grushevskaya HV, Krylov GG, Kruchinin SP, Vlahovic B (2018) Graphene quantum dots, graphene non-circular n-p-n-junctions: quasi-relativistic pseudo wave and potentials. In: Bonca J, Kruchinin S (eds) Proceedings NATO ARW "nanostructured materials for the detection of CBRN". Springer, pp 47–58
8. Radchenko TM, Tatarenko VA, Sagalianov IY, Prylutskyy YI, Szroeder P, Biniak S (2016) On adatomic-configuration-mediated correlation between electrotransport and electrochemical properties of graphene. Carbon 101:37–48
9. Kruchinin S, Nagao H, Aono S (2010) Modern aspect of superconductivity: theory of superconductivity. World Scientific, Singapore, p 232
10. Radchenko TM, Tatarenko VA, Sagalianov IY, Prylutskyy YI (2014) Effects of nitrogen-doping configurations with vacancies on conductivity in graphene. Phys Lett A 378(30–31):2270–2274
11. Radchenko TM, Shylau AA, Zozoulenko IV, Ferreira A (2013) Effect of charged line defects on conductivity in graphene: numerical Kubo and analytical Boltzmann approaches. Phys Rev B 87:195448-1-14
12. Radchenko TM, Shylau AA, Zozoulenko IV (2014) Conductivity of epitaxial and CVD graphene with correlated line defects. Solid State Commun 195:88–94

13. Los' VF, Repetsky SP (1994) A theory for the electrical conductivity of an ordered alloy. J Phys Condens Matter 6:1707–1730
14. Repetsky SP, Vyshyvana IG, Kruchinin SP, Bellucci S (2018) Influence of the ordering of impurities on the appearance of an energy gap and on the electrical conductance of grapheme. Sci Rep 8:9123
15. Slater JC, Koster GF (1954) Simplified LCAO Method for the periodic potential problem. Phys Rev 94(6):1498–1524
16. Repetsky SP, Vyshyvana IG, Kuznetsova EY, Kruchinin SP (2018) Energy spectrum of graphene with adsorbed potassium atoms. Int J Modern Phys B 32:1840030
17. Ducastelle F (1974) Analytic properties of the coherent potential approximation and its molecular generalizations. J Phys C Solid State Phys 7(10):1795–1816

Chapter 4
Multiferroics for Detection of Magnetic and Electric Fields

B. Dabrowski

Abstract Muliferroic materials are characterized by two or more primary ferroic orders: ferroelectric, ferromagnetic and ferroelastic. In multiferroics the coupling occurs between the magnetic (ferromagnetic or antiferromagnetic) and electric (ferroelectric) subsystems. This enables control of the dielectric polarization **P** by a magnetic field **H** and the manipulation of magnetization **M** by an electric field **E**, allowing design of a wide range of novel electronic devices for various sensing, memory, logical, energy, biomedical and other applications. I describe here our efforts to develop new class of single-phase muliferroic materials where a single element, the manganese, is responsible for both the ferroelectric and magnetic properties, which guarantees strong coupling which is necessary for practical applications.

Keywords Multiferroic · Detection of magnetic · Electric fields

4.1 Introduction

Search for muliferroic materials (or in general the magneto-electrics) has a long history because of the promise that if the coupling between the magnetic and electric subsystems is strong, the control of the dielectric polarization **P** by a magnetic field **H** and the manipulation of the magnetization **M** by an electric field **E** would allow design of a wide range of novel electronic devices. However, based on experimental evidence and the theoretical projections it was believed that while for technological applications the single-phase and single-element multiferroics are highly desirable, combining ferroelectric and magnetic order within one material, like in e.g. $BaTiO_3$, cannot be done. This belief, known as the "d^0-ness conjecture" assumed that whereas the magnetic propertied require at least partially filled *d*-

B. Dabrowski (✉)
Institute of Physics, Polish Academy of Sciences, Warsaw, Poland
e-mail: dabrowski@anl.gov

J. Bonča, S. Kruchinin (eds.), *Advanced Nanomaterials for Detection of CBRN*,
NATO Science for Peace and Security Series A: Chemistry and Biology,
https://doi.org/10.1007/978-94-024-2030-2_4

shells providing a non-zero magnetic moment, the usual displacive ferroelectricity requires empty d-shells [1]. Because of that conjecture, the considerable research effort was dedicated for development of bulk composites where two dissimilar materials, one magnetic and the other ferroelectric, were combined. Large range of different compounds were used in variety of 0-, 1- and 2-dimensional structures, however it was observed that the coupling between dissimilar phases was mostly based on indirect, extrinsic strain mediated coupling. Similarly, significant effort was spent on thin film composites where the improved interface coupling was achieved through direct bonding. Indeed, by miniaturization and multi-functionality in integrated devices, the best magneto-electric properties have been achieved to date for these structures [2–4].

Despite the d^0-ness conjecture, the search for single-phase multiferroics has been also pursued for decades. For example, considerable research was devoted to $BiFeO_3$ multiferroic [5]. However, it was observed that the coupling is vanishingly weak in this well-known compound because two dissimilar sublattices, the lone-pair active Bi–O network that orders ferro-electrically at 1140 K essentially do not interact with the cycloidal antiferromagnetic Fe–O network that orders at 640 K. We show here that it is possible to achieve displacive-type ferroelectric distortions in the single lattice magnetic d^3 system of the $Sr_{1-x}Ba_xMnO_3$ perovskite with the Mn–O bonds stretched beyond their equilibrium lengths, and the reason for the scarcity of similar multiferroic systems is caused by the difficulty of obtaining compounds with expected tolerance factor $t > 1$ near the room temperature. Obtaining such compounds is caused by stringed demands of the synthesis conditions brought about by the universal dependence of $t(T)$ that increases with temperature. Thus, we demonstrate that it is not the d^0-ness conjecture that limits the lack of the single-phase multiferroics where distortions of the single transition metal cations and oxygen are responsible for both ferroelectric and magnetic properties. Design and discovery of such compounds where the strong coupling between ferroelectric and magnetic properties is guaranteed due to presence of a single magnetic cation could lead to development of new class of very sensitive detection devices of magnetic and electric fields.

4.2 Synthesis Limitations of the Perovskites with Expected Tolerance Factor $t > 1$ Near 300 K

Since the early 2000' we have been searching for the displacive-type multiferroics in the antiferromagnetic perovskite system $Sr_{1-x}Ba_xMnO_3$ where the single Mn-ion would be responsible for both ferroelectricity and magnetism [6]. The guiding principle was based on the observation that ferroelectricity with the highest T_C of 400 K has been observed in the nonmagnetic perovskites $A^{2+}TiO_3$ for the largest ion A = Ba. When the tabulated equilibrium bond lengths of [Ba–O] and [Ti–O] are used [7], the expected tolerance factor $t = [Ba–O]/2^{1/2}[Ti–O]$ is larger than 1. Because the

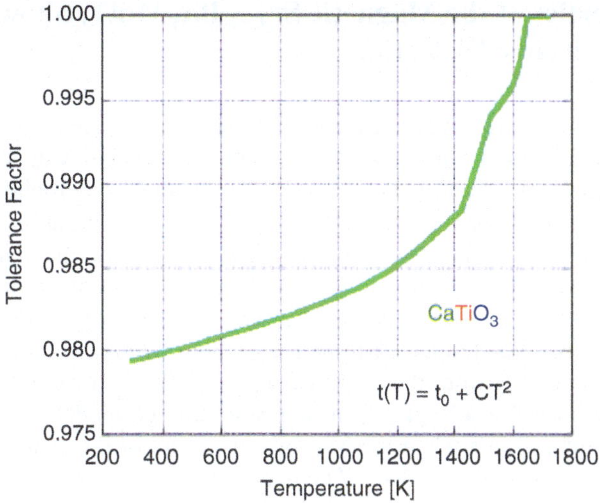

Fig. 4.1 Typical temperature dependence of the tolerance factor $t(T)$ on temperature for perovskites AMO_3 (M = Ti, Mn)

biggest possible experimentally measured t is 1 for the perovskite cubic structure, the [Ti–O] and [B–O] bonds have to be respectively under tension and compression when the expected $t > 1$ due to the 3-dimensional constraints of the structure. The [Ti–O] bonds are under the largest tension and are maximally elongated above the known equilibrium lengths for the largest available A^{2+} = Ba ion, explaining thus the highest T_C.

When the known tabulated equilibrium bond lengths are used, the expected tolerance factor t is also larger than 1 for the known cubic antiferromagnetic (T_N = 234 K) $SrMnO_3$ for which the [Mn–O] bonds are also under tension, and such tensions should increase with the substitution of a larger Ba for Sr [8]. However, initially we were not able to generate ferroelectricity in $Sr_{1-x}Ba_xMnO_3$ system because we could not substitute enough of Ba to induce sufficiently large tension on the [Mn–O] bonds due to the structural instability of perovskite structure at the synthesis temperatures of $T \sim 1700$ K. At these temperatures, the non-ferroelectric hexagonal structures formed for which the 3-dim constraints are not present and the [Mn–O] bonds are not under tension. Due to the ionic sizes of the A and Mn cations, the formation of the perovskite structure requires condition $t(T) < 1$ to properly fit cations and oxygen, which was not possible to satisfy due to the dependence of $t(T) \sim t_0 + \alpha T^2$ that increases with temperature as shown on Fig. 4.1 [9]. Only by using the optimized two-step synthesis procedure to initially decrease $t(T, d)$ below 1 by increasing the size of [Mn–O] bonds containing larger Mn^{3+} ions of the oxygen reduced material at high temperatures than the Mn^{4+} ions we were able to stabilize the perovskite $Sr_{1-x}Ba_xMnO_{3-d}$ phase.

4.3 Properties of the Magnetic $Sr_{1-x}Ba_xMnO_3$ Showing Ferroelectric Distortion

After oxidization of the oxygen reduced material at lower temperatures of 700 K, the oxygen stoichiometric and kinetically stable perovskites with large expected tolerance factor were achieved, and the new multiferroics were found for $x \geqslant 0.43$, which exhibited both robust ferroelectric distortions and the antiferromagnetism originating exclusively from a single-ion displacement of the magnetic Mn^{4+} (d^3) cations and oxygen [10]. Similar single-phase noncentrosymmetric *P4mm* space group structures were obtained for $x = 0.44$ and 0.45, see Fig. 4.2. Typical, displacive-type tetragonal ferroelectric phase with a polarization of $P_s \sim 14\ \mu C/cm^2$, as derived from the empirical relation $P_S^2 \propto (c/a - 1)$ based on X-ray diffraction (see Fig. 4.3) and from the point charge model based on neutron diffraction results, occurs when the Mn ions move out of the center of the MnO_6 octahedral units below $T_C \sim 350$ K. The required elongation of the [Mn–O] bonds was observed at 1.6% above the equilibrium length to split them into long and short bonds along the c-axis similar to the tetragonal distortions of tetragonal $BaTiO_3$. In addition, the Mn spins order below T_N of ~210 K into a simple G-type magnetic structure, while the displacive distortions decrease during the magnetic transition as seen on Fig. 4.3 causing reduction of P_s to $\sim 12\ \mu C/cm^2$, demonstrating that the two order parameters are strongly coupled [11]. We were not able, however, to measure directly the spontaneous polarization P_s due to considerable leaking currents of the obtained semiconducting materials. Based on the structural and magnetic studies the structure – properties phase diagram of the $Sr_{1-x}Ba_xMnO_3$ system was obtained as shown on Fig. 4.4.

The ideas presented here for the synthesis of multiferroic $Sr_{1-x}Ba_xMnO_3$ demonstrate that it should be possible to achieve similar displacive-type ferroelectric

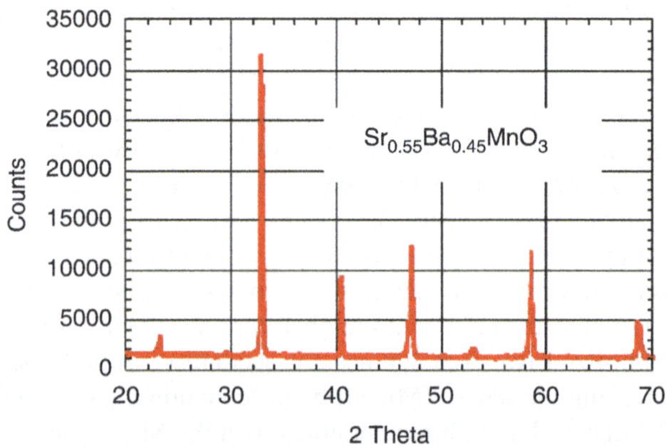

Fig. 4.2 Representative X-ray diffraction pattern for $Sr_{1-x}Ba_xMnO_3$ ($x = 0.45$) showing typical ferroelectric structure *P4mm* similar to tetragonal $BaTiO_3$

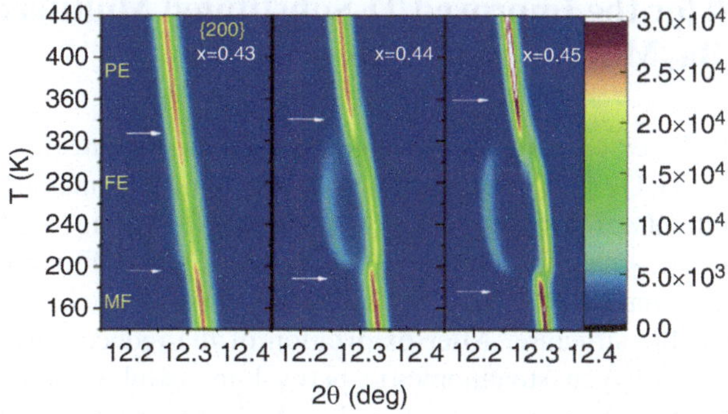

Fig. 4.3 Contour maps of the cubic {200} reflection from synchrotron X-ray powder diffraction for the $x = 0.43$, 0.44 and 0.45 samples. The tetragonal splitting of the peak is observed at T_c (top ar rows), which changes as a function of the Ba content. The splitting is partially suppressed below T_N (bottom arrows)

Fig. 4.4 Phase diagram of the structural, magnetic, and ferroelectric properties as a function of barium content (x) in the multiferroic $Sr_{1-x}Ba_xMnO_3$ system. PE, PM, FE, MF refer to paraelectric, paramagnetic, ferroelectric, and multiferroic phases, respectively

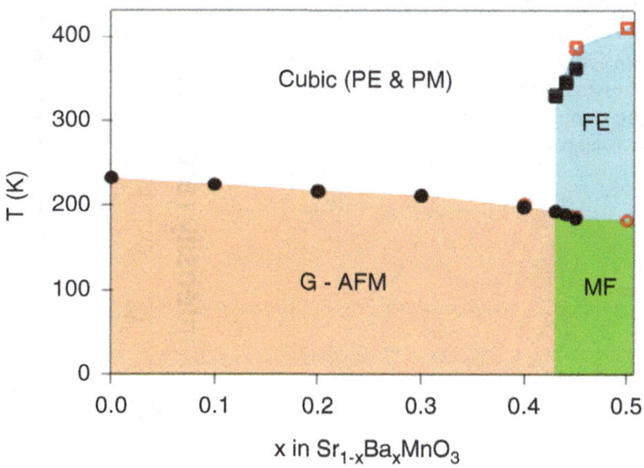

distortions also in the other magnetic d^n perovskite AMO_3 systems (A = Rare or Alkaline Earths; M = transition metals) with the M–O bonds stretched beyond their equilibrium lengths, and the reason for the scarcity of similar multiferroic systems is caused by the difficulty of obtaining compounds with expected tolerance factor $t > 1$ near the room temperature. It is, thus, not the d^0-ness conjecture that limits the lack of single-phase and single-ion multiferroics but the instability of such compounds. The design and discovery of such compounds is not trivial because several conditions regarding the ionic sizes, their valences, easiness of obtaining and stability of oxygen reduced and oxidized phases have to be satisfied simultaneously, but the expected strong coupling between ferroelectric and magnetic properties, which could lead to development of new classes of very sensitive detection devices of magnetic and electric fields makes search of them worthwhile

4.4 Search for the Improved Ti-Substituted Multiferroic $Sr_{1-x}Ba_xMnO_3$ Compounds

We have recently extended our investigations to the Ti-substituted system of $Sr_{1-x}Ba_xMn_{1-y}Ti_yO_3$ [12]. The similar two-step synthesis method was used to obtain the single-phase samples (see Fig. 4.5) within the range of $0.45 < x < 0.60$ and $0 < y < 0.12$ The room temperature X-ray diffraction data have shown a typical behavior of the oxygen reduced samples, which after synthesis in H_2/Ar display cubic structure. After oxidization of the reduced material at lower temperatures, the oxygen stoichiometric perovskites exhibit robust ferroelectric distortions. Magnetic measurements (see Fig. 4.6) confirmed by neutron powder diffraction showed transitions to antiferromagnetic phase below $T_N \sim 180$ K, which was further reduced from the $Sr_{1-x}Ba_xMnO_3$ ($x = 0.43$–0.55) system due to the

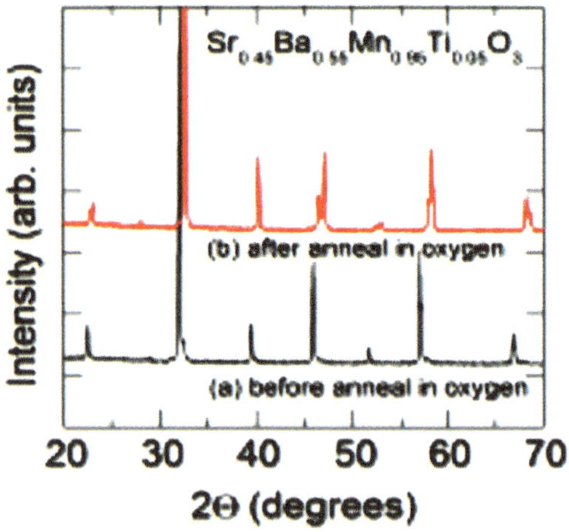

Fig. 4.5 Representative X-ray diffraction patterns for $Sr_{0.45}Ba_{0.55}Mn_{0.95}Tl_{0.05}O_{3-\delta}$ showing typical behavior of the samples obtained from the two-step synthesis method

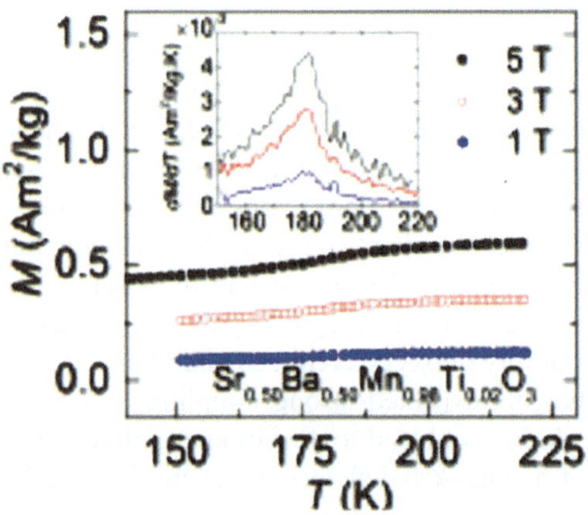

Fig. 4.6 Temperature derivatives of the dc magnetization data (see inset) showing determination of T_N for $Sr_{0.50}Ba_{0.50}Mn_{0.96}Tl_{0.02}O_3$ in applied magnetic fields of 0.1, 3.0, and 5.0 T

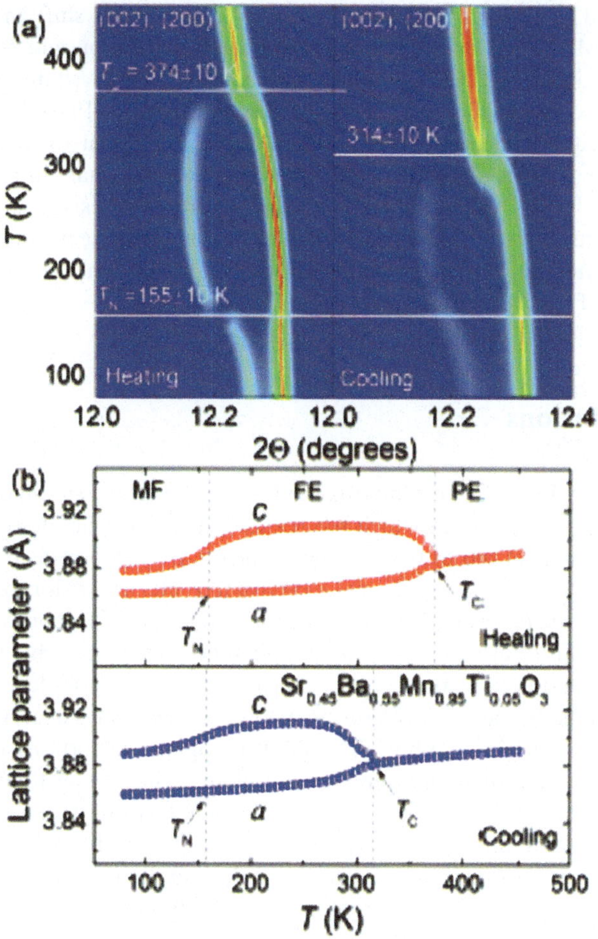

Fig. 4.7 Two dimensional contour maps of the X-ray data (**a**) and the refined lattice parameters (**b**) for $Sr_{0.45}Ba_{0.55}Mn_{0.95}Tl_{0.05}O_3$. The top panel shows the splitting of the cubic ⟨200⟩ reflection above and below T_N

higher Ba content and dilution of the magnetic Mn–O network by substitution of non-magnetic Ti.

The temperature dependent X-ray diffraction data shown on Fig. 4.7 was studied both on heating and cooling for the $x = 0.55$ and $y = 0.05$ sample. The displacive distortions significantly exceeding the size of distortions in $Sr_{1-x}Ba_xMnO_3$ as well as in the classical ferroelectric $BaTiO_3$ were observed, thus the polarization P_s was increased to $\sim 29\,\mu C/cm^2$ in the ferroelectric phase. Also, the ferroelectric transition temperatures were increased up to $T_C \sim 420$ K for several compositions. At all temperatures below T_C, the tetragonal distortions are caused by the (Mn, Ti) atoms shift along the c-axis away from the high symmetry positions, while the apical and equatorial oxygen atoms of the octahedra shift in the opposite direction to (Mn, Ti) atoms, thus, creating charge separation and ferroelectricity. The suppression of ferroelectric distortion below T_N was again observed. The equation $P_S^2 \propto (c/a - 1)$ suggested a P_S reduction to ~19 μC/cm^2 below T_N and strong magneto-electric coupling. The large tetragonal distortion surviving below T_N, proving robust multiferroic phase indicates important improvement over data observed for $Sr_{1-x}Ba_xMnO_3$. The temperature hysteresis properties seen on Fig. 4.7 indicate, however, the first-order character of the phase transition at T_C. This property, unfortunately, may limit application of $Sr_{1-x}Ba_xMn_{1-y}Ti_yO_3$ compounds for detection of magnetic and electric fields.

4.5 Conclusions

Development of the two-step synthesis method enabled the creation of a new series of multiferroic perovskites in which the Ba solubility at the Sr site of $SrMnO_3$ and the nonmagnetic Ti at the Mn site were extended. We demonstrated that it should be possible to achieve similar displacive-type ferroelectric distortions also in the other magnetic d^n perovskite AMO_3 systems when the M-O bonds are stretched beyond their equilibrium lengths. With Ti substitution, we were able to increase T_C up to ~420 K and polarization P_s to ~29 μC/cm^2 in the ferroelectric phase. The suppression of the ferroelectric distortion below T_N was always observed suggesting strong magneto-electric coupling. The large tetragonal distortions present below T_N, proved presence of robust multiferroic phase. The Ti substitution decreased T_N and suppressed antiferromagnetic interactions in favor of the ferroelectric order extending into the multiferroic state. The Ti substitution, however, enhanced the first-order character of the phase transition at T_C and did not sufficiently eliminate leakage currents to permit direct measurements of the spontaneous polarization by electrical measurements of the P-E curves.

Acknowledgments This work was supported by the Polish NCN through Grant No. 2018/31/B/ST5/03024.

References

1. Hill NA (2000) Why are there so few magnetic ferroelectrics? J Phys Chem B 104:6694
2. Palneedi H, Annapureddy V, Priya S, Ryu J (2016) Status and perspectives of multiferroic magneto-electric composite materials and applications. Actuators 5:9
3. Dzhezherya YI, Khrebtov AO, Kruchinin SP (2018) Sharp-pointed susceptibility of ferromagnetic films with magnetic anisotropy inhomogeneous in thickness. Int J Mod Phys B 32:1840034
4. Dzhezherya Y, Novak IY, Kruchinin S (2010) Orientational phase transitions of lattice of magnetic dots embedded in a London type superconductors. Supercond Sci Technol 23:1050111

5. Bibes M, Barthelemy A (2008) Towards a magnetoelectric memory. Nat Mater 7:425
6. Dabrowski B, Chmaissem O, Mais J, Kolesnik S, Jorgensen JD, Short S (2003) Tolerance factor rules for $Sr_{1-x-y}Ca_xBa_yMnO_3$ perovskites. J Solid State Chem 170:154
7. Shannon RD (1976) Revised effective ionic radii and systematic studies of interatomic distances in halides and chalcogenides. Acta Crystallogr A 32:751
8. Chmaissem O, Dabrowski B, Kolesnik S, Mais J, Brown DE, Kruk R, Prior P, Pyles B, Jorgensen JD (2001) Relationship between structural parameters and Néel temperature in $Sr_{1-x}Ca_xMnO_3$ ($0 \leq x \leq 1$) and $Sr_{1-y}Ba_yMnO_3$ ($y \leq 0.2$). Phys Rev B 64:134412
9. Baszczuk A, Dabrowski B, Avdeev M (2015) High temperature neutron diffraction studies of $PrInO_3$ and the measures of perovskite structure distortion. Dalt Trans 44:10817
10. Pratt DK, Lynn JW, Mais J, Chmaissem O, Brown DE, Kolesnik S, Dabrowski B (2014) Neutron scattering studies of the ferroelectric distortion and spin dynamics in the type-1 multiferroic perovskite $Sr_{0.56}Ba_{0.44}MnO_3$. Phys Rev B 90:140401
11. Somaily H, Kolesnik S, Mais J, Brown D, Chapagain K, Dabrowski B, Chmaissem O (2018) Strain-induced tetragonal distortions and multiferroic properties in polycrystalline $Sr_{1-x}Ba_xMnO_3$ ($x = 0.43$–0.45) perovskites. Phys Rev Mater 2:054408
12. Chapagain K, Brown DE, Kolesnik S, Lapidus S, Haberl B, Molaison J, Lin C, Kenney-Benson C, Park C, Pietosa J, Markiewicz E, Andrzejewski B, Lynn JW, Rosenkranz S, Dabrowski B, Chmaissem O (2019) Tunable multiferroic order parameters in $Sr_{1-x}Ba_xMn_{1-y}Ti_yO_3$. Phys Rev Mater 3:084401

Chapter 5
Many-Fermion Wave Functions: Structure and Examples

D. K. Sunko

Abstract Many-fermion Hilbert space has the algebraic structure of a free module generated by a finite number of antisymmetric functions called *shapes*. Physically, each shape is a many-body vacuum, whose excitations are described by symmetric functions (bosons). The infinity of bosonic excitations accounts for the infinity of Hilbert space, while all shapes can be generated algorithmically in closed form. The shapes are geometric objects in wave-function space, such that any given many-body vacuum is their intersection. Correlation effects in laboratory space are geometric constraints in wave-function space. Algebraic geometry is the natural mathematical framework for the particle picture of quantum mechanics. Simple examples of this scheme are given, and the current state of the art in generating shapes is described from the viewpoint of treating very large function spaces.

Keywords Many-fermion wave functions · Bosons · Hilbert space

5.1 Quantum Mechanics and Algebraic Geometry

The standard textbook picture of quantum mechanics is that one-body wave functions represent possible states of individual particles, while many-body wave functions are constructed from the one-body functions by respecting indistinguishability for a *given* number of particles, leading, in the case of fermions, to the well-known Slater determinants. This picture is called the *particle picture* of quantum mechanics. In advanced textbooks a *field picture* is introduced, which corresponds, as Dirac put it, to a "deeper reality," meaning it refers from the outset to an infinite number of degrees of freedom.

The particle picture has a deeper reality of its own. Any wave function of N identical fermions in d dimensions may be written [1]

D. K. Sunko (✉)
Faculty of Science, Department of Physics, University of Zagreb, Zagreb, Croatia
e-mail: dks@phy.hr

J. Bonča, S. Kruchinin (eds.), *Advanced Nanomaterials for Detection of CBRN*, NATO Science for Peace and Security Series A: Chemistry and Biology,
https://doi.org/10.1007/978-94-024-2030-2_5

$$\Psi = \sum_{i=1}^{D} \Phi_i \Psi_i, \tag{5.1}$$

where the Ψ_i are antisymmetric and Φ_i are symmetric functions of N particle coordinates. If the Φ_i were c-numbers, the Ψ in Eq. (5.1) would form a D-dimensional vector space. As it stands, it presents a finitely generated free module, where $D = N!^{d-1}$ is the dimension of the module (number of its generators). The Ψ still belong to the full infinite-dimensional Hilbert space spanned by all Slater determinants, because of the additional degrees of freedom in the symmetric functions Φ_i.

The scheme (5.1) has an important geometric interpretation, which was discovered by Menaechmos in his construction of the cube root $\sqrt[3]{a}$. He interpreted the original observation by Hippocrates of Chios, namely

$$y = x^2 \quad \& \quad y^2 = ax \implies x^4 = ax, \tag{5.2}$$

to mean that the solution of $x^3 - a = 0$ could be found by intersecting two parabolas. This insight is fundamental to algebraic geometry: to represent an unknown object as an intersection of known objects. Extending this idea, Omar Khayyam found 19 classes of cubics by constructing various intersections of conics to solve them. From a modern viewpoint, due principally to Hilbert, his classes may be presented as *ideals* generated by second-degree polynomials in x and y, e.g.

$$R = P \cdot (x^2 - y) + Q \cdot (ax - y^2), \tag{5.3}$$

where P and Q are arbitrary polynomials in x and y. This equation has the same structure as Eq. (5.1). Its defining characteristic is that simultaneous zeros of the generators are necessarily zeros of all members R of the ideal.

Shapes are to a fermion many-body wave function what conics are to cubics. They are generators of all solutions to the N-fermion wave equation which respect the Pauli principle. Like the conics, they are not arbitrary functions, and also like the conics, there is a finite number of them. The efficient generation of shapes is the subject of current research efforts, described in the second part of this chapter.

5.1.1 Technicalities

In order to implement the scheme (5.1) most simply, a technical step is necessary. The Bargmann transform [2] reads

$$\mathcal{B}[f](t) = \frac{1}{\pi^{1/4}} \int_{\mathbb{R}} dx\, e^{-\frac{1}{2}(t^2+x^2)+xt\sqrt{2}} f(x) \equiv F(t). \tag{5.4}$$

Here $f \in L^2(\mathbb{R})$ and $F \in \mathrm{F}(\mathbb{C})$, the *Bargmann space* of entire functions $F : \mathbb{C} \to \mathbb{C}$ such that

$$\int_{\mathbb{C}} |F(t)|^2 d\lambda(t) < \infty, \tag{5.5}$$

where

$$d\lambda(t) = \frac{1}{\pi} e^{-|t|^2} d\,\mathrm{Re}\,t\, d\,\mathrm{Im}\,t, \qquad \int_{\mathbb{C}} d\lambda(t) = 1. \tag{5.6}$$

The inverse Bargmann transform is then

$$\mathcal{B}^{-1}[F](x) = \frac{1}{\pi^{1/4}} \int_{\mathbb{C}} d\lambda(t)\, e^{-\frac{1}{2}\left(\bar{t}^2 + x^2\right) + x\bar{t}\sqrt{2}} F(t), \tag{5.7}$$

where the bar denotes complex conjugation.

Specifically, the Bargmann transform of Hermite functions $\psi_n(x)$ is

$$\mathcal{B}[\psi_n](t) = \frac{t^n}{\sqrt{n!}}. \tag{5.8}$$

It has the algebraically important property that quantum numbers (state labels) n add when single-particle wave functions are multiplied, $t^n t^m = t^{n+m}$. Because $n!m! \neq (n+m)!$, one must use unnormalized single-particle wave functions t^n, with scalar product

$$(t^n, t^m) = \int_{\mathbb{C}} \bar{t}^n t^m d\lambda(t) = n!\, \delta_{nm}. \tag{5.9}$$

In three dimensions, the Hermite functions in x, y, and z are mapped onto Bargmann-space variables t, u, and v, respectively. For N particles, the variables acquire indices, e.g. t_i, with $i = 1, \ldots, N$.

The technical advantage of Bargmann space is that the factorizations $\Phi_i \Psi_i$ in Eq. (5.1) can be interpreted literally, as factorizations of polynomials. The same would not be so easy in laboratory space, where quantum numbers are indices of special functions, which have opaque properties under multiplication. One should bear in mind, however, that the free-module structure (5.1) is an intrinsic feature of Hilbert space, irrespective of representation.

Another important technicality is that the generating function (Hilbert series [3]) which counts the shapes is known [1]. For N fermions in d dimensions, it is a polynomial $P_d(N, q)$, which satisfies *Svrtan's recursion*

$$N P_d(N, q) = \sum_{k=1}^{N} (-1)^{k+1} \left[C_k^N(q) \right]^d P_d(N-k, q), \tag{5.10}$$

where

$$C_k^N(q) = \frac{(1-q^N)\cdots(1-q^{N-k+1})}{(1-q^k)} \tag{5.11}$$

is a polynomial, and $P_d(0, q) = P_d(1, q) = 1$. For example, $P_2(3, q) = q^2 + 4q^3 + q^4$, meaning that, of the $D = 3!^{2-1} = 6$ shapes of $N = 3$ fermions in $d = 2$ dimensions, one is a second-degree polynomial, four are third-degree, and one is fourth-degree.

5.2 Simple Examples [4]

5.2.1 The Fractional Quantum Hall Effect

A particular example of the free-module structure (5.1) has been observed in the context of the $d = 2$ fractional quantum Hall effect (FQHE), albeit without noticing its generality. Adopting the notation of Ref. [5] for this subsection, one of the six shapes counted by $P_2(3, q)$ above is the (second-degree) ground-state Slater determinant,[1]

$$\begin{vmatrix} x_1 & x_2 & x_3 \\ y_1 & y_2 & y_3 \\ 1 & 1 & 1 \end{vmatrix}, \tag{5.12}$$

which is clearly not analytic in terms of the variables $z_j = x_j - iy_j$. One combination of the four third-degree shapes found in Ref. [1] is

$$\Psi_0 = \begin{vmatrix} z_1^2 & z_2^2 & z_3^2 \\ z_1 & z_2 & z_3 \\ 1 & 1 & 1 \end{vmatrix} = (z_1 - z_2)(z_1 - z_3)(z_2 - z_3), \tag{5.13}$$

while the other three involve terms with $\bar{z}_j$ on one or both rows of the determinant. The sixth (fourth-degree) shape goes into itself under the exchange $x_j \leftrightarrow y_j$, like the ground state (5.12), so it is not analytic in the z_j either.

On the other hand, Laughlin's $N = 3$ wave function for the FQHE, Eq. (7.2.12) of Ref. [5], contains a factor

[1]The localization terms $\exp(-x_1^2/2 - \ldots)$ are dropped for clarity.

$$(z_a + i z_b)^{3m} - (z_a - i z_b)^{3m} = \Phi_m (z_1 - z_2)(z_1 - z_3)(z_2 - z_3) = \Phi_m \Psi_0, \tag{5.14}$$

where

$$z_a = \frac{1}{\sqrt{6}}(z_1 + z_2 - 2z_3), \quad z_b = \frac{1}{\sqrt{2}}(z_1 - z_2), \tag{5.15}$$

and Φ_m is a symmetric polynomial in the z_j. The factorization (5.14) is well known and easy to prove directly, which brings Laughlin's wave function into the scheme (5.1). This correspondence proves, by enumeration, Laughlin's conjecture [5] that there is only one vacuum for $N = 3$ and $d = 2$ which satisfies the analyticity constraint.

5.2.2 *Two Electrons in a Quantum Dot*

Two identical fermions in a three-dimensional harmonic potential are the simplest model of a finite system. It is easy to show that the Bargmann-space angular momentum operator has the same form as in laboratory space,

$$L_z = -i(x\partial_y - y\partial_x) \xrightarrow{\mathcal{B}} -i(t\partial_u - u\partial_t) \equiv \mathfrak{L}_v, \tag{5.16}$$

and cyclically. Therefore, solid harmonics in Bargmann space are the same polynomials as in real space, with (x, y, z) simply replaced by (t, u, v).

The generating function for this case is $P_3(2, q) = 3q + q^3$. The ground-state triplet is a vector in Bargmann space just as in laboratory space,

$$\boldsymbol{\Psi} = (t_1 - t_2, u_1 - u_2, v_1 - v_2) = (\Psi_1, \Psi_2, \Psi_3), \tag{5.17}$$

while the fourth shape, appearing in the second-excited shell, is

$$\widetilde{\Psi} = \Psi_1 \Psi_2 \Psi_3, \tag{5.18}$$

geometrically a pseudoscalar (signed volume). Introducing a vector of symmetric functions along the same lines,

$$\boldsymbol{e} = (t_1 + t_2, u_1 + u_2, v_1 + v_2) = (\eta_1, \eta_2, \eta_3) \tag{5.19}$$

the first-excited shell is spanned by 9 vectors $\eta_i \Psi_j$, which is the scheme (5.1) again. Knowing the form of the solid harmonics, it is easy to recast the scheme in rotationally invariant combinations. E.g., $\boldsymbol{e} \cdot \boldsymbol{\Psi}$ is a scalar, while $(-\eta_1 + i\eta_2)(-\Psi_1 + i\Psi_2) = e_{11}\Psi_{11}$ is a state with total angular momentum and projection $l = m = 2$.

The second-excited shell is more interesting. In addition to squares η_i^2, there appear a new type of symmetric-function excitations, the discriminants

$$\Delta_i = \Psi_i^2, \quad i = 1, 2, 3, \tag{5.20}$$

which are excitations of *relative* motion. There is a total of 10 excitations involving relative motion alone, corresponding to the one-body oscillator states with three quanta. They are spanned by the fourth shape $\widetilde{\Psi}$ in addition to the nine states $\Delta_i \Psi_j$. Of the latter, one can easily construct a vector triplet,

$$(\Delta_1 + \Delta_2 + \Delta_3)\boldsymbol{\Psi}, \tag{5.21}$$

noting that the sum of discriminants is a rotational scalar like r^2. The remaining seven states constitute a rotational septiplet Ψ_{3m} with $l = 3$, where $\Psi_{33} = \Psi_{11}^3$ and $\widetilde{\Psi}$ is embedded in the $m = \pm 2$ states:

$$\widetilde{\Psi} \sim \Psi_{32} - \Psi_{3,-2}. \tag{5.22}$$

Even in this simplest possible example of a finite system, there appear two bands in the spectrum, because there are two shapes constraining the motion: the vector $\boldsymbol{\Psi}$, and the pseudoscalar $\widetilde{\Psi}$. All states in the spectrum can be classified according to whether they contain $\widetilde{\Psi}$ or not. The classification of finite-system spectra into bands is very like Omar Khayyam's classification of cubics by conics: bands are ideals generated by the shapes.

This example is quite revealing of the *kinematic* ("off-shell") nature of the constraint (5.1). The classification into bands is traditionally presented in the context of dynamics, i.e. some concrete equations of motion. Here one sees that bands are qualitative manifestations of geometric constraints in wave-function space, essentially many-body effects of the Pauli principle.

5.3 Large Spaces

Simulations of strongly correlated systems must contend with the well-known fermion sign problem [6]: it is not known in general how to update a many-fermion wave function consistently with an initial phase convention. Variational approaches avoid this problem, but at the price of limiting the possible range of wave functions to the form of the initial *ansatz*.

The shape approach has the potential to obviate both problems. Because the number of shapes is finite, and they can be generated algorithmically in closed form, the expression (5.1) is effectively a variational expression which is guaranteed to exhaust the whole wave-function space. It is natural to recast this program in probabilistic language, because the spaces involved are very large, so it is generally impossible to have a complete expression like (5.1) stored in memory. The principal line of research into shapes at present is to generate an arbitrary shape with equal a priori probability. In the remainder of this chapter, the current state of these efforts will be briefly presented. The principal results have not been published anywhere so far. In particular, the algorithm described in Ref. [7] is superseded here. From now on, the presentation is limited to the case of three dimensions.

5.3.1 The Size of the Space

The very large number of shapes $N!^2 \sim N^{2N}$ must be put in perspective. The highest shape is unique and has degree $G = 3N(N-1)/2$ [7]. The total number of one-particle states up to the G-th oscillator shell is the sum of the shell degeneracies up to it:

$$\sum_{n=0}^{G} \binom{n+2}{2} = \binom{G+3}{3} \sim N^6, \tag{5.23}$$

so for N particles in the first G shells the total number of states is

$$\binom{\sim N^6}{N} \sim N^{6N}. \tag{5.24}$$

Even though the number of shapes is unimaginably large, it is vanishingly small in comparison with the total number of states spanning the same range of oscillator shells. Beyond the G-th shell, no new shapes appear, so the fraction of shapes in the total space can be made small at will.

These considerations open the way to structured simulations of large spaces, in which one optimizes in the shape subspace before taking other states into account. It amounts to using Eq. (5.1) with $\Phi_i \in \mathbb{C}$ as a reduced variational expression in a first step. Such an approach makes physical sense because the nodal surface of the ground-state many-body function is expected to be an intersection of shapes alone: if the Φ_i introduced new nodes, these would correspond to excited states. [In Omar Khayyam's scheme (5.3), R can similarly have roots which are not solutions of the cubic, because of P and Q.] Thus one expects that corrections due to $\Phi_i \notin \mathbb{C}$ in the second step will change the geometry but not the topology of the ground-state nodal surface.

5.3.2 The Structure of Shapes in Three Dimensions

The highest-degree shape S_N of N identical fermions is a product of three 1D ground-state Slater determinants [7] $\tilde{\Delta}_N(t)$, cf. Eq. (5.18),

$$S_N = \tilde{\Delta}_N(t)\tilde{\Delta}_N(u)\tilde{\Delta}_N(v). \tag{5.25}$$

Slater determinants in Bargmann space are Vandermonde forms

$$\tilde{\Delta}_N(t) = \begin{vmatrix} t_1^{N-1}/(N-1)! & \cdots & t_N^{N-1}/(N-1)! \\ \vdots & & \vdots \\ t_1/1! & \cdots & t_N/1! \\ 1/0! & \cdots & 1/0! \end{vmatrix} = \prod_{1\le i<j\le N} \frac{t_i - t_j}{j-i}. \tag{5.26}$$

The normalization with factorials is for later convenience. Now it can be proven that symmetrized derivatives of S_N are shapes. Denoting a derivative with respect to a variable with a capital letter, e.g. T_i for d/dt_i, a symmetrized derivative is denoted by parentheses:

$$(T^a U^b V^c) = \sum_{i=1}^{N} T_i^a U_i^b V_i^c, \tag{5.27}$$

which is a *word* with $a+b+c$ commuting letters T, U, V, where $a, b, c \ge 0$ are integers. A product of words is a *sentence*:

$$(T^a U^b V^c)\cdots(T^x U^y V^z). \tag{5.28}$$

A sentence is a symmetric polynomial in the derivatives, call it $P(T, U, V)$. The defining characteristic of a shape is that it is orthogonal to states of the form $\Phi\Psi$, where Ψ is a shape, and $\Phi \neq 1$ a symmetric polynomial corresponding to a bosonic excitation of Ψ – these were called "trivial states" in Ref. [1].

The claim is that $P(T, U, V)S_N$ is a shape if S_N is a shape. The proof is standard [8]:

$$(P(T, U, V)S_N, \Phi\Psi) = (S_N, \Phi P(t, u, v)\Psi) = 0, \tag{5.29}$$

where $P(t, u, v)$ is the symmetric polynomial obtained by replacing all derivatives in $P(T, U, V)$ with the corresponding variables. The first equality follows because variables and derivatives are conjugate in the scalar product (5.9), and the second because the product $P(t, u, v)\Psi$ is multiplied by $\Phi \neq 1$, which gives a trivial state, no matter how $P(t, u, v)\Psi$ itself is resolved in the scheme (5.1).

For example, $(TU)S_2 = 2(v_1 - v_2)$, cf. Eq. (5.17). This approach is far more efficient than the originally published one [1], because it generates shapes directly,

Fig. 5.1 The number of shapes approaches the number of sentences asymptotically when the number of particles $N \to \infty$, as shown here for $N = 5, 10, 15$

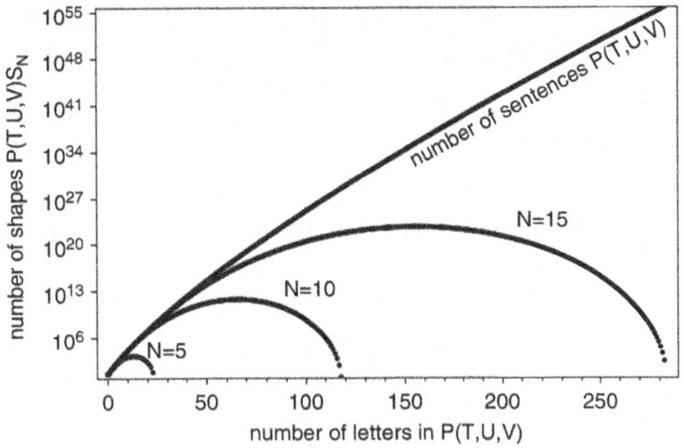

without the need to generate the much larger oscillator-shell spaces in which they are embedded. The main open problem at present is that the number of sentences is still much larger than the number of shapes. This redundancy is illustrated in Fig. 5.1. The number of sentences[2] is equal to the number of shapes only when the total number of letters in a sentence is equal or smaller than N, the number of particles. Higher and higher derivatives of S_N eventually reach the ground state for any finite N, after which the number of shapes is zero, as observed in the figure.

5.3.3 Shape Notation for a Given Number of Particles

The sentence notation is adapted to an arbitrary number of particles. Any sentence acting on S_N will give a shape (or zero), irrespective of the value of N. A notation focused on a given finite N is introduced for the redundancy problem now.

Because derivatives act on columns of $\tilde{\Delta}_N$, the factorials allow a compact expression for the outcome of such operations.

$$[n_1 \dots n_N] = \begin{vmatrix} t_1^{n_1}/n_1! & \cdots & t_N^{n_N}/n_N! \\ t_1^{n_1-1}/(n_1-1)! & \cdots & t_N^{n_N-1}/(n_N-1)! \\ \vdots & & \vdots \\ 0 \text{ or } 1 & \cdots & 0 \text{ or } 1 \end{vmatrix} . \tag{5.30}$$

For example, $\tilde{\Delta}_3(t) = [2\,2\,2]$ and $d\tilde{\Delta}_3(t)/dt_1 = [1\,2\,2]$. This determinant is a slight generalization of the so-called *confluent Vandermonde form*. Writing the three terms in the product (5.25) vertically in the order t, u, v, the variables can be left implicit, e.g.

[2]With all words having at least two distinct letters, because $(T^a)\Delta_N(t) = 0$ [7].

$$\begin{matrix}[0\,1\,2]\\ [1\,1\,2]\\ [1\,2\,2]\end{matrix} = \begin{vmatrix} 1 & t_2 & t_3^2/2 \\ 0 & 1 & t_3 \\ 0 & 0 & 1 \end{vmatrix} \cdot \begin{vmatrix} u_1 & u_2 & u_3^2/2 \\ 1 & 1 & u_3 \\ 0 & 0 & 1 \end{vmatrix} \cdot \begin{vmatrix} v_1 & v_2^2/2 & v_3^2/2 \\ 1 & v_2 & v_3 \\ 0 & 1 & 1 \end{vmatrix} =$$

$$1 \cdot (u_1 - u_2) \cdot (v_2 - v_3)(v_1 - v_2/2 - v_3/2). \qquad (5.31)$$

A symmetrized derivative is obtained by summing all possible arrangements of columns, e.g.

$$\begin{Bmatrix} 1\,1\,2 \\ 1\,1\,2 \\ 2\,2\,2 \end{Bmatrix} = \begin{matrix}[1\,1\,2]\\ [1\,1\,2]\\ [2\,2\,2]\end{matrix} + \begin{matrix}[1\,2\,1]\\ [1\,2\,1]\\ [2\,2\,2]\end{matrix} + \begin{matrix}[2\,1\,1]\\ [2\,1\,1]\\ [2\,2\,2]\end{matrix}. \qquad (5.32)$$

The curly brackets denote a *shape symbol*, or symbol for short. For a given N, sentences and symbols are related linearly, e.g.

$$(TU)^2 S_3 = \begin{Bmatrix} 0\,2\,2 \\ 0\,2\,2 \\ 2\,2\,2 \end{Bmatrix} + 2 \begin{Bmatrix} 1\,1\,2 \\ 1\,1\,2 \\ 2\,2\,2 \end{Bmatrix}, \qquad (5.33)$$

where the first term corresponds to the $i = j$ terms in the double sum $(TU)^2$, i.e. the word (T^2U^2), while the second comes from the parts with $i < j$ and $i > j$. The expansion of each sentence into symbols generates at least one distinct symbol (the second one in the example above), so the symbols are just as good a basis for the shapes as the sentences.

Constraints can be deduced for the symbols from the underlying determinants. For example, a symbol will be zero if all entries on any row are less than $N - 1$, because the corresponding determinant then has a row of zeros. These constraints reduce the number of symbols with respect to the number of sentences, which is practical enough for smaller problems, but there still remain many more symbols than shapes of a given degree in general.

If one could find a set of constraints which were both *efficient* in the sense that they allowed every distinct shape to be generated exactly once, and *operative* in the sense that they could be implemented in polynomial time in N, the problem of generating all shapes with equal a priori probability would be solved. Such progress is unlikely to happen by trial and error, because the redundancy problem has a structure and physical meaning of its own, described in the next section.

5.3.4 Syzygies and the Fermion Sign Problem

Sentences can be classified according to the total powers of T, U, and V appearing in them:

$$P(T, U, V) \simeq (T^a U^b V^c). \tag{5.34}$$

As an example, take $N = 4$ particles and generate all shapes $P(T, U, V)S_4$ with $(a, b, c) = (2, 2, 4)$. There are exactly five of them, so the whole subspace in the class of $(T^2U^2V^4)$ can be spanned by five distinct symbols, such as

$$\begin{Bmatrix} 2\,2\,3\,3 \\ 2\,2\,3\,3 \\ 1\,1\,3\,3 \end{Bmatrix}, \begin{Bmatrix} 2\,2\,3\,3 \\ 1\,3\,3\,3 \\ 1\,3\,2\,2 \end{Bmatrix}, \begin{Bmatrix} 2\,2\,3\,3 \\ 1\,3\,3\,3 \\ 1\,1\,3\,3 \end{Bmatrix}, \begin{Bmatrix} 1\,3\,3\,3 \\ 3\,1\,3\,3 \\ 1\,3\,2\,2 \end{Bmatrix}, \begin{Bmatrix} 1\,3\,3\,3 \\ 2\,2\,3\,3 \\ 1\,3\,2\,2 \end{Bmatrix}. \tag{5.35}$$

Practically, they are found as follows. The highest shape S_4 has [3 3 3 3] on each row, so the row-sum is 12. A derivative such as $(T^2U^2V^4)$ subtracts 2 from the first and second rows, and 4 from the third, so all shapes in the same class can be generated by symbols whose row-sums are 10, 10, and 8 respectively. When all so-far known constraints and symmetries are implemented, 21 symbols are allowed. One expands them one by one in the underlying variables t_i, u_i, v_i until five distinct ones are found. The dimension of the subspace (five) is known in advance from a refinement of Svrtan's recursion: in the formula (5.10), drop the $(-1)^{k+1}$ and replace (for $d = 3$)

$$\left[C_k^N(q)\right]^3 \to C_k^N(T) \cdot C_k^N(U) \cdot C_k^N(V). \tag{5.36}$$

With these modifications, the recursion gives a polynomial $B(N, T, U, V)$, such that the coefficient of $T^aU^bV^c$ is the dimension of the shape subspace of class $(T^aU^bV^c)$. In this particular example,

$$B(4, T, U, V) = \ldots + 5 \cdot T^2U^2V^4 + \ldots, \tag{5.37}$$

so the algorithm can stop as soon as five distinct symbols are found. Among all the 21 symbols, one finds relations such as

$$\begin{Bmatrix} 2\,2\,3\,3 \\ 1\,3\,3\,3 \\ 3\,1\,2\,2 \end{Bmatrix} + \begin{Bmatrix} 2\,2\,3\,3 \\ 1\,3\,3\,3 \\ 1\,3\,2\,2 \end{Bmatrix} - \begin{Bmatrix} 2\,2\,3\,3 \\ 1\,3\,3\,3 \\ 1\,1\,3\,3 \end{Bmatrix} = 0. \tag{5.38}$$

In the theory of invariants, such relations are called *syzygies* [3]. They appear characteristically as polynomial expressions in some determinants, which vanish when the determinants are expanded in the underlying variables. When the determinants express geometrical constraints, a syzygy shows which constraints imply one another. An example is Desargues' theorem: adjusting some lines to intersect is the same as adjusting some points to be collinear. The syzygy (5.38) is a challenge to begin developing such geometric insights for constraints induced by the Pauli principle in wave-function space. Notice, for example, that this particular syzygy is one-dimensional, because the first two rows in all three symbols are the same.

Syzygies are algebraic expressions of the fermion sign problem. It appears in Eq. (5.38) as three different wave functions which interfere destructively because there are actually only two distinct functions. Generating only distinct symbols is solving the sign problem, because it fixes the expression (5.1) as a variational *ansatz*.

Syzygies of shape symbols have physical meaning. Sentences represent deexcitations (loss of quanta) of the high-lying special state S_N. In a fast deexcitation cascade starting with that state, expressions like (5.38) are physical interference effects between different branches of the cascade: not to include them would get the branching ratios wrong. Syzygies become a problem only when one wants to model the *equilibrium* state, requiring that all distinct wave functions be included with equal a priori probabilities.

Algebraically, syzygies generate their own ideal, the *syzygy ideal* of the invariant ring. Calculations in the quotient ring *modulo* the syzygy ideal are real in the sense that one is not inadvertently manipulating zeros. Methods developed for such computations are of two kinds. One requires "unpacking" the determinantal expressions in terms of the underlying variables, such as the algorithm described above, which discovered the basis (5.35) and the syzygy (5.38). The other is symbolic, in the sense that some rules at the level of the symbols themselves determine which expressions are allowed. This second type of method is the goal of the present research, because expanding the symbols in terms of the underlying variables rapidly becomes prohibitive when the number of fermions increases. It amounts to finding efficient constraints on the symbols, as discussed in the previous section.

5.4 Connection to Field Theory

It has been pointed out before [1] that the excitations Φ_i in Eq. (5.1) are counted in a manner strictly analogous to the quantization of the electromagnetic field. Namely, the excitations in the three directions in space are mutually independent. Every Φ_i can be expanded in monomials of the form $\Phi_j^x \Phi_k^y \Phi_l^z$, where each $\Phi^{x,y,z}$ is a symmetric one-dimensional function *by itself*. A 3D excitation is just an iteration of 1D excitations.

The connection of these excitations to field theory is straightforward: just let the number of particles $N \to \infty$. This limit is implicit in the theory of symmetric functions, where one assumes by default that N is fixed but may be arbitrarily large, as if the "supply of variables" were inexhaustible [9]. Physically, the heat capacity of the electromagnetic field increases without bound with temperature because the field has infinitely many degrees of freedom in a finite volume, so the supply of degrees of freedom (wave-numbers) is inexhaustible, even if not all are excited.

The important difference between the excitations Φ_i and field degrees of freedom is that the former have no zero-point energy, which resides in the shapes [1]. In the limit $N \to \infty$, the mapping of sentences to shapes (Fig. 5.1) becomes one-to-one. Notably, all words in a sentence must have at least two distinct letters, because one-dimensional words (T^a) give zero when acting on S_N [7]. While S_N is a product of 1D forms, one needs to "tie together" the spatial directions in a word in order to obtain a shape in general. This property is in sharp contrast with the excitations, which factorize into one-dimensional functions $\Phi^{x,y,z}$ *only*. A field-theoretic description of fermionic vacua should map to the space of (infinitely long) sentences (5.28), not to the excitations Φ_i. It remains to be seen whether such a connection is possible.

5.5 Discussion

The success of the Slater determinant basis rests on the physical fact that the ground-state Slater determinant is a good starting point for the description of the ground state and low-lying excitations of many real systems. The simple algebraic structure of this determinant then provides the formal advantage, that one can manipulate a very large expression – a Slater determinant of $N = 10^{23}$ particles – without ever having to "unpack" it in terms of the one-particle labels. This advantage is realized through the well-known formalism of second quantization.

The advantage disappears in strongly correlated cases, which are formally characterized as those where the initial *ansatz* needs to be more complex than a single Slater determinant. These are known as configuration-interaction approaches in quantum chemistry, going back to the Heitler-London wave function. One can still use the second-quantized notation, but its cost-effectiveness is ultimately compromised by the *physical* necessity to calculate in terms of the individual orbitals. The end point of such a deconstruction are quantum Monte Carlo (QMC) simulations, which rapidly become prohibitive with a larger number of particles.

The present work puts back some structure into the latter efforts, motivated by the desire to write a strongly correlated variational *ansatz* without prejudice. A QMC calculation deals with a very large Hilbert space. However, not all states in that space are born equal. A minority, the shapes, are distinguished by the ability to act as vacuum states, or algebraically, as generators of the Hilbert space as a free module. These are natural generalizations of the ground-state Slater determinant. Learning to manipulate these states at the formal level, using only the shape symbols as labels, is the corresponding generalization of the second-quantized formalism. It promises to mobilize the apparatus of algebraic geometry and classical invariant theory for many-body problems with strong correlations.

The interpretation of the fermion sign problem in terms of syzygies is a case in point. The syzygies are a nuisance in equilibrium calculations, which is the sign problem, but they also have a real meaning. Physically, the sentences and symbols represent chains of de-excitation of the initial (highest) shape S_N. A syzygy like (5.38) means that the same wave function may be constructed by apparently different laboratory preparations, e.g. photon-loss sequences. In chemists' language, it indicates different synthetic pathways to the same entangled state.

The daunting size of the shape space, $N!^2$ in three dimensions, has two mitigating factors. One is that current simulations work in a much larger space by default, so one who is not afraid of N^{6N} can hardly complain that N^{2N} is too much. However, that leaves open the possibility that such heavy work for very large N may be unnecessary, which leads to the second mitigating factor. Experience with actual strongly correlated systems – the FQHE [5], large molecules [10, 11], and modern functional materials [12, 13] – indicates a hierarchical organization of the wave function, in which comparatively few electrons create a correlated template, which is then extended across the system. With the developments described above, the shape formalism is already operational for $N \sim 5$, which is the number of identical fermions in a d shell, so it is more applicable to real-world problems than appears from the present article.

Investigating the geometry of wave-function space seems worthy both as an end in itself and as a practical tool. It is a new open frontier of fundamental quantum mechanics.

Acknowledgments I thank J. Bonča and S. Kruchinin for the invitation to present these results at the NATO Advanced Research Workshop in Odessa. Conversations with M. Primc and D. Svrtan are gratefully acknowledged. This work was supported by the Croatian Science Foundation under Project No. IP-2018-01-7828 and University of Zagreb Support Grant 20283207.

References

1. Sunko DK (2016) Natural generalization of the ground-state Slater determinant to more than one dimension. Phys Rev A 93:062109. https://doi.org/10.1103/PhysRevA.93.062109
2. Bargmann V (1961) On a Hilbert space of analytic functions and an associated integral transform part I. Commun Pure Appl Math 14(3):187. https://doi.org/10.1002/cpa.3160140303
3. Sturmfels B (2008) Algorithms in invariant theory, 2nd edn. Springer, Wien
4. Rožman K, Sunko DK (2020) Generic example of algebraic bosonisation. Eur Phys J Plus 135:30 . https://doi.org/10.1140/epjp/s13360-019-00015-0
5. Laughlin RB (1990) Elementary Theory: the Incompressible Quantum Fluidge. In: Prange RE, Girvin SE (eds) The quantum Hall effect, 2nd edn. Springer, New York, pp 233–301
6. Hirsch JE (1985) Two-dimensional Hubbard model: numerical simulation study. Phys Rev B 31:4403. https://doi.org/10.1103/PhysRevB.31.4403
7. Sunko DK (2017) Fundamental building blocks of strongly correlated wave functions. J Supercond Nov Magn 30:1. https://doi.org/10.1007/s10948-016-3799-1
8. Bergeron F, Préville-Ratelle LF (2012) Higher trivariate diagonal harmonics via generalized Tamari posets. J Comb 3(3):317. https://doi.org/10.4310/JOC.2012.v3.n3.a4
9. Stanley RP (1999) Enumerative combinatorics. Cambridge University Press, Cambridge
10. Nakatsuji H, Nakashima H (2015) Solving the Schrödinger equation of molecules by relaxing the antisymmetry rule: inter-exchange theory. J Chem Phys 142(19):194101. https://doi.org/10.1063/1.4919843
11. Kruchinin S (2016) Energy spectrum and wave function of electron in hybrid superconducting nanowires. Inter J Mod Phys B 30(13):1042008

12. Soldatov AV, Bogolyubov NN Jr, Kruchinin SP (2006) Method of intermediate problems in the theory of Gaussian quantum dots placed in a magnetic field. Condens Matter Phys 9(1):1
13. Park H, Millis AJ, Marianetti CA (2014) Computing total energies in complex materials using charge self-consistent DFT + DMFT. Phys Rev B 90:235103. https://doi.org/10.1103/PhysRevB.90.235103

Chapter 6
Factors and Lattice Reactions Governing Phase Transformations in Beta Phase Alloys

O. Adiguzel

Abstract Metals and alloys have different phases at different conditions in crystallographic basis, and these phases are described in phase diagrams as alloy composition-temperature, or composition-pressure dependent. Beta phases of alloys mainly have bcc-based ordered or disordered structure and these phases are very sensitive to external conditions, and phase structures turn into other crystal structures by lowering temperature and stressing material, by means of dual thermal and stress induced martensitic transformations. Lattice vibrations and interatomic interactions play an important role in the processing of transformation. Thermal induced martensitic transformation occurs in atomic scale in the material on cooling from parent phase region, and interatomic interactions govern this transition. These interactions are described by pair-wise potential function between all of the atom pairs, and embedded atom electron cloud potential functions. Thermal induced martensitic transformation occurs along with lattice twinning on cooling and ordered parent phase structures turn into twinned martensite structures, these structures turn into detwinned martensite structures by stressing material by means of strain induced martensitic transformation. Thermal induced martensitic transformation occurs as a lattice reaction with the cooperative movement of atoms on two opposite two opposite directions, ⟨110⟩-type directions on the {110}-type planes of austenite matrix which is basal plane of martensite. Copper based alloys exhibit this property in metastable β-phase region, which has bcc-based structures at high temperature parent phase field. Lattice invariant shear and twinning is not uniform in these alloys and gives rise to the formation of complex layered structures. In the present contribution, X-ray diffraction and transmission electron microscopy (TEM) studies were carried out on two copper based CuAlMn and CuZnAl alloys, and reached results were interpreted.

O. Adiguzel (✉)
Department of Physics, Firat University, Elazig, Turkey
e-mail: oadiguzel@firat.edu.tr

J. Bonča, S. Kruchinin (eds.), *Advanced Nanomaterials for Detection of CBRN*,
NATO Science for Peace and Security Series A: Chemistry and Biology,
https://doi.org/10.1007/978-94-024-2030-2_6

Keywords Martensitic transformations · Shape memory effect · Lattice twinning · Detwinning

6.1 Introduction

Metals and alloys have different phases at different conditions in crystallographic basis, and these phases are described in phase diagrams as alloy composition-temperature, or composition-pressure dependent. Beta phases of alloys mainly have bcc-based ordered or disordered structure and these phases are very sensitive to external conditions, and phase structures turn into other crystal structures by lowering temperature and stressing material, by means of crystallographic transformation, thermal and stress induced martensitic transformations. Thermal treatments and phase transformations are important factor in the processing of materials, like strengthening the steel and shape changes with variation of temperature. Lattice vibrations and interatomic potentials play an important role in the processing of transformation. Thermal induced martensitic transformation occurs in atomic scale in the material on cooling from parent phase region, and interatomic interactions govern this transition. Martensitic transformations have displacive character, and movements are confined to interatomic lengths smaller than the lattice parameters of the material crystal. Interatomic interactions occur as pair-wise interaction or embedded atom electron cloud interactions. These interactions are described by pair-wise potential function between all of the atom pairs, and embedded atom electron cloud potential functions. Martensitic transformation is based on the location of atoms in the crystal; and structural analysis can be done by using the atomic positions in the simulation studies.

Thermal and stress induced martensitic transformations initiate an event called shape memory effect mainly in β-phase region of certain alloys, called shape memory alloys. These alloys take place in a class of functional materials with this peculiar property, shape memory effect. This property is characterized by the recoverability of desired shape on the material at different temperatures. Shape memory effect is performed in only thermal manner on heating and cooling in a temperature interval after deformation at low temperature martensitic condition, and this behavior can be called thermoelasticity. Shape memory effect is based on successive thermal and stress induced martensitic transformations, and microstructural mechanisms are responsible for shape memory behavior. In particular, the twinning and detwinning processes are essential as well as martensitic transformation in reversible shape memory **effect** [1, 2]. Thermal induced martensite occurs on cooling along with lattice twinning by means of a shear-like mechanism, and ordered parent phase structure turns into twinned martensite as multivariant martensite in self-accommodating manner. Twinned martensitic structures turn into detwinned or reoriented martensitic structures by means of stress induced martensitic transformation by stressing material in low temperature condition. Thermal induced martensitic transformations occur with cooperative movement

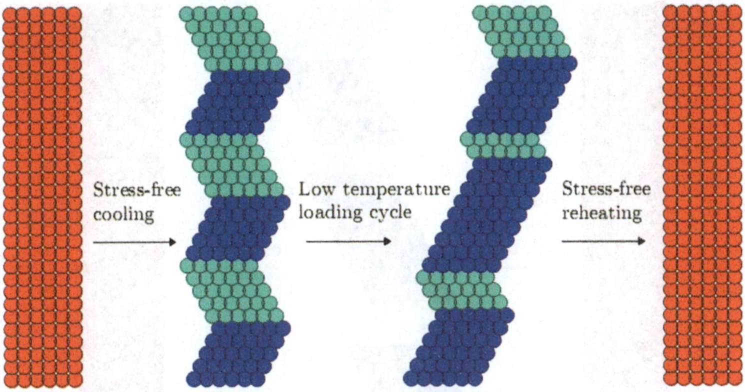

Fig. 6.1 Schematic illustration of transformation mechanism in shape-memory effect [2]

of atoms by means of lattice invariant shears on a {110}-type plane of austenite matrix which is basal plane of martensite. The lattice invariant shears occurs, in two opposite directions, ⟨110⟩-type directions on the {110}-type basal planes. The {110}-plane family has 6 certain lattice planes; {110}, {1 −1 0}, {101}, {10−1}, {011}, {0 1 −1}. This kind of shear can be called as {110} ⟨110⟩-type mode and possible 24 martensite variants occur. Shape memory alloys can be deformed plastically in low temperature martensitic condition, and recover the original shape on heating over the austenite finish temperature. The material cycles between the deformed and original shapes on cooling and heating in reversible shape memory case. By applying external stress, martensite variants are forced to reorient into a single variant leading inelastic strains, and deformation of shape memory alloys in martensitic state proceeds through a martensite variant reorientation or detwinning of **twins** [1, 2]. The twinning occurs with internal stresses, while detwinning occurs with the external stresses. The basic mechanism of phase transformations and shape memory effect in crystallographic level is schematically illustrated in Fig. 6.1 [2]. The deformed material recovers the original shape in bulk level, and crystal structure turns into the parent phase structure on first heating.

Shape memory alloys exhibit another property called superelasticity (SE), which is performed by mechanical stress in a constant temperature in parent phase region. These alloys can be deformed just over austenite finish temperature and recover the original shape on releasing the stress in superelastic manner. Deformation at different temperature exhibits different behavior beyond shape memory effect and superelasticity.

Shape memory effect is performed in a temperature interval depending on the forward (austenite → martensite) and reverse (martensite → austenite) transformation, on cooling and heating, respectively. Superelasticity is performed in the parent austenite phase region, just over austenite finish temperature. Superelastic materials are deformed in the parent phase region and, shape recovery is carried out instantly and simultaneously upon releasing the applied stress. This property

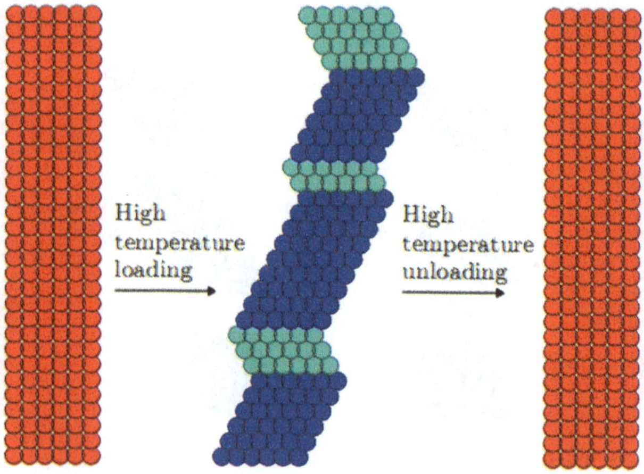

Fig. 6.2 Schematic illustration of phase transformation in superelasticity [2]

exhibits classical elastic material behavior, but stressing and releasing paths are different. Stress-strain behavior is different in two cases, shape memory effect and super-elasticity. Deformation is performed plastically in product martensitic condition in shape memory case. Meanwhile, the material is deformed in parent phase region in superelasticity, and recovers the original shaper after releasing the external stress. Superelasticity is also result of martensitic transformation; stress induced martensitic transformation, which is induced by applying external stress only in mechanical manner. With this stress, ordered parent austenite phase structures turn into the fully detwinned martensite. The basic mechanism of phase transformation in superelasticity is schematically illustrated in Fig. 6.2 [2].

Superelasticity is performed in non-linear way, unlike normal elastic materials and exhibits rubber like behavior. Loading and unloading paths in stress-strain diagram are different in superelasticity, and hysteresis loop refers to the energy dissipation. Deformation at different temperature exhibits different behavior beyond shape memory effect and superelasticity [3–7]. The hysteresis loop refers to the absorption of strain energy and these alloys are mainly used as deformation absorbent materials in damping devices and buildings, due to the absorbance of strain energy during any disaster or seismic events, like earthquake. This hysteresis is one of the available characteristics for potential dissipative applications [8]. The potential applications of shape memory alloys (SMA) in damping devices for civil structures, like buildings and bridges, to smooth out the oscillations produced by earthquakes and winds has been a subject of increasing interest in recent years. The superelasticity and the hysteresis cycle associated to the martensitic transformation in shape memory alloys are used to dissipate the energy of oscillations [8, 9] With these properties, shape memory alloys with successive thermally and stress induced martensitic transformations are potential candidates for highly effective, nonstandard mechanical damping systems [9]. The loading and releasing paths

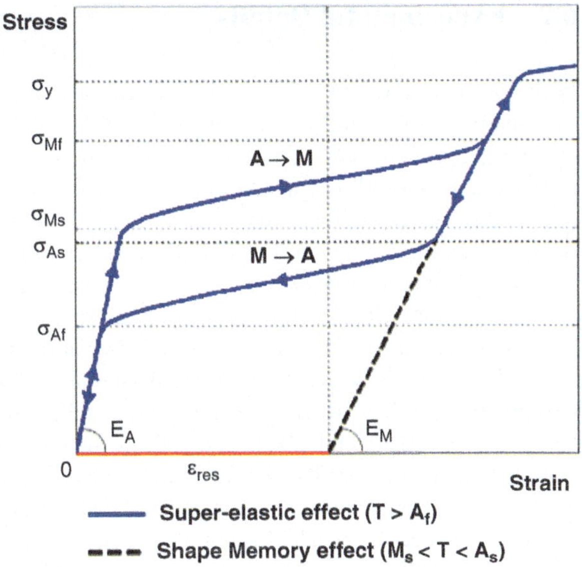

Fig. 6.3 Stress-strain diagram of shape memory effect and superelasticity [10]

demonstrating the shape memory and superelastic effects and stress-strain diagram is shown in Fig. 6.3 [10].

Copper based alloys exhibit this property in metastable β-phase region, which has B2 or DO_3 – type ordered lattice at high temperature, and these structures martensitically turn into layered complex structures with lattice twinning process, on cooling from high temperature austenitic phase region. Martensitic transformations occur mainly in two steps in copper based ternary alloys. First one is Bain distortion, and second one is lattice invariant shear. Bain distortion consists of an expansion of 26% parallel to the ⟨001⟩-type axes, and compression of 11% parallel to the ⟨110⟩ and ⟨1$\bar{1}$0⟩-type directions [3]. Lattice invariant shears occur with cooperative movement of atoms less than interatomic distances on {110}-type close packet plans of austenite matrix.

The lattice invariant shears occur, in two ⟨110⟩-type opposite directions on the {110}-type basal planes and this type of shear can be called as {110} ⟨110⟩-type mode and has 24 variants in self-accommodating manner [11–14]. These lattice invariant shears are not uniform in copper alloys and give rise to the formation of unusual complex layered structures called long period layered structures such as 3R, 9R or 18R depending on the stacking sequences on the close-packed planes of the ordered lattice. The complicated long-period stacking ordered structures mentioned above can be described by different unit cells. All of these martensite phases are long-period stacking ordered structures that is the underlying lattice is formed by stacks of close-packed planes. In case the parent phase has a B2-type superlattice, the stacking sequence is ABCBCACAB(9R) [11, 13].

6.2 Experimental Details

In the present paper, two copper based ternary shape memory alloys were selected for investigation; Cu-26.1%Zn 4%Al and Cu-11%Al-6%Mn (in weight). The martensitic transformation temperatures of these alloys are over the room temperature and both alloys are entirely martensitic at room temperature. Specimens obtained from these alloys were solution treated for homogenization in the β-phase field (15 min at 830 °C for CuZnAl alloy and 20 min at 700 °C for CuAlMn alloy), then quenched in iced-brine to retain the β-phase and aged at room temperature after quenching (both alloys). Powder specimens for X-ray examination were prepared by filling the alloys. Specimens for TEM examination were also prepared from 3 mm diameter discs and thinned down mechanically to 0.3 mm thickness. These specimens were heated in evacuated quartz tubes in the β-phase field (15 min at 830 °C for CuZnAl and 20 min at 700 °C for CuAlMn) for homogenization and quenched in iced-brine. These specimens were also given different post-quench heat treatments and aged at room temperature. TEM and X-ray diffraction studies carried out on these specimens. TEM specimens were examined in a JEOL 200CX electron microscope, and X-ray diffraction profiles were taken from the quenched specimens using Cu-K_α radiation with wavelength 1.5418 Å.

6.3 Results and Discussion

X-ray powder diffractograms were taken from CuZnAl and CuAlMn samples.

An x-ray powder diffractogram taken from the long term aged CuAlMn alloy sample is shown in Fig. 6.4. This diffractogram has been indexed on the monoclinic M18R basis. X-ray diffraction profile of CuZnAl alloy also exhibit similar

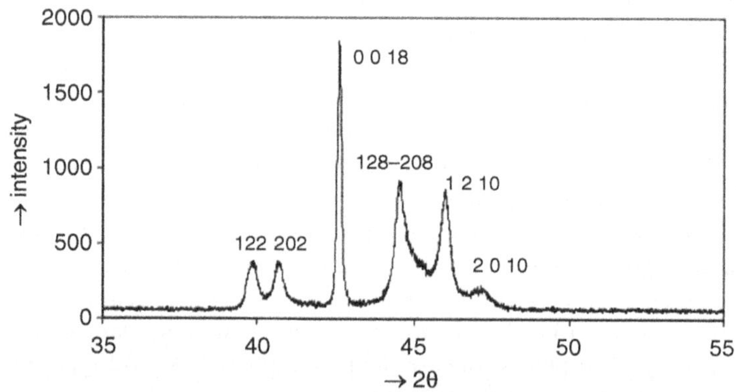

Fig. 6.4 An X-ray diffractogram taken from the long term aged CuAlMn alloy sample

Fig. 6.5 Two electron diffraction patterns taken from the CuZnAl and CuAlMn alloy samples

configuration. Two electron diffraction patterns taken from CuZnAl alloy sample are also shown in Fig. 6.5, respectively.

X-ray powder diffractograms and electron diffraction patterns reveal that these alloys exhibit superlattice reflections. X-ray diffractograms and electron diffraction patterns reveal that both alloy have the ordered structure in martensitic condition, and exhibit superlattice reflections. A series of X-ray powder diffractograms and electron diffraction patterns were taken from both CuZnAl and CuAlMn alloy samples in a large time interval and compared with each other. It has been observed that electron diffraction patterns exhibit similar characteristics, but some changes occur at the peak locations and intensities on the X-ray diffractograms with aging duration. These changes occur as rearrangement or redistribution of atoms in the material, and attribute to new transitions in diffusive manner [3, 13, 15]. The ordered structure or super lattice structure is essential for the shape memory quality of the material. In the shape memory alloys, homogenization and releasing the external effect is obtained by ageing at β-phase field for adequate duration.

On the other hand, post-quench ageing and service processes in devices affect the shape memory quality, and give rise shape memory losses. These kinds of results lead to the martensite stabilization in the reordering or disordering manner. In order to make the material satisfactorily ordered and to delay the martensite stabilization, copper-based shape memory alloys are usually treated by step-quenching after homogenization. Metastable phases of copper-based shape memory alloys are very sensitive to the ageing effects, and any heat treatment can change the relative stability of both martensite and parent phases [8, 15]. Martensite stabilization is closely related to the disordering in martensitic state. Structural ordering is one of the important factors for the formation of martensite, while atom sizes have important effect on the formation of ordered structures [8, 16, 17].

Although martensitic transformation has displacive character, martensite stabilization is a diffusion controlled phenomena, and this result leads to redistribution of atoms on the lattices sites. Stabilization is important factor and causes to memory losses, and changes in main characteristics of the material; such as, transformation temperatures, and diffracted angles and peak intensities. Martensitic transformation in copper-based β-phase alloys is based on one of the $\{110\}_\beta$ planes of parent phase called basal plane for martensite. The (110) basal plane which has a rectangular shape in parent phase is subjected to hexagonal distortion and undergoes a hexagon.

The powder specimens were aged at room temperature after quenching process, and many x-ray diffractograms have been taken from both of the alloy samples in a large time interval. Although all of the diffractograms exhibit similar characteristics, some changes have been observed at diffraction angles and intensities of diffraction peaks on the diffractograms with aging duration. These changes are attributed to new transitions which have diffusive character. It means that some neighbor atoms change locations. In particular, some of the neighbor peak pairs have moved toward each other. It is interesting that miller indices of these plane pairs provide a special relation:

$$\left(h_1^2 - h_2^2\right)/3 = \left(k_2^2 - k_1^2\right)/n,$$

where $n = 4$ for 18R martensite [3]. These plane pairs can be listed as follow; (122)–(202), (128)–(208), (1 2 10)–(2 0 10), (040)–(320). This result can be attributed to a relation between interplane distances of these plane pairs and rearrangement of atoms on the basal plane. In these changes, atom sizes play important role. The different sizes of atomic sites lead to a distortion of the close-packed plane from an exact hexagon and thus a more close-packed layered structure may be expected. In the disordered case, lattice sites are occupied randomly by the atoms, and atom sizes can be taken nearly equal, and martensite basal plane becomes an ideal hexagon, whereas, the lattice sites are occupied regularly by different atoms which have different sizes.

6.4 Conclusion

It can be concluded from the above results that the copper-based shape memory alloys are very sensitive to the ageing treatments. X-ray diffraction angles and peak intensities in X-ray diffractograms change with the ageing time in martensitic condition. In particular, some successive peak pairs come close each other. These changes lead to the martensite stabilization in the redistribution or disordering manner, and stabilization proceeds by a diffusion-controlled process. The martensite stabilization is a diffusion controlled phenomena and leads to redistribution of atoms on the lattices sites, although martensitic transformation has displacive character. The basal plane of martensite turns into a hexagon by means of Bain distortion with martensite formation on which atom sizes have important effect. In case the atoms occupying the lattice sites have the same size, the basal plane of martensite becomes regular hexagon; otherwise the deviations occur from the hexagon arrangement of the atoms. The above mentioned peaks come close each other in the disordered case, and occur separately in the ordered case. The changes in the diffraction angles of the selected plane pairs can be a measure of the ordering degree in martensite.

References

1. Ma J, Karaman I, Noebe RD (2010) High temperature shape memory alloys. Int Mater Rev 55:257–315
2. Richards AW (2013) Interplay of martensitic transformation and plastic slip in polycrystals. Ph.D. thesis, California Institute of Technology
3. Adiguzel O (2013) Phase transitions and microstructural processes in shape memory alloys. Mater Sci Forum 762:483–486
4. Ermakov V, Kruchinin S, Fujiwara A (2008) Electronic nanosensors based on nanotransistor with bistability behaviour. In: Bonca J, Kruchinin S (eds) Proceedings NATO ARW "Electron transport in nanosystems". Springer, pp 341–349
5. Dzhezherya Y, Novak IY, Kruchinin S (2010) Orientational phase transitions of lattice of magnetic dots embedded in a London type superconductors. Supercond Sci Technol 23:1050111–105015
6. Ermakov V, Kruchinin S, Pruschke T, Freericks J (2015) Thermoelectricity in tunneling nanostructures. Phys Rev B 92:115531
7. Repetsky SP, Vyshyvana IG, Nakazawa Y, Kruchinin SP, Bellucci S (2019) Electron transport in carbon nanotubes with adsorbed chromium impurities. Materials 12:524
8. De Castro F, Sade M, Lovey F (2012) Improvements in the mechanical properties of the 18R ↔ 6R high hysteresis martensitic transformation by nanoprecipitates in CuZnAl alloys. Mater Sci Eng A 543:88–95
9. de Castro Bubani F., Lovey F, Sade M, Cetlin P (2016) Numerical simulations of the pseudoelastic effect in CuZnAl shape-memory single crystals considering two successive martensitic transitions. Smart Mater Struct 25:1. https://doi.org/10.1088/0964-1726/25/2/025013
10. Barbarino S et al (2014) A review on shape memory alloys with applications to morphing aircraft. Smart Mater Struct 23:1–19
11. Zhu JJ, Liew KM (2003) Description of deformation in shape memory alloys from DO3 austenite to 18R martensite by group theory. Acta Mater 51:2443–2456
12. Sutou Y et al (2005) Effect of grain size and texture on pseudoelasticity in Cu–Al–Mn-based shape memory wire. Acta Mater 53:4121–4133
13. Adiguzel O (2017) Thermoelastic and pseudoelastic characterization of shape memory alloys. Int J Mater Sci Eng 5(3):95–101
14. Casati R et al (2014) Thermal cycling of stress induced martensite for high performance shape memory effect. Scr Mater 80:13–16
15. Li Z, Gong S, Wang MP (2008) Macroscopic shape change of $Cu_{13}Zn_{15}Al$ shape memory alloy on successive heating. J Alloys Compd 452:307–311
16. Guo YF et al (2007) Mechanisms of martensitic phase transformations in body-centered cubic structural metals and alloys: molecular dynamics simulations. Acta Mater 55:6634–6641
17. Aydogdu A, Aydogdu Y, Adiguzel O (2004) Long-term ageing behaviour of martensite in shape memory Cu–Al–Ni alloys. J Mater Process Technol 153–154:164–169

Chapter 7
Quantum-Chemical Calculations of Pure and Phosphorous Doped Ultra-small Silicon Nanocrystals

Sh. Makhkamov, F. Umarova, A. Normurodov, N. Sulaymonov, O. Ismailova, A. E. Kiv, and M. Yu. Tashmetov

Abstract The development of nanotechnology has led to the use of various nanostructures in order to modify and produce new materials. In this regard, there is an increasing interest in studying the properties of nanostructures, methods for their preparation, and mechanisms for their incorporation into various matrices.The diamond-like hydrogen-passivated ultra-small $Si_{29}H_{36}$, $Si_{34}H_{36}$, $Si_{59}H_{60}$, $Si_{71}H_{60}$ nanocrystals were studied by the first-principle method. In this ultra-small length scale, the dependence of properties with gradually increasing sizes does not practically investigate. A detailed analysis of the structural deformation and charge distribution has been conducted to understand how structures change occurs within this length scale, where quantum confinement effects become predominant.

Keywords Ultra-small silicon nanocrystals · Phosphorous doping · Quantum-chemical calculation · Structure · Electronic properties

---

Sh. Makhkamov · A. Normurodov · N. Sulaymonov · M. Yu. Tashmetov
Institute of Nuclear Physics Academy of Sciences, Tashkent, Uzbekistan

F. Umarova (✉)
Institute of Nuclear Physics Academy of Sciences, Tashkent, Uzbekistan

Institute of Ion-Plasma and Laser Technologies Academy of Sciences, Tashkent, Uzbekistan
e-mail: umarova_inp@mail.ru

O. Ismailova
Turin Polytechnic University in Tashkent, Tashkent, Uzbekistan

Uzbek-Japan Innovation Center of Youth, Tashkent, Uzbekistan

A. E. Kiv
South-Ukrainian National Pedagogical University, Odessa, Ukraine

J. Bonča, S. Kruchinin (eds.), *Advanced Nanomaterials for Detection of CBRN*, NATO Science for Peace and Security Series A: Chemistry and Biology,
https://doi.org/10.1007/978-94-024-2030-2_7

7.1 Introduction

The study of ultra-small silicon nanoparticles is a very active area of research because of the interesting fundamental physical properties of these mesoscale objects and of promising applications in advanced electronic devices and optoelectronic devices. The non-toxicity of silicon nanoparticle allows it to be used in gas sensors and biological objects. In order to realize the optimal functioning of the nanoparticles during their actual applications, their thermal stability, compatibility in different media and other properties are needed to investigate. A clear understanding of the underlying mechanism is necessitated to uncover the relationship between these complicated issues, such as size, shape, and surface composition, thermal and chemical stability.

Doping of nanostructured silicon allows to be controlled of its properties. For example, the mechanisms of ion-induced creation of heavy clusters have been intensively studied. This refers to clusters of heavy atoms, the masses of which are large in comparison with the masses of atoms that make up the compound. The purposeful creation of such clusters can lead to a significant improvement in the properties of materials, such as polymers, materials for reactor engineering, thermoelectric applications, etc. [1, 2].

In this work, by carrying out ab initio calculations, we investigated the stability of Si nanocrystals with P atom depending on their size. The possibility of doping of nanostructured silicon with a diameter of up to 6 nm has been shown in numerous experimental and theoretical investigations published up to the present [3–15]. The theoretical analysis of simulations indicates that doping of P in Si nanoparticles is energetically favourable for nanocrystals with the diameters of more than 2 nm, and in nanocrystals with a diameter of less than 2 nm P atoms tends to substitute Si near the surface [11]. However, ultra-small Si nanostructures doped with P, having mean diameters in the range of 2 nm or below, are obtained using ultra-thin P-SiO_2 films in multilayer structures [16]. Such nanocrystals, embedded in a SiO_2 matrix, are formed during high temperature treatment, and it has been shown that P atoms move in the Si-rich region. In connection with that, we systematically investigated structural deformations, charge distributions, P formation energies and electronic properties of ultra-small particles with silicon core sizes of 0.88 - 1.41 nm.

The paper is organized as follows: in Sect. 7.2, we represent the computational method used, the structural properties, charge distributions, the energies of phosphorous formation, electronical properties of doped nanocrystals. Our conclusion is given in Sect. 7.3.

7.2 Results and Discussion

7.2.1 Computational Method

For calculations we used the first-principle ORCA software package [17] based on the density functional theory (DFT). A density functional theory (DFT) has been used with a def2-SVP basis set The exchange-correlation functional is parameterized by the Becke–Lee–Yang–Parr (BLYP) scheme.The simple basis set with combination of BLYP have been successfully used to predict the stability of lowest lying energy of silicon nanocrystals [18] and the experimentally-measured thermal conductivities for isotopically pure silicon [19]. The spherical diamond-like $Si_{29}H_{36}$, $Si_{34}H_{36}$, $Si_{59}H_{60}$, $Si_{71}H_{60}$ nanocrystals with the diameters of silicon cores of 0.88, 1.03, 1.26, and 1.41 nm, respectively, are cut out from the bulk crystal with taking all atoms within a sphere of a given radius. Each nanocrystal is centred on a silicon atom. The surface dangling bonds of the initial nanocrystals are saturated by hydrogen and then a full geometry optimization without symmetry constraints was performed. Spin-restricted and unrestricted Hartree–Fock calculations have been performed for pure and doped clusters correspondingly. The ground state structure of pure nanocrystal with a spin multiplicity $M = 1$ ($M = 2S + 1$) was used. Once the ground state structure of doped structure was obtained, the central or subsurface silicon atoms were replaced by P atom and again optimized. The energy of the doped crystal was calculated with a multiplicity $M = 2$. The Mulliken atomic population analysis is used in calculations.

For doped structures, the optimized geometries of initial nanostructures were used. Both structural and electronic properties have been investigated as a function of the size and position of the impurity within nanoparticles.

7.2.2 Structural Properties

The calculated parameters of the initial undoped nanoparticles show the structural changes after optimization. The diameters of silicon cores of nanocrystals change in various directions and nanoparticles become asymmetrical. However, the central parts of nanoparticles retain their tetrahedral symmetry. The lengths of Si–Si bonds are ~2.39–2.40 Å, the widths of the band gap (HOMO-LUMO gap) of undoped nanoparticles are significantly expanded in comparison with bulk silicon.

Figures 7.1 and 7.2 show the optimized geometries of doped nanoparticles with the phosphorous atom replacing the Si atom in the centre (P_c) and on the subsurface sphere (P_{ss}), respectively. In Tables 7.1 and 7.2, the calculated parameters of pure and doped nanocrystals are presented. The structural changes induced by the presence of a single P atom depend on the sizes of nanoparticles. In the $Si_{29}P_cH_{36}$, the symmetry is strongly deviated from optimized initial one. The phosphorous atom is displaced from the central position; it is possible, Jahn–Teller distortion of the

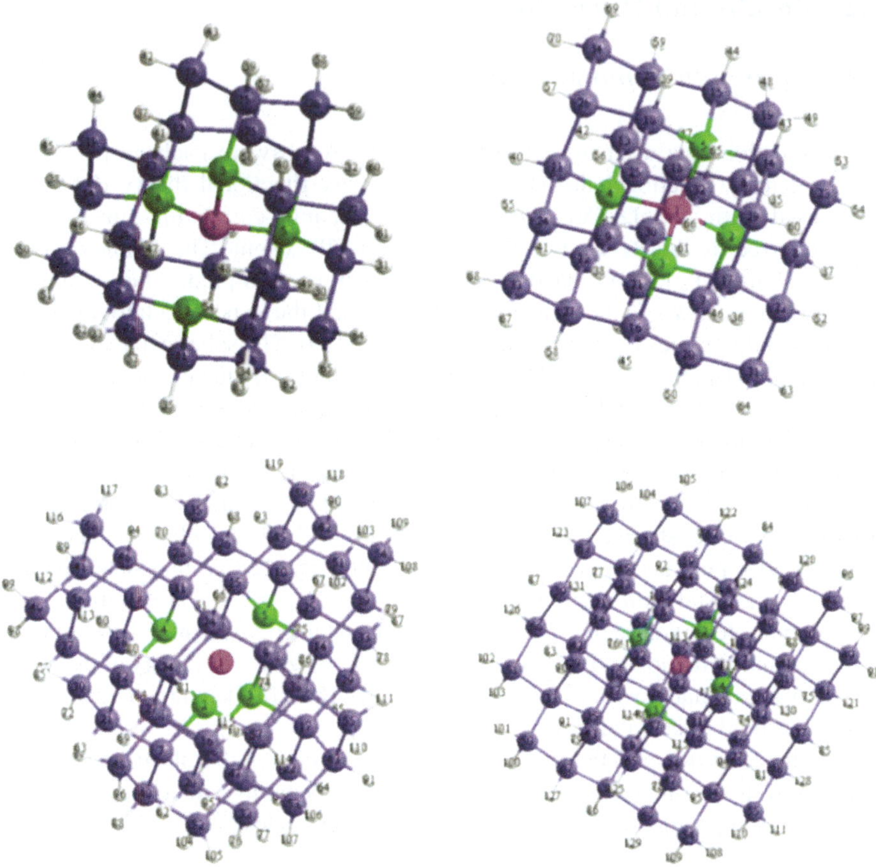

Fig. 7.1 Structure of optimized geometries of $Si_{28}P_cH_{36}$, $Si_{33}P_cH_{36}$, $Si_{58}P_cH_{60}$, $Si_{70}P_cH_{60}$ nanocrystals. The silicon and hydrogen atoms denote as violet and grey. P_c is the phosphorous atom located in the centre denotes by red colour. The atoms on the first coordination sphere denote with green colour

central P atom occurs [20]. The P atom with Si has three shorter (2.345 Å each) and one longer (2.91 Å) bonds, also the diameter of the silicon core increases. In the rest nanoparticles some reconstruction occurs around the P atom. Of greatest interest is the structure of a larger nanocrystal $Si_{70}P_cH_{60}$. The lengths of all four Si–P_c bonds are the same and increase up to 2.444 Å. The following Si–Si bonds between the first and second spheres are also the same and equal to 2.387 Å. Thus, the spherical surrounding of P atom is retained, the silicon core diameter does not change as compared with the undoped one.

Structural deformation of nanostructures is accompanied by changes in the charge distribution on the spheres. In all nanostructures, a charge of the central phosphorous atom is negative and increases relative to undoped structures. In

Table 7.1 Diameters of the silicon core and charges on atoms of undoped and doped silicon nanocrystals

			Average charges on coordination spheres of nanocrystals				
Nano-crystal	Diameter of the silicon core, Å	Charge of the central atom	1st sphere	2nd sphere	3rd sphere	4th sphere	H atoms
$Si_{29}H_{36}$	7.98–8.81	−0.07	−0.11	−0.004	0.02	–	0.01
$Si_{28}P_cH_{36}$	8.03–8.85	−0.21	−0.06	−0.01	0.02	–	0.010
$Si_{28}P_{ss}H_{36}$	8.80–8.88	0.03	−0.13	−0.01	0.02	–	0.001
$Si_{34}H_{36}$	9.90–11.08	−0.07	−0.09	0.01	0.02	–	0.004
$Si_{33}P_cH_{36}$	9.90–11.08	−0.15	−0.08	0.01	0.02	–	0.01
$Si_{33}P_{ss}H_{36}$	9.83–11.16	0.03	−0.13	0.006	0.01	–	0.01
$Si_{59}H_{60}$	12.02–12.57	−0.06	−0.06	−0.09	0.04	0.003	0.01
$Si_{58}P_cH_{60}$	12.02–12.57	−0.13	−0.03	−0.08	0.04	0.003	0.015
$Si_{58}P_{ss}H_{60}$	12.03–12.60	0.03	−0.06	−0.1	0.04	0.003	0.01
$Si_{71}H_{60}$	11.73–14.09	−0.04	−0.07	−0.07	−0.14	0.036	0.15
$Si_{70}P_cH_{60}$	11.73–14.09	−0.14	−0.04	−0.06	−0.14	0.036	0.025
$Si_{70}P_{ss}H_{60}$	11.76–14.19	0.03	−0.04	−0.08	−0.14	0.037	0.003

Table 7.2 Lengths (Å) of Si–P bonds in doped nanocrystals

Nano-crystals	Lengths of Si-P_c bonds	Nano- crystals	Lengths of Si–P_{ss} bonds	Nano- crystals	Lengths of Si–P_c bonds	Nano- crystals	Lengths of Si–P_{ss} bonds
$Si_{28}P_cH_{36}$	2.91	$Si_{28}P_{ss}H_{36}$	2.332	$Si_{58}P_cH_{60}$	2.451	$Si_{58}P_{ss}H_{60}$	2.356
	2.345		2.50		2.47		2.402
	2.345		2.50		2.464		2.337
	2.345		2.357		2.464		2.649
$Si_{33}P_cH_{36}$	2.40	$Si_{33}P_{ss}H_{36}$	2.399	$Si_{70}P_cH_{60}$	2.443	$Si_{70}P_{ss}H_{60}$	2.35
	2.399		2.397		2.443		2.34
	2.40		2.396		2.443		2.44
	2.399		2.397		2.443		2.44

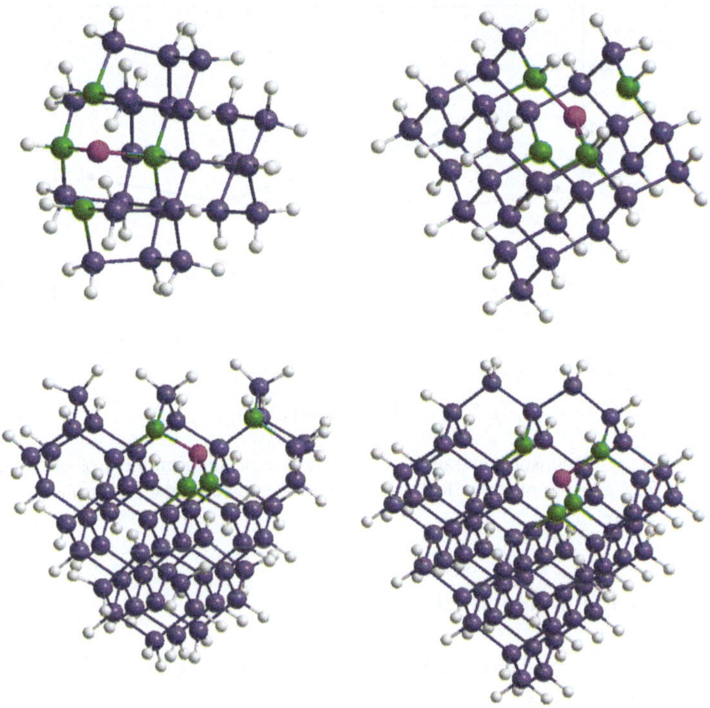

Fig. 7.2 Structure of optimized geometries of $Si_{28}P_{ss}H_{36}$, $Si_{33}P_{ss}H_{36}$, $Si_{58}P_{ss}H_{60}$, $Si_{70}P_{ss}H_{60}$. The silicon and hydrogen atoms denote as violet and grey. P_{ss} is the phosphorous atom replacing the Si atom on the subsurface sphere. The silicon atoms nearby phosphorous denote with green colour

addition, an average negative charge on neighbouring sphere decreases. The outer spheres of the silicon core are positively charged. As a result, a "spherical dipole" with negative core and positively charged surface is formed.

Calculations of structures doped by P atom on subsurface sphere show that ultra-small $Si_{28}P_{ss}H_{36}$ nanocrystal is strongly deformed as well, the lengths of Si-P_{ss} bonds change differently and the diameter of silicon core increases. With the size increases, the deformation decreases Because of local deformation around the impurity in the subsurface region a slight increase in the core diameter is observed. The charge of the central atom is positive, and its value does not change with the increase in the size of the nanocrystal.

7.2.3 Formation energy

The energies of phosphorous formation are calculated as the difference between the total energies of the undoped and doped structures. The calculation was carried out

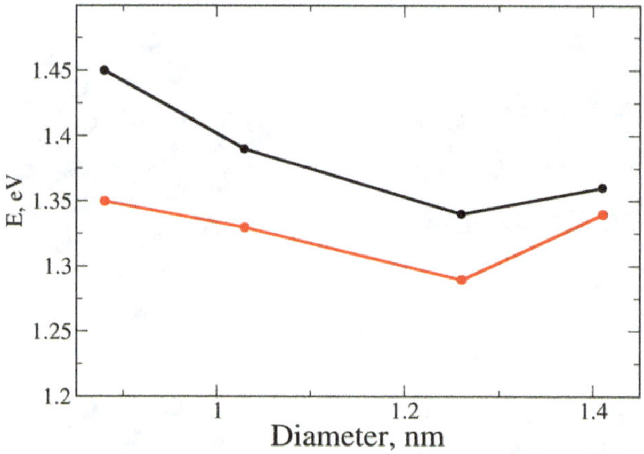

Fig. 7.3 The energy of phosphorus impurity formation depending on the size of Si nanocrystallites: upper curve – phosphorus atom located in the centre, lower curve – phosphorous atom is on the subsurface sphere

according to the following formula [6, 7]

$$E_f = E(\mathrm{Si}_{n-1}\mathrm{PH}_m) - E(\mathrm{Si}_n\mathrm{H}_m) + \mu_{\mathrm{Si}} - \mu_{\mathrm{P}}, \tag{7.1}$$

here, μ_{Si} is total energy per atom of bulk silicon, μ_{P} is total energy per atom of phosphorous.

The calculated formation energies presented in Fig. 7.3 show that the incorporation of phosphorous in ultra-small silicon crystals is more preferred in the largest ones. As can be seen in Fig. 7.3 also, the energies of phosphorous formation in central and subsurface positions begin draw nearly in $Si_{70}P_cH_{60}$ and $Si_{70}P_{ss}H_{60}$ nanocrystals. The energy difference between these states become insignificantly. To verify this fact, four positions of P atom in different direction in subsurface sphere were defined. Calculated formation energies of P atom in these positions have almost equal values ($\sim$1.34 $\div$ 1.35 eV). We conclude that both the centre and subsurface positions of the P atom in the nanocrystals $Si_{70}P_cH_{60}$ and $Si_{70}P_{ss}H_{60}$ can be energetically stable. The structural parameters of the nanocrystal $Si_{70}P_cH_{60}$ make it possible to place the atom P in the centre, since it has a tetrahedral geometry, the first sphere expands, the energy of formation of the atom P in the central and subsurface positions differs insignificantly and the P atom can be located in a favourable position in the centre or closely.

The authors [11], studying the energies of P location in Si freestanding nanocrystals, predicted that there is a critical size of 2 nm, below which the P atom will be expelled to the surface. The possibility of formation of doped Si nanocrystals with a diameter of 1.5–2.0 nm, embedded in SiO_2 matrix, has been experimentally shown in [13], in accordance with the results of TEM [21, 22]. The phosphorous peak in ToF-SIMS indicates that the phosphorous in silicon-rich range may exist inside oxide film. The probability of P atom to leave the Si region is limited by the lower diffusivity of P in surrounding matrix.

7.2.4 *Electronic Properties*

Electronic properties of undoped Si nanocrystals have changed as compared with bulk silicon due to quantum confinement, and their band gaps (the energy difference between the top of valence band and bottom of conductive band) have considerably extended and controlled by the sizes of the nanocrystals (Fig. 7.4).

As can be seen from Fig. 7.5, the edges of valence and conductive bands in doped nanocrystals of the same size with undoped ones only insignificantly changes. After doping of silicon nanocrystal with P, a spin splitting level is created in the band gap, and its location depends on the size of the nanocrystal. Figure 7.5 shows the calculated energy levels for structures investigated, when the impurity is located in the centre and near the surface. Phosphorous levels are located below the conduction band In this case, the HOMO-LUMO gap strongly decreases and equalled the energy difference between the phosphorous levels and the bottom of conducting band.

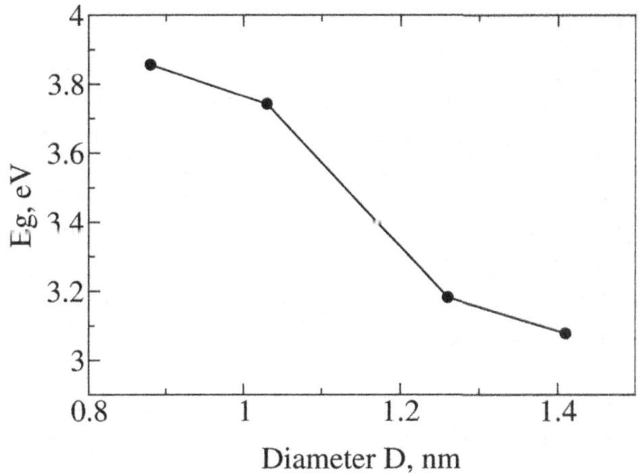

Fig. 7.4 The band gap of undoped nanocrystals as a function of the size of the silicon core. For the other curves difference between top of valence band and bottom of conducting band has been shown

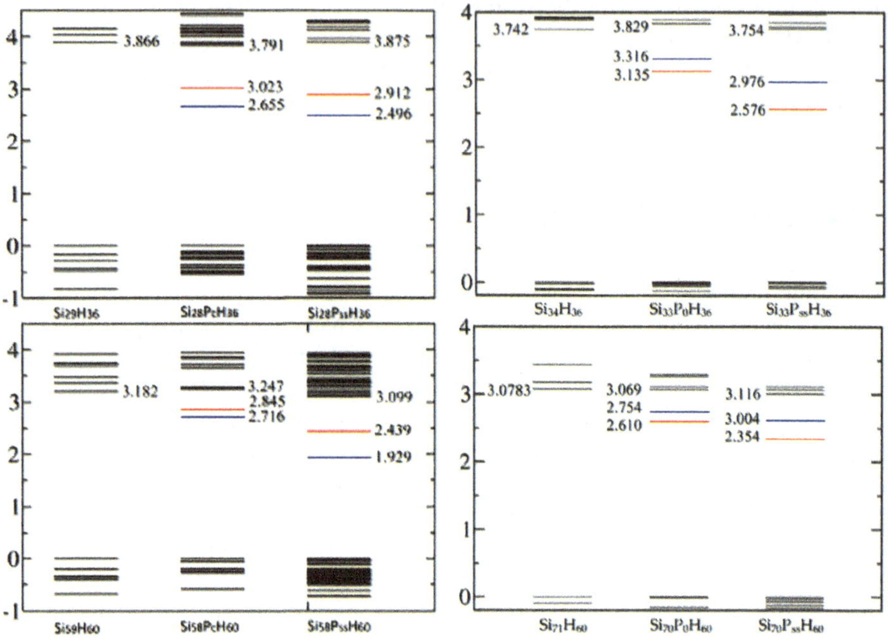

Fig. 7.5 Calculated energy levels for nanocrystals of different sizes. From left to right – initial undoped nanocrystals, with the phosphorous atom located in the centre and in the subsurface region

7.3 Conclusion

The structural and electronic properties of ultra-small nanocrystals $Si_{29}H_{36}$, $Si_{34}H_{36}$, $Si_{59}H_{60}$, $Si_{71}H_{60}$ have been studied by the first-principle method. The structural analysis has shown that all nanocrystals undergo rearrangements initiated by the presence of phosphorous. In ultra-small $Si_{29}H_{36}$ nanocrystals, deformation covers all the spheres. The phosphorous atom is displaced from the central position, it is possible, Jahn-Teller distortion of central P atom occurs due to spin splitting. Structural parameters of the nanocrystal $Si_{70}P_cH_{60}$ make it possible to place the atom P in the favourable position in the centre or closely. Changes in the HOMO-LUMO gap of undoped nanocrystals are related to the effects of quantum confinement. The width of the band gap of doped structures decreases strongly as phosphorous introduces the additional levels below the conduction band.

References

1. Kavetskyy T et al (2019) Formation of heavy clusters in ion-irradiation compounds. Vacuum 164:149–152
2. Kirievsky K et al (2019) Ion-induced n-p inversion of conductivity in TiNiSn compound for thermoelectric applications. J Appl Phys 126:155106. https://doi.org/10.1063/1.5121825

3. Mimura A, Fujii M, Hayashi S, Kovalev D, Koch F (2000) Photoluminescence and free-electron absorption in heavily phosphorus-doped Si nanocrystals. Phys Rev B62(19):12625
4. Kovalev D, Heckler H, Polisski G, Koch F (1999) Optical properties of Si nanocrystals. Phys Status Solidi B 215(2):871–932
5. Linnros J (2005) Optoelectronics: nanocrystals brighten transistors. Nat Mater 4:117
6. Zhou Z, Friesner RA, Brus L (2003) Electronic structure of 1 to 2 nm diameter silicon core/shell nanocrystals: surface chemistry, optical spectra, charge transfer, and doping. J Am Chem Soc 125:15599
7. Melnikov DV, Chelikowsky JR (2004) Quantum confinement in phosphorus-doped silicon nanocrystals. Phys Rev Lett 92:046802
8. Pawlak BJ, Gregorkiewicz T, Ammerlaan CAJ, Takkenberg W, Tichelaar FD, Alkemade PFA (2001) Experimental investigation of band structure modification in silicon nanocrystals. Phys Rev B 64:115308
9. Chelikowsky JR, Alemany MMG, Chan T-L, Dalpian GM (2011) Computational studies of doped nanostructures. Rep Prog Phys 74(4):046501
10. Ossicini S, Bisi O, Degoli E, Marri I, Iori F, Luppi E, Magri R et al (2008) First-principles study of silicon nanocrystals: structural and electronic properties, absorption, emission, and doping. J Nanosci Nanotechnol 8(4):479–492
11. Chan TL, Tiago ML, Kaxiras E, Chelikowsky JR (2008) Size limits on doping phosphorus into silicon nanocrystals. Nano Lett 8(2):596–600
12. Iori F, Degoli E, Luppi E, Magri R, Marri I, Cantele G, Ninno D, Trani F, Ossicini S (2006) Doping in Silicon nanocrystals: an ab-initio study of the structural, electronic and optical properties. J Lumin 121:335–339
13. Repetsky SP, Vyshyvana IG, Kruchinin SP, Bellucci S (2018) Influence of the ordering of impurities on the appearance of an energy gap and on the electrical conductance of graphene. Sci Rep 8:9123
14. Kruchinin SP, Zolotovsky A, Kim HT (2013) Band structure of new ReFe AsO superconductors. J Modern Phys 4:608–611
15. Repetsky SP, Vyshyvana IG, Kuznetsova EY, Kruchinin SP (2018) Energy spectrum of graphene with adsorbed potassium atoms. Int J Mod Phys B 32:1840030
16. Perego M, Bonafos C, Fanciulli M (2010) Phosphorus doping of ultra-small silicon nanocrystals. Nanothechnology 21(2):025602
17. Neese F (2012) The ORCA program system. Wiley Interdiscip Rev Comput Mol Sci 2:73–78
18. Yoo S, Shao N, Zeng XC (2009) Reexamine structures and relative stability of medium-sized silicon clusters: low-lying endohedral fullerene-like clusters Si_{30}–Si_{38}. Phys Lett A 373:3757–3760
19. Jain A, McGaughey AJH (2015) Effect of exchange–correlation on first-principles-driven lattice thermal conductivity predictions of crystalline silicon. Comput Mater Sci 110:115–120
20. Jahn HA (1938) Stability of polyatomic molecules in degenerate electronic states. II. Shin degenerate. Proc R Soc A164:117–131
21. Cho E-C, Park S, Hao X, Song D, Conibeer G, Park S-C, Green MA (2008) Silicon quantum dot/crystalline silicon solar cells. Nanotechology 19(24):245201
22. Zacharias M, Heitmann J, Scholz R, Kahler U, Schmidt M, Blasing J (2002) Size-controlled highly luminescent silicon nanocrystals: a SiO/SiO_2 superlattice approach. Appl Phys Lett 80(4):661

Chapter 8
Theoretical Aspects of Nanosensors for Radiation Hazards Detecting

G. Tatishvili, T. Marsagishvili, M. Matchavariani, and Z. Samkharadze

Abstract Theoretical aspects of using the modern opportunities of nanotechnology for creation of new type sensors for determination of lowest concentration radiation sources are under consideration.

Taking into account abilities of modern high sensitive facilities for analytical signals registration, are suggested sensors based on changing of electroconductivity of the system while the adsorption of impurity particle on the semiconductor electrode under the radiation influence. It is recommended to use such sensors for determination of radioactive substances in biological liquids and in water ecosystems.

The Green function technique makes possible to analyze the kinetics of the elementary act of charge transfer processes in detail.

Analytical expressions for kinetic parameters of charge transfer processes in such systems are obtained.

Keywords Nanosensor · Radiation · Electroconductivity · Adsorption · Charge transfer · Green function

8.1 Introduction

Continuous monitoring to identify the sources of CBRN (chemical, biological, radiation, nuclear) hazards is of the highest importance for ensuring the safety of the population and requires multi-point monitoring in real time.

Traditional environmental control measures, involving periodic sampling followed by laboratory testing, of course, cannot satisfy the CBRN monitoring requirements, since due to the length of time between sampling and obtaining

G. Tatishvili (✉) · T. Marsagishvili · M. Matchavariani · Z. Samkharadze
R. Agladze Institute of Inorganic Chemistry and Electrochemistry of Iv. Javakhishvili Tbilisi State University, Tbilisi, Georgia
e-mail: tati@iice.ge

J. Bonča, S. Kruchinin (eds.), *Advanced Nanomaterials for Detection of CBRN*, NATO Science for Peace and Security Series A: Chemistry and Biology,
https://doi.org/10.1007/978-94-024-2030-2_8

analysis results, the risks of technological, environmental, and terrorist disasters increase sharply.

The solution of this problem is the development and creation of miniature detectors with different operating principles, converting information about the state of the medium into electrical and optical signals, with a subsequent signal transfer at different distances. In addition, the energy consumption of these sensors is negligible. They can receive power, for example, from solar panels, thermoelectric generators, kinetic energy converters, etc. Accordingly, such energy efficiency makes them necessary for any autonomous control systems.

The development of nanotechnology opens up great opportunities for the development of a large variety of such sensors. Ultra-small sizes of them contribute to bringing their sensitivity practically to theoretically possible limits. Such nanosensors are very suitable for detecting extremely low concentrations of particles of any type. In particular, in case of radiation sources, it is the very important issue, because of very low concentrations of the radionuclides their detection is connected with some difficulties. At the same time, long-term exposure of them, even at their low concentration, brings to essential changes in biological processes.

Different processes occur in substances under the influence of high energy radiation (heavy charged particles, neutrons, electrons, X-ray beams, etc.). As a result of such influence different short-lived formations occur in the substance, such as excited molecules, free radicals, solvated electrons, etc.

Dosimetry based on registration of the products created as a result of the influence of radiation, has essential value when investigation of radiation processes [1–3].

Possibilities of modern technical registration facilities may be used during the investigation of radiation sources, when their concentration is low. For example, modern electronic equipment makes it possible to note even insignificant changes in the electronic structure of materials that can be caused by various phenomena, for example, heating, irradiation, adsorption, etc. At the same time, one can increase the concentration of radionuclides in liquid medium with the help of synthetic or natural sorbents [4–6].

So, application of sensitive methods and increase of the concentration by use of any sorbents allows usage of the sensors based on registration of electroconductivity. It is suggested to use such sensors for determination of some substances existent in biological liquids (with radioactive substances) and for determination of radioactive detrimental impurities in water ecosystems. The analytical signal of these sensors may be transmitted for long distances. That allows installation of these devices in places which are inaccessible or dangerous for the personnel.

This paper shows theoretical evaluations for processes that can occur in a system that describes the operation principle of certain types of nano-sensors to identify radiation sources with very low concentrations.

Systems for electrochemical nanosensors with electrodes made of semiconductor materials with thin-film coatings operating in an irregular condensed medium, in particular, the liquid polar phase, are considered.

A convenient apparatus for describing the processes under study is the Hamiltonian formalism and the method of delayed Green functions [7–10].

Representation of the medium using the temperature Green's functions constructed on the polarization operators of the molecules of the medium makes it possible to take into account the effects of frequency and spatial dispersion [11, 12].

Here we will investigate the sensors based on registration of electroconductivity.

The principle of effective mechanism may be based on the following: under the effect of radioactive sources impurity particles of solution are adsorbed on the electrode surface. At this time, according to the concepts of chemisorption, a charge transfer takes place from the conduction band of semiconductor on an adsorbed impurity particle in a condensed medium [13], which causes a change of electroconductivity and correspondingly the magnitude of current transmitted between electrodes of the sensor changes.

Among the components, used for creation of sensors' electrodes, there are no substances harmful to the environment (materials with repeated processing and regeneration possibilities will be used).

8.2 Description of the Investigated System and Processes

The system under study consists of semiconductor with a thin-film coating, nonregular polar condensed medium (electrolyte) with polar impurity molecules of low concentration.

The processes passing in the investigated system may be divided into several groups. First group – charge transfer processes between particles and between electrode and particle in condensed medium. Second group – electron excitation processes of some particles and processes of electron excitation energy transfer between particles in condensed medium.

Consequently, the process may be divided into two types. First type processes – charge transfer form electrode on particle takes place, i.e. current, which passes between the electrodes of the sensor changes. Second type processes – molecule excitation occurs in the film on the surface of the electrode under the influence of radiation and transfer of excitation energy takes place.

Since in this paper we consider theoretical aspects concerning the operating mechanism of only a nanosensor based on the principle of determining changes in the electrical conductivity of a system, we will not touch on the issue of describing the processes of charge transfer between particles with photon radiation. We also will not describe electron transfer between two impurity particles due to their low concentration in the electrolyte.

8.3 Charge Transfer Processes from Semiconductor Electrode on Adsorbed Impurity Particle in Electrolyte

Charge transfer processes from an electrode on adsorbed impurity particle may be described analytically, at selection of a particular system, for which system's Hamiltonian will be written correctly [14, 15]. The principal complicity of research of electron transfer processes is connected with necessity of quantum approach. In frames of this approach, there are no accepted models for charge transfer and energy transfer processes in non-regular condensed medium. Usage of the mathematical physics apparatus of temperature Green function permits the description of complex condensed systems and charge transfer processes in such systems. This method permits to unify theoretical approaches and to use different models for the description of spatial and frequency effects of dispersion.

8.3.1 System Hamiltonian

The Hamiltonian of the initial state can be written in the form:

$$H^{\mathrm{i}} = H^{\mathrm{i}}_{\mathrm{sc}} + H^{\mathrm{i}}_{\mathrm{m}} + H^{\mathrm{i}}_{\mathrm{p}} + H^{\mathrm{int}}_{\mathrm{p,m}} + H^{\mathrm{int}}_{\mathrm{sc,m}} + H^{\mathrm{int}}_{\mathrm{sc,p}}, \tag{8.1}$$

where $H^{\mathrm{i}}_{\mathrm{sc}}$ is the Hamiltonian of semiconductor electrode in initial state; $H^{\mathrm{i}}_{\mathrm{p}}$ is the Hamiltonian of impurity particles in initial state; $H^{\mathrm{i}}_{\mathrm{m}}$ is the Hamiltonian of condensed medium; $H^{\mathrm{int}}_{\mathrm{sc,m}}$ is the interaction of electrode with the medium; $H^{\mathrm{int}}_{\mathrm{p,m}}$ is interaction of particle with the medium; $H^{\mathrm{int}}_{\mathrm{sc,p}}$ is the interaction Hamiltonian of a particle with electrode. The particle is assumed polarized and "solvated".

Analogously the Hamiltonian in final state has the form (the particle is in electron-excited state):

$$H^{\mathrm{f}} = H^{\mathrm{f}}_{\mathrm{sc}} + H^{\mathrm{f}}_{\mathrm{m}} + H^{\mathrm{f*}}_{\mathrm{p}} + H^{\mathrm{int}}_{\mathrm{p,m}} + H^{\mathrm{int}}_{\mathrm{sc,m}} + H^{\mathrm{int}}_{\mathrm{sc,p}} + e\eta, \tag{8.2}$$

where η is overvoltage in point of particle location.

Electron distribution by energetic levels of electrode is described by a Fermi distribution for semiconductor electrodes.

$$f(\varepsilon) = (1 + \exp((\varepsilon - \varepsilon_{\mathrm{F}})/kT))^{-1}, \tag{8.3}$$

where ε_{F} is Fermi energy, k is Boltzmann constant, T is temperature.

However, if the energy of level essentially differs from the energy of Fermi level, then distribution of energy levels transfers into Boltzmann distribution.

The second question, which is necessary to solve for semiconductor electrodes is the number of quantum states in unit of volume of a substance $\rho(\varepsilon)$. It is difficult to conduct exact calculations for such values, however, it is possible to make some evaluations for crystals.

8.3.2 *Current Density for Electron Transfer Processes from a Semiconductor Electrode on Adsorbed Impurity Particles*

In accordance with the general theory of charge transfer processes, the cathode current density is written in the form:

$$i_c = e\pi \int d\varepsilon \rho(\varepsilon) f(\varepsilon) W_c(\varepsilon), \tag{8.4}$$

where $\rho(\varepsilon)$ is the density of single-electron states in electrode, $f(\varepsilon)$ is an electron Fermi distribution function, $W_c(\varepsilon)$ is probability of the elementary act of electron transfer from energy level with ε energy on a particle in unit time. Integration by $\varepsilon$is carried out over the whole energy spectrum of the electrode.

The probability of an elementary act of electron transfer from an electrode to a particle has the form:

$$W_c(\varepsilon) = \frac{\beta}{i} e^{\beta F_i} \int\limits_{C_\theta} d\theta \mathrm{Sp}\left[e^{-\beta(1-\theta)H_i} L_{fi} e^{-\beta\theta H_f}\right]; \quad \beta = 1/kT. \tag{8.5}$$

Here, H_i and H_f are the Hamiltonians of initial and final states of the system; F_i is free energy of initial state, including electron energy on the electrode and ion energy in electrolyte solution; L_{fi} is electron-resonance integral of electron transfer from an electrode on a particle. Integration contour passes parallel to the imaginary axis in a band $0 < \mathrm{Re}\theta < 1$.

Let's separate electron energy ε, counted out from Fermi energy ε_F, from the Hamiltonian of the initial state $H_i$and corresponding free energy F_i:

$$H_i = \varepsilon - \varepsilon_F + H_{iF}; \quad F_i = \varepsilon - \varepsilon_F + F_{iF} \tag{8.6}$$

and the expression for the current density will be presented in the form:

$$i_c = e\pi \int\limits_{C_\theta} d\theta \mathrm{Sp}\left[e^{-\beta(1-\theta)H_{iF}} L_{fi} e^{-\beta\theta H_f} L_{fi}\right] \frac{\beta}{i} \int\limits_{-\infty}^{\infty} d\varepsilon e^{\beta\theta(\varepsilon-\varepsilon_F)} f(\varepsilon). \tag{8.7}$$

When conducting analytical calculations of current density, the integration by ε energy is carried out first. For metals, the dependence of the density of single-electron states $\rho(\varepsilon)$ as a function of energy behaves as a slowly varying function (for example, in the model of free electrons the dependence of ρonε is of a root nature).

Suppose that the energy dependence of the electron-resonance integral L from energy ε is weak. As a rule, for metals this approximation is performed quite well. Moreover, integration over ε in (8.7) can be carried out approximately by the saddle point method. In this case, it can be shown that:

$$i_{\mathrm{c}} = e\frac{\beta}{i}\mathrm{e}^{\beta F_{\mathrm{iF}}} \int\limits_{\mathrm{C}_\theta} \mathrm{d}\theta \rho(\varepsilon^*) \mathrm{Sp}\left[\mathrm{e}^{-\beta(1-\theta)H_{\mathrm{iF}}} L_{\mathrm{fi}}(\varepsilon^*) \mathrm{e}^{-\beta\theta H_{\mathrm{f}}} L_{\mathrm{fi}}(\varepsilon^*)\right] \times$$

$$\int\limits_{-\infty}^{\infty} \mathrm{d}\varepsilon^* \mathrm{e}^{\beta\theta(\varepsilon^*-\varepsilon_{\mathrm{F}})} f(\varepsilon^*), \tag{8.8}$$

where ε^* is the point at which the expression by ε reaches a maximum. Integration by θ is carried out by the saddle point method.

Omitting intricate calculations, we'll adduce quantum expression for the density of cathode current of heterogeneous process with the participation of semiconductor electrode with Fermi distribution

$$i_{\mathrm{c}} = e\pi \left|L_{\mathrm{fi}}(\boldsymbol{R}^*, \psi^*)\right|^2 \int d\varepsilon \cdot kT\rho(\varepsilon) \exp(-\ln(\sin \pi\theta^*)) \times$$

$$\Phi(\boldsymbol{R}^*, \psi^*) U(\boldsymbol{R}^*, \Psi^*) \exp\left\{ - e\eta - \beta\theta^* \Delta F - \Psi^m(\boldsymbol{R}^*, \psi^*; \theta) - \right.$$

$$\beta \sum_{n=1}^{N} E_{rn} \frac{\theta^*(1-\theta^*)\omega_n^{\mathrm{i}}\omega_n^{\mathrm{f}}}{(1-\theta^*)(\omega_n^{\mathrm{i}})^2 + \theta^*(\omega_n^{\mathrm{f}})^2} - \theta \sum_{k=1}^{N} \ln(\omega_n^{\mathrm{f}}/\omega_n^{\mathrm{i}}) -$$

$$\left. \theta^* \ln\left[1 + \beta \sum_{k=1}^{N} (G_k(\boldsymbol{R}^*, \psi^*) + \frac{\omega_k^{\mathrm{f}}}{\omega_k^{\mathrm{i}}} \bar{G}_k(\boldsymbol{R}^*, \psi^*)) \frac{\sqrt{2E_{rk}\omega_k^{\mathrm{i}}}}{\omega_k^{\mathrm{f}}} \right] \right\} (2\pi)^{1/2} \times$$

$$\left[\prod_{s=1}^{N} \frac{\omega_s^{\mathrm{i}}}{\sqrt{(1-\theta^*)(\omega_s^{\mathrm{i}})^2 + \theta^*(\omega_s^{\mathrm{f}})^2}} \right] \left\{ 1 + \sum_{k=1}^{N} \left[G_k(\boldsymbol{R}^*, \psi^*) + \frac{\omega_k^{\mathrm{f}}}{\omega_k^{\mathrm{i}}} \bar{G}_k(\boldsymbol{R}^*, \psi^*) \right] \times \right.$$

$$\left. (\beta\theta^* - F_\omega(\theta^*)) \sqrt{2E_{rk}\omega_k^{\mathrm{i}}} \theta^* (\omega_k^{\mathrm{f}})^2 \left[(1-\theta^*)(\omega_k^{\mathrm{i}})^2 + \theta^*(\omega_k^{\mathrm{f}})^2 \right]^{3/2} \right\}. \tag{8.9}$$

Here ΔF is the free energy of the process, including particle's electron excitation energy in final state; E_{rn} is the reorganization energy of n-th intramolecular degree of freedom of the particle, ω_n^{i} and ω_n^{f} correspondingly are frequencies of intramolecular oscillation at the beginning and at the end of the process. Functions $G_k(\boldsymbol{R}^*, \psi^*)$ and $\bar{G}_k(\boldsymbol{R}^*, \psi^*)$ are connected with the interaction of intramolecular oscillations of impurity particles with fluctuations of molecule polarization of electrolyte.

$$G(\boldsymbol{R}, \psi) = -\frac{1}{2} \int drdr' \frac{\partial E_i}{\partial Q} (\boldsymbol{r}; \boldsymbol{R}, \psi)\, g_{ik}^R(\boldsymbol{r}, \boldsymbol{r}'; \omega = 0) \Delta E_k(\boldsymbol{r}'; \boldsymbol{R}, \psi), \tag{8.10}$$

$$\bar{G}(\boldsymbol{R}, \psi) = -\frac{1}{2} \int drdr' \Delta E_i (\boldsymbol{r}; \boldsymbol{R}, \psi)\, g_{ik}^R(\boldsymbol{r}, \boldsymbol{r}'; \omega = 0) \frac{\partial E_k}{\partial Q} (\boldsymbol{r}'; \boldsymbol{R}, \psi). \tag{8.11}$$

ΔE is change of electric field intensity of impurity particle and electrode during charge transfer process. Green function g^R is a function of polarization fluctuations' operators of amorphous solid and liquid at finite temperature $T = 1/\beta$.

In factorization approximation for function g^R we have:

$$g_{ik}^R(\boldsymbol{r}, \boldsymbol{r}'; \omega) = g_{ik}^R(\boldsymbol{r}, \boldsymbol{r}'; \omega = 0) f(\omega). \tag{8.12}$$

The reorganization energy of the medium for charge transfer processes may be determined by the expression:

$$E_r^m(\boldsymbol{R}, \psi) = -\frac{1}{2} \int drdr' \Delta E_i (\boldsymbol{r}; \boldsymbol{R}, \psi)\, g_{ik}^R(\boldsymbol{r}, \boldsymbol{r}'; \omega = 0) \Delta E_k(\boldsymbol{r}'; \boldsymbol{R}, \psi). \tag{8.13}$$

Here $\Delta E_i (\boldsymbol{r}; \boldsymbol{R}, \psi)$ is the change of system's electrostatic field strength during the transfer process.

Star in designations of coordinates signifies the value of this coordinate in point of maximum at taking the corresponding integral by the method of saddle points as a rule.

Saddle point θ^* may be found from the equation:

$$e\eta + \beta \Delta F + \frac{\partial \psi^m(R^*, \psi^*; \theta)}{\partial \theta} + \frac{\partial}{\partial \theta} \sum_{k=1}^{N} \beta E_{rk} \frac{\theta(1-\theta)\omega_k^{\mathrm{i}}\omega_k^{\mathrm{f}}}{(1-\theta)(\omega_k^{\mathrm{i}})^2 + \theta(\omega_k^{\mathrm{f}})^2} + \ln \prod_{k=1}^{N} \frac{\omega_k^{\mathrm{f}}}{\omega_k^{\mathrm{i}}} +$$

$$\ln\left\{1 + \beta \sum_{k=1}^{N} \left[G_k\left(\boldsymbol{R}^*, \psi^*\right) + \frac{\omega_k^{\mathrm{f}}}{\omega_k^{\mathrm{i}}} \bar{G}_k\left(\boldsymbol{R}^*, \psi^*\right)\right] \frac{\sqrt{2E_{rk}\omega_k^{\mathrm{i}}}}{\omega_k^{\mathrm{f}}}\right\} + \pi \operatorname{ctg}(\pi\theta) = 0;$$

$$\beta = 1/kT. \tag{8.14}$$

In this formula L_{fi} is resonance integral of particle interaction with surface of semiconductor electrode.

Matrix element is calculated by application of wave functions in framework of concrete model. Resonance integral L_{fi} can be considered as some phenomenological parameter. Arguments of this resonance integral characterize geometric characteristics of the process, distance up to surface, space orientation of particle during charge transfer. Function $U(R^*, \psi^*)$ is calculated for concrete processes with consideration of the geometry of the particles and electrodes. Function $\Phi(\boldsymbol{R}^*, \psi^*)$ is the distribution function of reduced particles. Model function can be selected as this function. Function $\Psi^m(\boldsymbol{R}^*, \psi^*; \theta)$ is function of medium reorganization. Its formal expression is:

$$\Psi^m(\boldsymbol{R}^*, \psi^*, \theta) = \frac{1}{\pi}\int d\boldsymbol{r}d\boldsymbol{r}'\Delta E_i(\boldsymbol{r}, \boldsymbol{R}^*, \psi^*)\Delta E_k(\boldsymbol{r}', \boldsymbol{R}^*, \psi^*)\times$$

$$\int_{-\infty}^{\infty} d\omega \mathrm{Im} g_{ik}^R(\boldsymbol{r}, \boldsymbol{r}'; \omega)\frac{\mathrm{sh}\frac{\beta\omega(1-\theta)}{2}\mathrm{sh}\frac{\beta\omega\theta}{2}}{\omega^2\mathrm{sh}\frac{\beta\omega}{2}}. \tag{8.15}$$

In factorization approximation for function g^R, reorganization function of the medium may be presented as:

$$\Psi^m(\boldsymbol{R}^*, \psi^*, \theta) = E_r^m\frac{2}{\hbar}\int_{-\infty}^{\infty} d\omega f(\omega)\frac{\mathrm{sh}\frac{\beta\omega(1-\theta)}{2}\mathrm{sh}\frac{\beta\omega\theta}{2}}{\omega^2\mathrm{sh}\frac{\beta\omega}{2}}. \tag{8.16}$$

At taking of integrals by $\boldsymbol{r}$ and $\boldsymbol{r}'$ except of the structure of the medium it is necessary to consider, that as effects of spatial dispersion of medium (function $g(r, r')$), so effects of its frequency dispersion (function $f(\omega)$) must be described by different model functions with consideration of existence of definite modes of medium polarization. At calculation of reorganization functions G and $\bar{G}$ everything that is concerned with medium reorganization must be taken into account.

The calculation of such a process parameter as the reorganization energy can also be carried out in the framework of various models. An important issue, at that, is the knowledge of the dielectric constant near the particle and electrode. It should be considered, that near the particle and the electrode, the value of the dielectric constant will be significantly different (below the value in the solution volume).

As conducted estimations show, realization of analytical calculation completely is not possible and it is necessary to carry out numerical integration.

The activation energy of the process can be determined by formula:

$$E_a = -\ln(\sin\pi\theta^*) + \theta^*\left(1-\theta^*\right)E_r^m - \Delta F\theta^* +$$

$$\sum_{k=1}^{N} E_{rk}\theta^*(1-\theta^*)\omega_k^{\mathrm{i}}\omega_k^{\mathrm{f}}\left[\left(1-\theta^*\right)(\omega_k^{\mathrm{i}})^2 + \theta^*(\omega_k^{\mathrm{f}})^2\right]^{-1} +$$

$$kT\theta^* \ln\left[1+\beta\sum_{k=1}^{N}\left(G_k+(\omega_k^{\mathrm{f}}/\omega_k^{\mathrm{i}})\bar{G}_k\right)\right]\times$$

$$\left(2E_{rk}\omega_k^{\mathrm{i}}\right)^{1/2}(\omega_k^{\mathrm{f}})^{-1}+\theta^*\sum_{k=1}^{N}\ln(\omega_k^{\mathrm{f}}/\omega_k^{\mathrm{i}}). \tag{8.17}$$

As can be seen from the analysis of the expressions obtained for the current density of the process of particle adsorption on the electrode, in the study of specific processes, a detailed analysis of the particle properties is required (the vibrational spectrum of the particle in the liquid volume and the vibrational spectrum of the particle after adsorption). The parameters of the vibrational spectra of particles can be obtained both by direct measurements (spectra in a liquid), and by using model calculations from the data of infrared spectra of particles in vacuum and spectra of liquids (electrolytes).

Simplified Models

If the interaction of intramolecular oscillations of particle with fluctuations of medium polarization will be neglected, then the expression for activation energy will be appreciably simplified:

$$E_a=-\ln(\sin\pi\theta^*)+\theta^*\left(1-\theta^*\right)E_r^m-e\eta-\Delta F\theta^*+$$

$$\sum_{k=1}^{N}E_{rk}\theta^*(1-\theta^*)\omega_k^{\mathrm{i}}\omega_k^{\mathrm{f}}\left[\left(1-\theta^*\right)(\omega_k^{\mathrm{i}})^2+\theta^*(\omega_k^{\mathrm{f}})^2\right]^{-1}. \tag{8.18}$$

Correspondingly, equation for determination of θ^* become:

$$e\eta+\beta\Delta F-\ln(\sin\pi\theta^*)+(1-2\theta)E_r^m+$$

$$\frac{\partial}{\partial\theta}\sum_{k=1}^{N}\beta E_{rk}\frac{\theta(1-\theta)\omega_k^{\mathrm{i}}\omega_k^{\mathrm{f}}}{(1-\theta)(\omega_k^{\mathrm{i}})^2+\theta(\omega_k^{\mathrm{f}})^2}=0. \tag{8.19}$$

If, besides this, the intramolecular reorganization of the particle can be neglected, or if the particle is monoatomic, then:

$$E_a=-\ln(\sin\pi\theta^*)+\theta^*\left(1-\theta^*\right)E_r^m-e\eta-\Delta F\theta^*. \tag{8.20}$$

And the equation for determining θ^* has the form:

$$e\eta+\beta\Delta F-\ln(\sin\pi\theta)+(1-2\theta)E_r^m=0. \tag{8.21}$$

Adduced expressions allow simple evaluation of kinetic parameters in the framework of strongly simplified models.

8.4 Conclusions

Charge transfer processes from an electrode on an adsorbed impurity particle are described analytically.

Analytical expressions for rate, for cathode current density and kinetic parameters of charge transfer processes between electrode and adsorbed impurity particle are obtained.

References

1. Marsagishvili T, Machavariani M, Tatishvili G (2013) Conversion of substances by the application of laser radiation. Eur Chem Bull 3(2):127–132
2. Marsagishvili T, Machavariani M, Tatishvili G, Khositashvili R, Tskhakaia E, Ananiashvili N, Metreveli J, Kikabidze-Gachechiladze M (2015) Theoretical investigations of nanosensors for radiation processes. In: International conference on nanotechnologies and biomedical engineering. IFMBE proceedings, vol 55. Springer, pp 528–532
3. Ananiashvili N, Metreveli J, Kikabidze-Gachechiladze M (2015) Nano-sensors for modeling of radiation processes. Reports of enlarged session of the 29th seminar, vol 29. I. Vekua Institute of Applied Mathematics
4. Marsagishvili T, Machavariani M, Tatishvili G (2013) Khositashvili R, Lekishvili N, Ion-exchange processes in the channels of zeolites. Asian J Chem 25(10):5605–5606
5. Marsagishvili T, Ananiashvili N, Metreveli J, Kikabidze-Gachechiladze M, Machavariani M, Tatishvili G, Tskhakaya E, Khositashvili R (2014) Applications of Georgian zeolites for the extraction of useful components from natural and waste waters. Eur Chem Bull 3(1):102–103
6. Marsagishvili T, Machavariani M, Tatishvili G, Ckhakaia E (2014) Thermodynamic analysis of processes with the participation of zeolites. Bulg Chem Commun 46(2):423–430
7. Abrikosov AA, Gorkov LP, Dzyaloshinski IE (1963) Methods of quantum field theory in statistical physics. Dover Publ., INC, New York, 341p
8. Ermakov V, Kruchinin S, Fujiwara A (2008) Electronic nanosensors based on nanotransistor with bistability behaviour. In: Bonca J, Kruchinin S (eds) Proceedings NATO ARW "Electron transport in nanosystems". Springer, pp 341–349
9. Ermakov V, Kruchinin S, Hori H, Fujiwara A (2007) Phenomena in the resonant tunneling through degenarate energy state with electron correlation. Int J Mod Phys B 11:827–835
10. Repetsky P, Vyshyvana IG, Nakazawa Y, Kruchinin SP, Bellucci S (2019) Electron transport in carbon nanotubes with adsorbed chromium impurities. Materials 12:524
11. Dogonadze RR, Marsagishvili TA (1985) The chemical physics of solvation, part 1. Elsevier, Amsterdam, pp 39–76
12. Dogonadze RR, Kuznetsov RR, Marsagishvili TA (1980) The present state of the theory of charge transfer processes in condensed phase. Electrochim Acta 25(1):1–28
13. Marsagishvili T, Machavariani M, Tatishvili G, Khositashvili R, Marsagishvili T, Machavariani M, Tatishvili G, Tskhakaia E (2015) Theoretical models for photocatalysis process. Int J Res Pharm Chem 5(1):215–221
14. Marsagishvili TA, Tatishvili GD (1993) To the theory of electromagnetic radiation in the radio-wave range in non-stationary heterogeneous systems. Poverkhnost 5:9
15. Marsagishvili TA, Tatishvili GD (1993) The theory of radio wave emission in electrochemical systems. Russ J Electrochem 29:1278

Part II
Nanosensor

Chapter 9
Chemoelectrical Gas Sensors of Metal Oxides with and Without Metal Catalysts

G. A. Mousdis, M. Kompitsas, G. Petropoulou, and P. Koralli

Abstract The interest for gas sensors is uprising due to the increased demand for many applications such as: Security, Environment, Industry, Medicine etc. Therefore, the development of new, cheap, easy to prepare, simple, low energy consuming and reliable sensors have become of great significance. A category of gas sensors with these characteristics is the metal oxide chemoelectrical sensors that are the most common ones used today. The sensing properties of these sensors depend mainly on the surface properties such as roughness, porosity, crystallinity etc. that are strongly depended on the preparation method. Moreover, the addition of metal nanoparticles and especially nanoparticles of noble metals with catalytic properties differentiate (mostly improve) the sensing efficiency. In this work we used the PLD technique to prepare thin films of Cu_xO ($1 < x < 2$) and surface decorated some of them with gold nanoparticles to be used as sensors. These sensors consisted of the sensing thin film on the surface of a quartz tube surrounding a ceramic heater with low thermal capacity. The films are characterized by AFM and SEM techniques. A significant response to several concentrations of the hydrogen and acetone at relative low temperatures was demonstrated.

Keywords Cu_xO thin films · Au · Nanoparticles · Hydrogen sensor · Acetone sensor

9.1 Introduction

Gas sensors have a great impact on many areas such as security, environment, industry, medicine, automotive applications, space houses, sensors networks etc. Moreover, process and manufacturing industries find extensive applications of

G. A. Mousdis (✉) · M. Kompitsas · G. Petropoulou · P. Koralli
NHRF-National Hellenic Research Foundation, Theoretical and Physical Chemistry Institute-TPCI, Athens, Greece
e-mail: gmousdis@eie.gr

J. Bonča, S. Kruchinin (eds.), *Advanced Nanomaterials for Detection of CBRN*, NATO Science for Peace and Security Series A: Chemistry and Biology, https://doi.org/10.1007/978-94-024-2030-2_9

various toxic and combustible gases, such as hydrogen sulfide (H_2S) [1], NH_3 [2] CO [3], H_2 [4], flammable volatile organic compounds [5] etc. Gas sensors can monitor the concentration of these gases detecting and avoiding gas leaks.

There is a huge need in the market for new cheap small, stable reliable and low power consuming gas sensors. This need in combination with the large applications range triggered the last decades a huge worldwide research for new sensor materials with improved properties. Gas sensor manufacturing technologies have been further improved by the development of nanotechnology [6–10], low-power and low-cost microelectronic circuits [11], chemometrics [12, 13], and microcomputing [14, 15]. The global gas sensor market was valued at USD 2.05 billion in 2018 and an increase of 7.8% from 2019 to 2025 is expected (http://www.grandviewresearch.com/industry-analysis/gas-sensors-market).

A gas sensor is a transducer that converts the presence and measures the concentration of a gas into a physical property [16]. There are many categories of gas sensors such as electrochemical, optical, semiconducting, capacitance, calorimetric, ultrasonic etc. The type of sensor to be used depends on the application demands and characteristics. The factors that have great importance in the sensor type decision are: gas type and concentration range, humidity, temperature, pressure, gas velocity, chemical poisons and/or interfering species, power consumption, response time, maintenance interval, fixed or portable, point or open path etc.

The most common category of gas sensors are the resistivity sensors (also known as chemiresistors). The transduction mechanism of resistive gas sensors is based on the change in resistance of the sensing layer upon adsorption and reaction with the target gas molecules. The sensing layer usually determines the sensitivity and the selectivity. In most cases, the sensing material is an organic polymer [17, 18], an organic [19] or hybrid semiconductor [20] or a metal oxide (MO_x) [21]. Although the number of chemiresistors based on polymers is increasing [22], those based on metal oxides are the most popular. Their popularity is due to: simple construction, low cost, compact size, high compatibility with microelectronic processing high accuracy, big variety of detectable gases, and their ability to monitor on a real-time basis. Gas-sensing characteristics such as gas response, selectivity, stability, and short response/recovery time are closely dependent upon the preparation method and condition of the sensor. But there are two main disadvantages, their poor selectivity and the elevated operating temperatures. The addition of a catalyst can decrease the operating temperature and even increase the selectivity and the sensitivity in respect to a specific analyte gas.

The MO_x films can detect, reductive (e.g., H_2, CO, Hydrocarbons), oxidative (e.g., O_2, O_3) or neutral (e.g., CO_2, H_2O) gases. Their main use is the detection of leaks of toxic or explosive gases at chemical and industrial processes [23]. Lately their use is expanded into many other activities concerning human health [24], drinks and food quality monitoring [25], traffic safety [26] etc. A big variety of transition or post transition metals have been used for the preparation of metal oxide sensors. The main reason of choosing these metals is the small energy difference of their cations (d^n configuration) with the d^{n+1} or d^{n-1} configuration [14]. The most effective metal oxide sensors are those with cation configurations d^0 (e.g. TiO_2) or d^{10} (e.g. SnO_2, ZnO).

Copper is a transition metal with the Cu^+ and Cu^{+2} configuration [Ar]-$3d^{10}$, -d^9 respectively. The difference in energy between these two states is small because of the primarily antibonding character of these two high-lying d orbitals [27]. Copper oxide (CuO) is a p-type semiconductor with a direct bandgap of 1.2–1.9 eV [28]. Gas sensors based on CuO nanostructures have attracted considerable attention because of their excellent sensitivity and selectivity in detecting a range of gases and vapors: water vapor, HCOH, C_2H_5OH, NO_2, H_2S, and CO [29, 30]. A range of different techniques such as wet chemical methods [31], sol-gel [32], spray pyrolysis [33], sputtering [34] etc. have been used to prepare CuO thin films.

As a part of research to understand and configure the parameters that influence the sensing properties of MO_x films, we have prepared copper oxide thin films with and without dispersed gold nanoparticles on quartz tubes substrates using the Pulsed Laser Deposition technique (PLD). The thin films were studied as gas sensors and the effect of the Au nanoparticles as a catalyst on gas sensing was investigated. Structural, morphological, and compositional properties were studied by atomic force microscopy (AFM), Scanning Electron Microscopy (SEM), and Energy Dispersive X-ray (EDS) Spectroscopy, respectively. The sensing properties of CuO and CuO:Au towards H_2, and acetone were investigated at different concentrations and operating temperatures up to 345 °C.

9.2 Experimental Techniques

9.2.1 Film Growth

The CuO thin film sensor has been grown by the PLD method (Fig. 9.1a) [35, 36]. The laser beam was delivered by a Mo. Nano S 130-10 Q-switched Nd:YAG laser (Litron Lasers) 532 nm and with pulse duration 10 ns. The repetition rate was 10 Hz and a total of 4 h was employed to complete a deposition. The beam was focused on the Cu target resulting to 14 J/cm^2 fluence. This foil target was mounted on a XY-translation stage that was driven by a microprocessor and performed a meander-like movement. Thus, target drilling was avoided. Before deposition, the high vacuum chamber was pre-evacuated down to a base pressure of 10^{-5} mbar. Then, oxygen reactive gas was inserted and kept at a 20 Pa dynamic pressure during film growth. The CuO thin film was deposited under RT on a Pyrex tube substrate ($D_{in} = 8$ mm, thickness 1 mm) that was rotating vertically by an electrical motor with a few revolutions/s. After deposition, the CuO thin film was annealed at 500 °C for 1 h. For the CuO:Au compound thin film sensor, a gold target was ablated for 15 min in vacuum by the same technique as above to partially cover the CuO thin film surface with Au nanoparticles and all measurements were repeated.

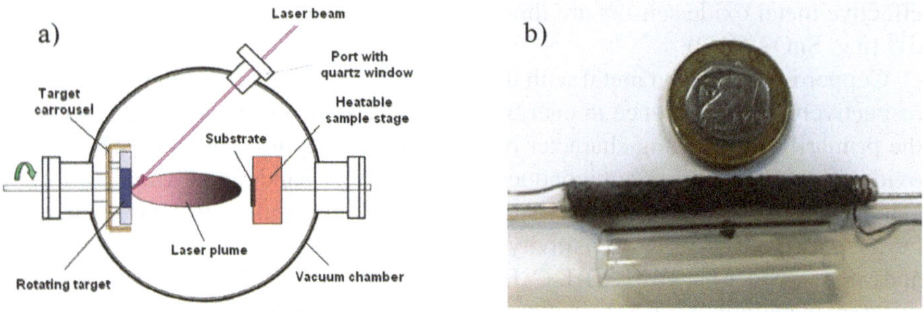

Fig. 9.1 (**a**) Schematic diagram of the PLD set-up for the preparation of Cu_xO and Cu_xO–Au doped films. (**b**) Photo of the high temperature ceramic cylindrical heater and the CuO thin film on a quartz tube

9.2.2 Gas Testing Apparatus

Gas testing was performed inside an aluminum chamber with controllable pressure. For sensing, the tube substrate was heated by a home-made cylindrical oven: this consisted of an alumina tube (4 mm in Dia.) including a Ni–Ni/Cr thermocouple for temperature measurement, a Ni–Cr wire wrapping up the tube that served as heating element (Fig. 9.1b) with the current from a stabilized power supply to set the desired operating temperature. A high-temperature paste covered the oven and transferred the heat to the quartz tube substrate efficiently. Prior to gas testing, the chamber was evacuated down to 10^{-4} mbar by a mechanical pump. Then the chamber was filled with atmospheric pressure from a bottle (80% N_2, 20% O_2), by using a Bronkhorst gas flowmeter and the pressure was controlled by a Baratron gauge. Analyte gas concentrations were calculated from partial pressure measurements displayed on the Baratron gauge. The film sensor was connected in series with a Keithley pico-Ameter in an electrical circuit, biased by a voltage of 1 V_{dc}. Upon inserting the analyte gas, the resistivity change of the sensor resulted to a current change through the circuit that was recorded in real time, digitized and displayed on the computer screen.

9.3 Films Characterization

9.3.1 Scanning Electron Microscopy (SEM/EDS) Results

Figure 9.2 show SEM images of the prepared CuO films with different magnification. As we can observe (Fig. 9.2a), a smooth and uniform surface of CuO film is formed with small particles (size <5 μm) on it. It is known from the literature that PLD grown thin films exhibit on their surface a large number of such droplet-like

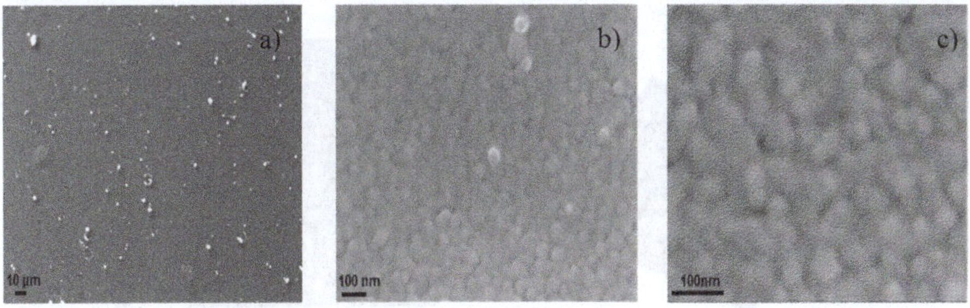

Fig. 9.2 SEM images of CuO film

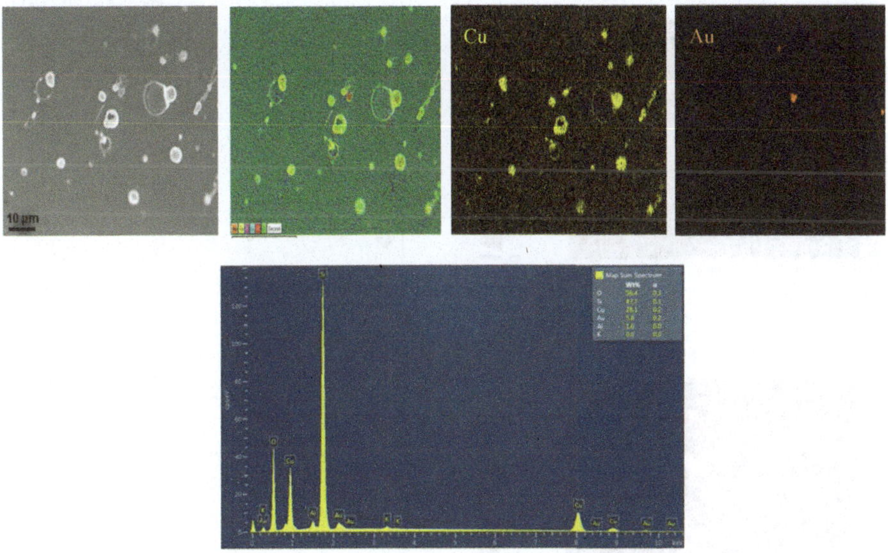

Fig. 9.3 SEM and EDS images of CuO:Au film

particles that consist of pure target material (here copper) which is ejected from the target during laser ablation. Images of higher magnification (Fig. 9.2b and c) reveal that the surface of CuO is granular, highly dense and compact, consisting of a homogeneous distribution of CuO grains with sizes 50–80 nm.

The SEM/EDS analysis (Fig. 9.3) confirmed that the droplet-like particles consist indeed of Cu (the yellow colour corresponds to Cu). Moreover, we can see on the surface of the CuO/Au films small particles of Au with sizes ~1 µm (orange colour is the Au).

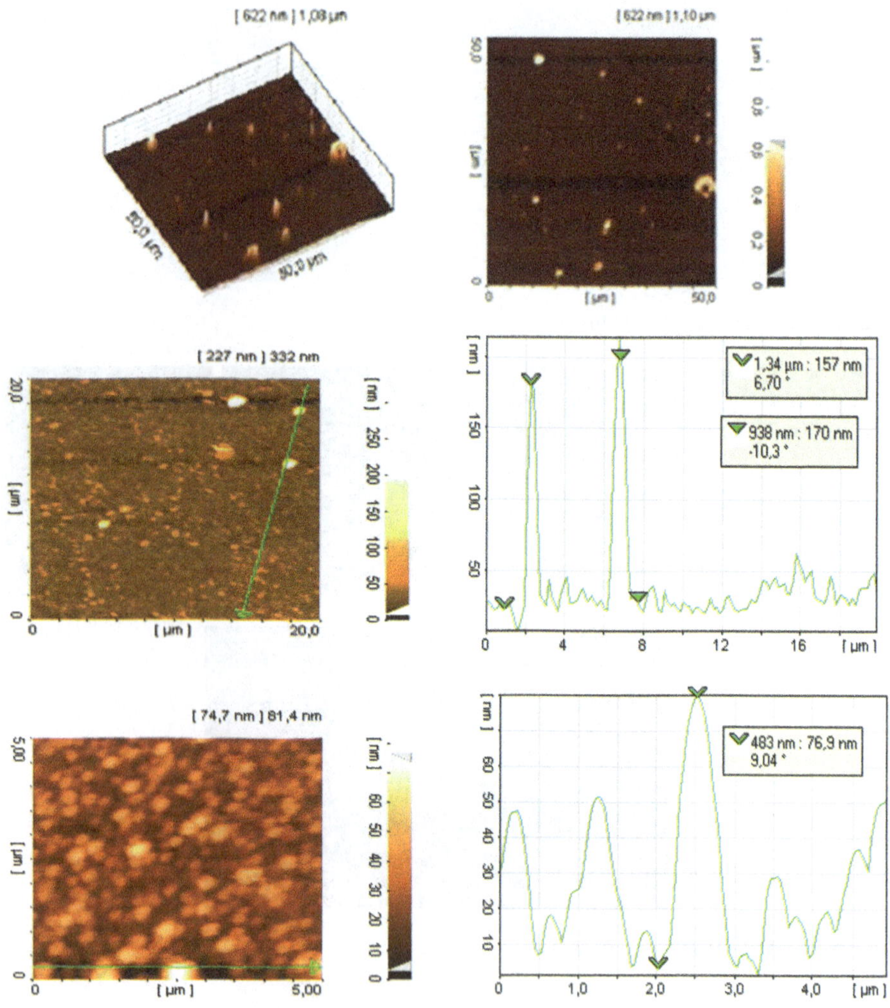

Fig. 9.4 3D, 2D and profile AFM images of CuO films

9.3.2 Atomic Force Microscopy (AFM) Results

From the AFM images (Fig. 9.4) we can see that the surface of the films is uniform with some Cu particles on it with sizes heights less than 700 nm. From the two-dimensional AFM (contact mode) images, as well as from the profile images, we observe that the surface of the films is very smooth and exhibit hillock morphology, including columnar grains growing along the c-axis, perpendicular to the substrate surface. The grain boundaries of CuO are clear and the grains seem to have similar dimensions and round shape in plane.

9.4 Gas Sensing Results

9.4.1 Sensing Mechanism

The fundamental sensing mechanism of metal oxide based gas sensors relies on the change in electrical resistivity of thin films on exposure to a target gas. The gas molecules interact with the metal oxide surface, acting as donors or acceptors of electrons thus altering the resistivity of the metal oxide film. Initially, O_2 air molecules are adsorbed on the film surface, bind to it by film electrons, and subsequently dissociates and transforms to charged ions according to the following reactions:

$$O_2\ (\text{gas-adsorb}) + e^- \longleftrightarrow O_2^-\ (T \leqslant 200\,^\circ\text{C})$$

$$1/2O_2\ (\text{gas-adsorb}) + e^- \longleftrightarrow O^-\ (T \geqslant 200\,^\circ\text{C})$$

$$1/2O_2\ (\text{gas-adsorb}) + 2e^- \longleftrightarrow O^{-2}\ (T \geqslant 200\,^\circ\text{C})$$

The reactions are endothermic and the equilibrium is moved to the right by increasing the temperature [37]. The gas molecules to be detected react with the adsorbed oxygen are oxidized and as a result the bonding electron is released back to the film. This induces a change of the number of charge carriers of the semiconducting metal oxide and consequently its resistivity (Fig. 9.5). Depending of the semiconductor type (n- or p-type) of the sensor and the type of analyte gas (oxidizing or reducing) the resistivity increases or decreases respectively [38] (Table 9.1).

Besides the type of the analyte, the magnitude of the resistivity change for a specific MO_x film to an analyte depends on the number of the surface complexes, the proximity to the gas, and the sensor temperature. For these reasons the morphology of the sensing layer (e.g. porosity, crystallinity, and surface roughness etc.) are

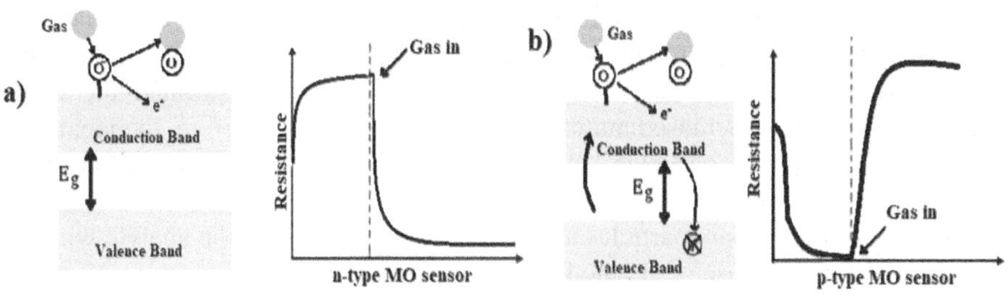

Fig. 9.5 Schematic representation of band models of conductive mechanism and resistance change for (**a**) an n- or (**b**) p-type metal oxide, without and with exposure to analyte gas, for reducing gases

Table 9.1 Dependence of resistance change from the type of semiconductor and the gas

Classification	Oxidizing gases	Reducing gases
n-type	Resistance increase	Resistance decrease
p-type	Resistance decrease	Resistance increase

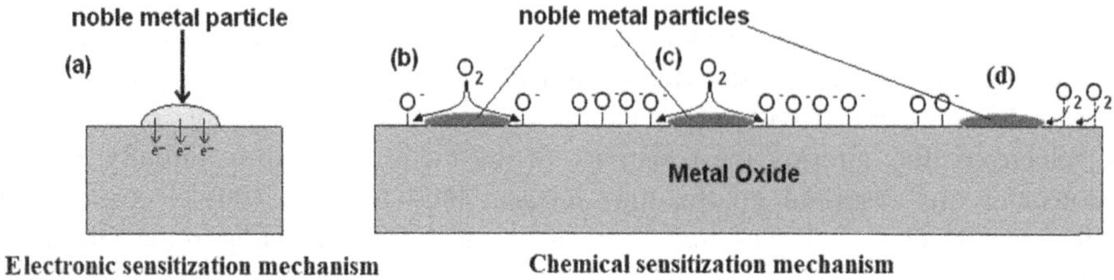

Fig. 9.6 Schematic depiction of the electronic sensitization mechanism (**a**) and the chemical sensitization mechanism (**b**, **c**, **d**)

crucial for the sensor efficiency [39]. As a consequence, the properties of the sensor strongly depend on the preparation method and processing (e.g. annealing).

The sensing properties can be improved by the presence of metal nanoparticles (mostly noble metals) on the surface of the sensor. This improvement can be attributed to two different sensitization mechanisms: (a) the electronic sensitization, caused by interfacial electronic redistribution and (b) the chemical sensitization [40]. In the electronic sensitization mechanism, there is a transfer of electrons between the noble metal and the MO_x beneath, due to the work function difference (Fig. 9.6a). In the chemical sensitization mechanism, the noble metals facilitate the chemical reactions between target gas and metal oxide surface through the spill-over phenomenon [4, 41]. The results depend strongly on the type and the concentration of the noble metal.

More specifically, the action of noble metal nanoparticles is based manly in the production of charged oxygen species due to highly effective dissociation catalytic ability (Fig. 9.6b) [42]. The noble metals are effective oxygen dissociation catalysts. They react with oxygen molecules and dissociate them to charged species (Fig. 9.6b). The charged oxygen diffuses onto the surface of the metal oxide (Fig. 9.6c). In addition, the noble metal nanoparticles attract and dissociate the O_2 molecules adsorbed in the neighboring metal oxide area before desorption (Fig. 9.6d). As a result, they create an "effective area" [43] around the metal nanoparticle with an increased number of charged species. Therefore, the ideal case would be that the nanoparticles are evenly dispersed on the surface so that these "effective areas" cover the whole surface of the film.

In many cases the nanoparticles also catalyze the reaction of the analyte with the O species e.g. in the case of H_2 the following reactions occur:

$$\text{(a) } H_2(g) \rightarrow 2H_{ads}, \quad \text{(b) } H_{ads} \rightarrow H^+_{ads} + e^-,$$
$$\text{(c) } H^+_{ads} + O^-_{ads} \rightarrow OH, \quad \text{(d) } 2OH \rightarrow H_2O + O_{ads},$$

where the first 2 reactions (a and b) were catalyzed by Pd.

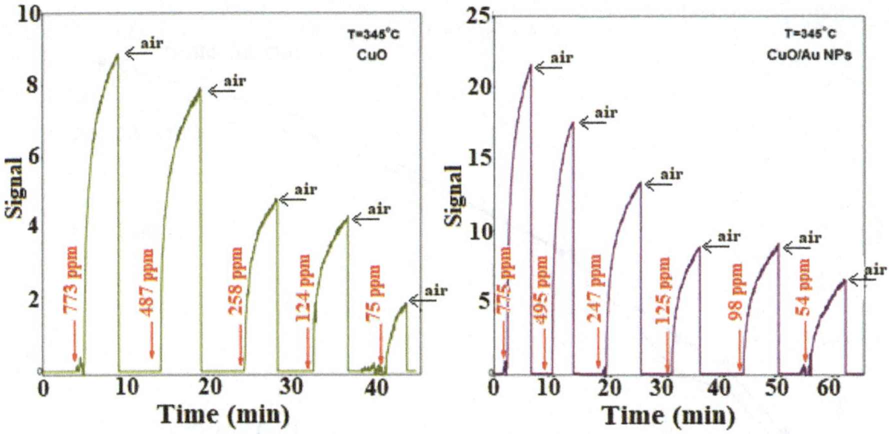

Fig. 9.7 Response of the CuO and CuO:Au sensors at 345 °C for different concentrations of H_2

9.4.2 *H_2 Sensing*

Hydrogen sensing was studied at two different temperatures (305 °C and 345 °C). In all cases there was a clear and fast response to the presence of H_2 (rise time less than 1 min). A typical set of real-time measurements is shown in Fig. 9.7 which presents the transient response of a CuO and a CuO:Au sensor to various concentrations of H_2 in air, for an operating temperature of 345 °C.

In the case of the Au doped sample the signal at 305 °C was more than three times bigger than that of the pure CuO (Fig. 9.8) resulting at a lower detection limit (Table 9.2). By increasing the temperature the signal improved significantly in the case of pure CuO and slightly in the case of CuO:Au sensor.

The enhancement of the sensor sensitivity towards hydrogen has been also observed previously in the case of the NiO:Au nanocomposites [44]. Au nanoparticles acting as catalysts are reducing the dissociation barrier for both hydrogen and atmospheric oxygen in their vicinity. The produced highly active atomic species interact with each other very fast, in particular at this high operating temperature, and it results to an additional increase of the film resistance [45].

9.4.3 *Acetone Sensing*

Acetone sensing was studied at two different temperatures (305 °C and 340 °C). In both cases, we can observe again a clear and fast response to the presence of acetone. A typical set of real-time measurements is shown in Fig. 9.9, which presents the transient response of a CuO and a CuO:Au sensor to various concentrations of acetone in air, for an operating temperature of 340 °C. On Fig. 9.10 we can see

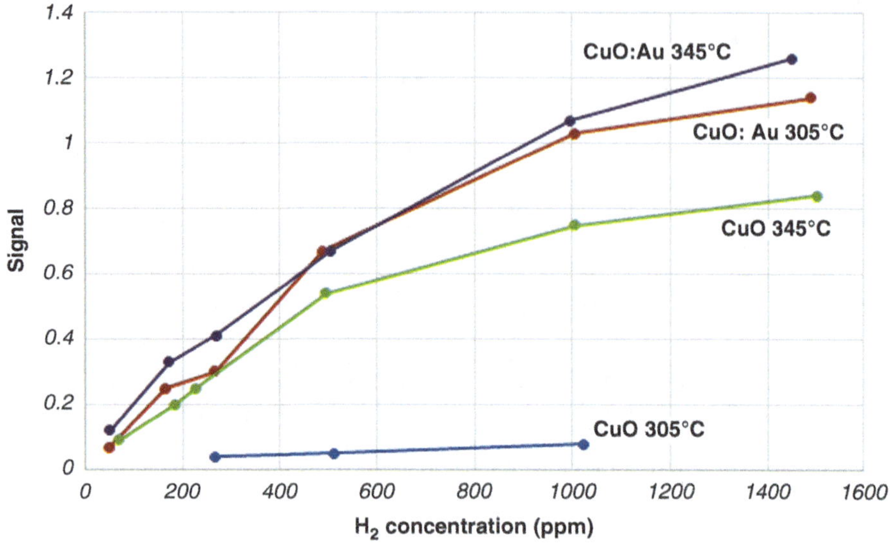

Fig. 9.8 Diagrams of the CuO and CuO:Au sensors response at 305 °C and 345 °C for different concentrations of H_2

Table 9.2 Detection Limits of CuO and CuO: Au sensors at 305 °C and 345 °C

Sample		Lower detection limit of H_2	Lower detection limit of acetone
CuO	At 305 °C	267 ppm	56 ppm
CuO:Au		50 ppm	128 ppm
CuO	At 345 °C	70 ppm	7 ppm
CuO:Au		50 ppm	12 ppm

the response of CuO and CuO:Au sensors to various concentrations of Acetone in air at 305 °C and 345 °C.

We can notice from this graph that partial surface coverage with Au particles deteriorates the sensor sensitivity towards acetone at both temperatures, in contrast with what was observed in the case of H_2. A possible explanation would be as follows: Au metallic nanoparticles act as catalysts again, promoting the dissociation of adsorbed oxygen only, but not the dissociation of acetone that would lead to signal enhancement, as for H_2. On the other side, the Au nanoparticles cover a part of the film surface, therefore less film surface is available to acetone to oxidize. To prove this hypothesis, more CuO:Au nano-composites with different concentrations of Au nanoparticles are necessary to be tested as acetone sensors. Such experiments are now under consideration.

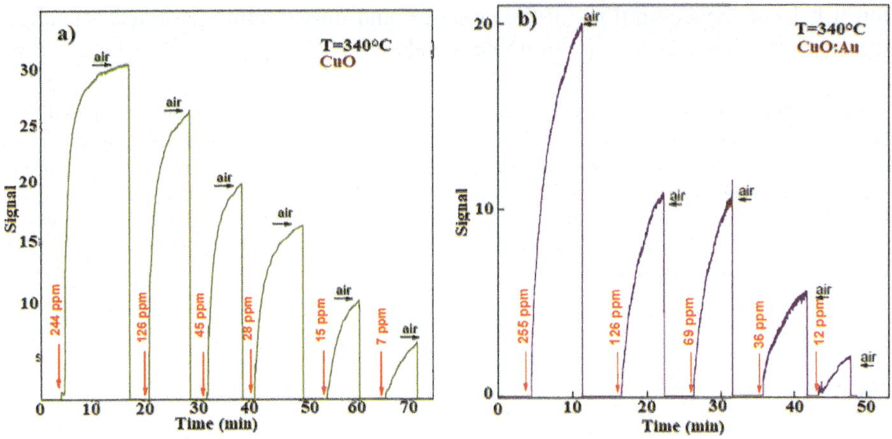

Fig. 9.9 Response of the CuO and CuO:Au sensors at 340 °C for different concentrations of Acetone

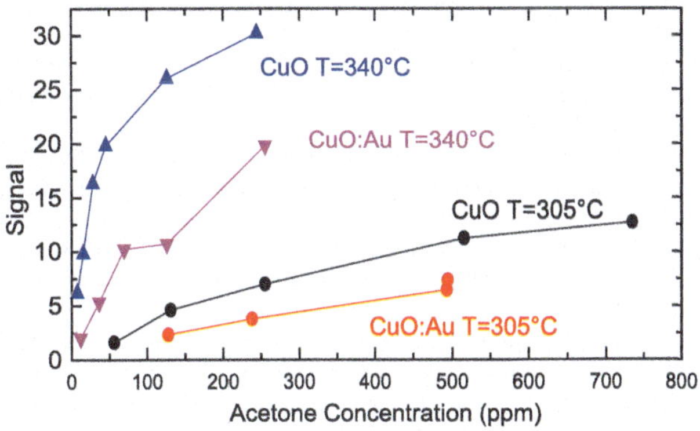

Fig. 9.10 Diagrams of the CuO and CuO:Au sensors response at 305 °C and 340 °C for different concentrations of Acetone

9.5 Conclusions

We have deposited CuO and nanocomposite CuO:Au thin films on quartz tubes by the PLD technique. A new type high temperature ceramic cylindrical heater was inserted into the tubes to prepare low cost gas sensors. AFM and SEM/ADX images have shown the granular film surface with a large effectiveness, appropriate to be used as gas sensors. They were tested as H_2 and acetone sensors with satisfactory results. The presence of Au nanoparticles on the film surface enhances in the case of H_2 detection the sensing efficiency significantly, but in the case of acetone deteriorates it. The experimental results showed that CuO and CuO:Au films prepared by PLD can be used as cost effective sensors. Experiments are in progress towards lower concentrations of the analytes, and investigating parameters such as the concentration and the type of noble metals.

Acknowledgments We acknowledge support of this work by the project "Advanced Materials and Devices" (MIS 5002409) which is implemented under the "Action for the Strategic Development on the Research and Technological Sector", funded by the Operational Program "Competitiveness, Entrepreneurship and Innovation" (NSRF 2014–2020) and co-financed by Greece and the European Union (European Regional Development Fund). One of the authors (P. Koralli) would like to thank the "IKY Fellowships of Excellence for Postgraduate Studies in Greece – SIEMENS PROGRAM" for a scholarship.

References

1. Gong J, Chen Q, Lian M-R, Liu N-C, Stevenson RG, Adami F (2006) Micromachined nanocrystalline silver doped SnO_2 H_2S sensor. Sens Actuators B Chem 114:32
2. Aspiotis N, El Sachat A, Athanasekos L, Vasileiadis M, Mousdis G, Vainos N, Riziotis C (2013) Diffractive ammonia sensors based on Sol–Gel nanocomposites materials. Sens Lett 11:1415
3. Stamataki M, Mylonas D, Tsamakis D, Kompitsas M, Tsakiridis P, Christoforou E (2013) CO-sensing properties of Cu_xO-based nanostructured thin films grown by reactive pulsed laser deposition. Sens Lett 11:1964
4. Fasaki I, Suchea M, Mousdis G, Kiriakidis G, Kompitsas M (2009) The effect of Au and Pt nanoclusters on the structural and hydrogen sensing properties of SnO_2 thin films. Thin Solid Films 518:1109
5. Alexiadou M, Kandyla M, Mousdis M, Kompitsas M (2017) Pulsed laser deposition of ZnO thin films decorated with Au and Pd nanoparticles with enhanced acetone sensing performance. Appl Phys A 123:262
6. Lima T-C, Ramakrishna S (2006) A conceptual review of nanosensors. Z Naturforsch 61a:402
7. Ermakov V, Kruchinin S, Fujiwara A (2008) Electronic nanosensors based on nanotransistor with bistability behaviour. In: Bonca J, Kruchinin S (eds) Proceedings NATO ARW "Electron transport in nanosystems". Springer, pp 341–349
8. Ermakov V, Kruchinin S, Hori H, Fujiwara A (2007) Phenomena in the resonant tunneling through degenarate energy state with electron correlation. Int J Mod Phys B 11:827–835
9. Ermakov V, Kruchinin S, Pruschke T, Freericks J (2015) Thermoelectricity in tunneling nanostructure. Phys Rev B 92:115531
10. Repetsky SP, Vyshyvana IG, Nakazawa Y, Kruchinin SP, Bellucci S (2019) Electron transport in carbon nanotubes with adsorbed chromium impurities. Materials 12:524
11. Chiu S-W, Tang K-T (2013) Towards a chemiresistive sensor-integrated electronic nose: a review. Sensors 13:14214
12. Lavine B, Workman J (2008) Chemometrics. Anal Chem 80:4519
13. Pravdova V, Pravda M, Guilbault GG (2007) Role of chemometrics for electrochemical sensors. Anal Lett 35:2389
14. Hagleitner C, Hierlemann A, Lange D, Kummer A, Kerness N, Brand O, Baltes H (2001) Smart single-chip gas sensor microsystem. Nature 414:293
15. Feng S, Farha F, Li Q, Wan Y, Xu Y, Zhang T, Ning H (2019) Review on smart gas sensing technology. Sensors (Basel, Switzerland) 19:3760
16. Mousdis G, Kompitsas M, Fasaki I (2011) Electrochemical sensors for the detection of hydrogen prepared by pld and sol-gel chemistry. In: Reithmaier JP, Paunovic P, Kulisch W, Popov C, Petkov P (eds) Proceedings NATO ARW "Nanotechnological basis for advanced sensors". Springer, pp 401–407

17. Indarit N, Kim Y-H, Petchsang N, Jaisutti R (2019) Highly sensitive polyaniline-coated fiber gas sensors for real-time monitoring of ammonia gas. RSC Adv 46:26773
18. Cichosz S, Masek A, Zaborski M (2018) Polymer-based sensors: a review. Polym Test 67:342
19. Garcia-Breijo E, Gómez-Lor Pérez B, Cosseddu P (eds) (2016) Organic sensors: materials and applications. Institution of Electrical Engineers, India, SciTech Publishing, Inc. ISBN: 978-1-84919-985-8
20. Wang S, Kang Y, Wang L, Zhang H, Wang Y (2013) Organic/inorganic hybrid sensors: a review. Sens Actuators B Chem 182:467
21. Adhikari B, Majumdar S (2004) Polymers in sensor applications. Prog Polym Sci 29:699
22. Lee J-H (2019) Technological realization of semiconducting metal oxide-based gas sensors. In: Barsan N, Schierbaum K (eds) Metal oxides, gas sensors based on conducting metal oxides. Elsevier, Amsterdam/Oxford/Cambridge, p 167–216, ISBN: 9780128112243
23. Oyabu T (1991) A simple type of fire and gas leak prevention system using tin oxide gas sensors. Sens Actuators B 5:221
24. Xing R, Xu L, Song J, Zhou C, Li Q, Liu D, Wei Song H (2015) Preparation and gas sensing properties of In_2O_3/Au nanorods for detection of volatile organic compounds in exhaled breath. Sci Rep 5:10717
25. Ghasemi-Varnamkhasti M, Mohtasebi SS, Siadat M, Lozano J, Ahmadi H, Razavi SH, Dicko A (2011) Aging fingerprint characterization of beer using electronic nose. Sens Actuators B 159:51
26. Seetha M, Mangalaraj D (2012) Nano-porous indium oxide transistor sensor for the detection of ethanol vapours at room temperature. Appl Phys A 106:137
27. Schneider WF, Hass KC, Ramprasad R, Adams JB (1996) Cluster models of Cu binding and CO and NO adsorption in Cu-exchanged zeolites. J Phys Chem 100:6032
28. Comini E, Faglia G, Sberveglieri G, Pan Z, Wang ZL (2002) Stable and highly sensitive gas sensors based on semiconducting oxide nanobelts. Appl Phys Lett 81:1869
29. Stamataki M, Mylonas D, Tsamakis D, Kompitsas M, Tsakiridis P, Christoforou E (2013) CO-sensing properties of Cu_xO-based nanostructured thin films grown by reactive pulsed laser deposition. Sens Lett 11:1964
30. Liao L, Zhang Z, Yan B, Zheng Z, Bao QL, Wu T, Li CM, Shen ZX, Zhang JX, Gong H (2009) Multifunctional CuO nanowire devices: P-type field effect transistors and CO gas sensors. Nanotechnology 20:085203
31. Wang W, Liu Z, Liu Y, Xu C, Zheng CC, Wang G (2003) A simple wet-chemical synthesis and characterization of CuO nanorods. Appl Phys A 76:417
32. Armelao L, Barreca D, Bertapelle M, Bottaro G, Sada C, Tondello E (2003) A sol-gel approach to nanophasic copper oxide thin films. Thin Solid Films 442:48
33. Moumen A, Hartiti B, Comini E, El Khalidi Z, Arachchige HMMM, Fadili S, Thevenin P (2019) Preparation and characterization of nanostructured CuO thin films using spray pyrolysis technique. Superlattice Microstruct 127:2
34. Samarasekara P, Kumara NTRN, Yapa NUS (2006) Sputtered copper oxide (CuO) thin films forgas sensor devices. J Phys Condens Matter 18:2417
35. Fasaki I, Giannoudakos A, Stamataki M, Kompitsas M, Gyorgy E, Mihailescu IN, Roubani-Kalantzopoulou F, Lagoyannis A, Harissopulos S (2008) Nickel oxide thin films synthesized by reactive pulsed laser deposition: characterization and application to hydrogen sensing. Appl Phys A 91:487
36. Fasaki I, Kandyla M, Tsoutsouva MG, Kompitsas M (2013) Optimized hydrogen sensing properties of nanocomposite NiO:Au thin films grown by dual pulsed laser deposition. Sens Actuators B 176:103
37. Williams DE (1999) Semiconducting oxides as gas-sensitive resistors. Sens Actuators B 57:1
38. Shankar P, Bosco J, Rayappan B (2015) Gas sensing mechanism of metal oxides: The role of ambient atmosphere, type of semiconductor and gases a review. Sci Lett J 4:126
39. Barsan N, Weimar U (2001) Conduction model of metal oxide gas sensors. J Electroceramics 7:143

40. Liu C, Kuang Q, Xie Z, Zheng L (2015) The effect of noble metal (Au, Pd and Pt) nanoparticles on the gas sensing performance of SnO_2-based sensors: a case study on the {221} high-index faceted SnO_2 octahedra. Cryst Eng Commun 17:6308
41. Müller SA, Degler D, Feldmann C, Türk M, Moos R, Fink K, Studt F, Gerthsen D, Bârsan N, Grunwaldt J-D (2018) Exploiting synergies in catalysis and gas sensing using noble metal-loaded oxide composites. Chem Cat Chem 10:864
42. Kolmakov A, Klenov DO, Lilach Y, Stemmer S, Moskovits M (2005) Enhanced gas sensing by individual SnO_2 nanowires and nanobelts functionalized with Pd catalyst particles. Nano Lett 5:667
43. Tsu K, Boudart M (1961) Recombination of atoms at the surface of thermocouple probes. Can J Chem 39:1239
44. Kandyla M, Chatzimanolis-Moustakas C, Koumoulos EP, Charitidis C, Kompitsas M (2014) Nanocomposite NiO:Au hydrogen sensors with high sensitivity and low operating temperature. Mater Res Bull 49:552
45. Morrison SR (1987) Selectivity in semiconductor gas sensors. Sens Actuators 12:425

Chapter 10
Ion Track Etching Revisited: Influence of Aging on Parameters of Irradiated Polymers as Required for Advanced Devices

Y. Bondaruk, A. Kiv, L. Alfonta, H. Garcia-Arrellano, J. Vacik, G. Muñoz Hernández, V. Hnatowicz, I. Donchev, and D. Fink

Abstract Aging of polymers might contribute to possible malfunction of devices based on them, such as biosensors and others. As hitherto still little is known about the influence of aging on pristine and swift heavy ion-irradiated polymer foils, we had recently started with the re-examination of the etching behaviour and the electronic properties of such materials, which had been subjected to ageing for up

---

Y. Bondaruk
South-Ukrainian K.D. Ushynsky National Pedagogical University, Odessa, Ukraine

A. Kiv (✉)
South-Ukrainian K.D. Ushynsky National Pedagogical University, Odessa, Ukraine

Department of Materials Engineering, Ben-Gurion University of the Negev, Beer-Sheva, Israel
e-mail: kiv@bgu.ac.il

I. Donchev
Department of Materials Engineering, Ben-Gurion University of the Negev, Beer-Sheva, Israel

L. Alfonta
Department of Life Sciences and Ilse Katz Institute for Nanoscale Science and Technology, Ben-Gurion University of the Negev, Beer-Sheva, Israel

H. Garcia-Arrellano
Departamento de Ciencias Ambientales, División de Ciencias Biológicas y de la Salud, Universidad Autónoma Metropolitana-Lerma, Lerma de Villada, Estado de México

G. Muñoz Hernández
Departamento de Fisica, Universidad Autónoma Metropolitana-Iztapalapa, México, DF, México

D. Fink
Departamento de Fisica, Universidad Autónoma Metropolitana-Iztapalapa, México, DF, México

Nuclear Physics Institute, ASCR, Řež, Czech Republic
e-mail: fink@xanum.uam.mx

J. Vacik · V. Hnatowicz
Nuclear Physics Institute, ASCR, Řež, Czech Republic

J. Bonča, S. Kruchinin (eds.), *Advanced Nanomaterials for Detection of CBRN*, NATO Science for Peace and Security Series A: Chemistry and Biology,
https://doi.org/10.1007/978-94-024-2030-2_10

to ~20 years. For this sake, we correlated the as-measured and normalized onset times of etching and etching speeds of the various etching stages with each other. As expected, the sample aging affects the magnitude of these parameters. Surprisingly, this aging effect also differs on both sides of a given sample. The etching parameters were found to correlate not only with both the thickness and the surface roughness of the examined foil samples, but also with the fluence during the ion irradiation. All these effects can be understood easily within a consistent model.

As for some examinations, very long etching times were required, the question arose how interruptions during the track etching procedure might affect the final result. In the appendix it is shown that transient sample drying is the safest way to avoid disturbances or falsifications of the measurements.

Keywords Ion track · Irradiated polymers · Nanopores

10.1 Introduction: The Present State-of-the-Art

Aged polymers become an increasing matter of concern, as they often contribute to environmental damage. Aging might also render polymeric devices such as biosensors and others useless after some time. Therefore it is important to study such effects in greater detail. This is done here as well on pristine as on ion-irradiated polymers.

The impact of swift heavy ions (SHI) into polymers creates straight zones of radiation damage, the so-called "latent ion tracks" [1], which are rich in excess free volume and radiochemical products and hence can be easily etched towards nanopores, the so-called "etched ion tracks". Due to their peculiar properties, such as micrometric to nanometric sizes which are useful for filtration and ultrafiltration [2–5]; an electronic behaviour which is useful for establishing an emerging polymeric electronics [6, 7], and large aspect ratios and easy transformation to biosensors by cladding their walls with e.g., enzymes [8–18], they gained special interest in recent times.

As the latter application made it necessary to compare the results of different biosensors (made of foils with different polymers, irradiation conditions and ages) with each other, a detailed study was recently performed to quantify their track etching parameters in dependence of the etching progress [19]. As etching is known to proceed in i ($i = 1, 2, 3$ and eventually even 4) different stages (named the track core C, penumbra P, halo H (if existing), and bulk B), the most important parameters are the onset times t_i of these stages and their etching speeds v_i. Most of these parameters correlate well with each other according to power laws, either using the as-measured data or after normalizing them via $\tau_i = t_i/t_B$ and $\nu_i = v_i/v_B$. The latter approach enables one to compare different biosensors with each other.

It was also found that these parameters correlate with the polymer ages and foil sides. Polymeric aging initiates different diffusion-driven decomposition, segregation and disintegration processes (briefly summarized in e.g., [7]) and thus leads to enhanced etching onset times and speeds with increasing age, as well of the un-irradiated polymers themselves as of the embedded SHI tracks. The pronounced side dependence of etching of both the polymer itself and the tracks embedded there seems to be essentially a consequence of the segregation and surface precipitation of polymeric additives, which often differs slightly from each other on both sample sides. As these layers act as diffusion barriers against etchant attack, the etching parameters exhibit some foil side asymmetry which is enhanced the older the samples are, i.e. the longer the segregation and precipitation lasts.

Here we will add some more experimental findings that give us a better hint about the nature of the side asymmetry of track etching. Furthermore, we show that there exist additional correlations between the track etching parameters and both the foil thickness and the ion track density (which is equal to the SHI fluence). We also examine possible correlations with the surface roughness. All these findings can be understood within a consistent model.

10.2 Experimental

10.2.1 The Samples and Measuring Techniques

From our earlier decade-long research work, a wide range of different SHI-irradiated polymer foils was still available which were re-examined between 2014 and 2016; the results are presented in this work. These samples were partly up to ~20 years old and they had meanwhile been stored in the dark under ambient lab conditions (~18 to 22 °C). The specific data of these samples were compiled in the first paper of this series [19], together with some representative etching results of both foil sides (as recorded by the electrical currents transmitted through the SHI-irradiated polymer foils as a function of etching time, while applying a test voltage across the foils). In all cases, the SHI irradiation energy was chosen so that the range of the SHIs exceeded the foil thickness by far so that the transferred energy density was roughly the same (usually within ~1% or less) on both foil sides. In all experiments PET, polycarbonate and cellulose nitrate samples were etched in 2014–2016 with 4 M NaOH (and the polyimide samples with 5% NaOCl) at room temperature in a specially designed etching chamber (made after the example of Ref. [20]). As it had been found that these aged foils exhibited a pronounced side dependence of track etching, we always etched the foils from both sides, and we always defined the side of faster etchant breakthrough as the "side A" and the opposite side as "side B".

The low etching concentration and temperature were chosen here intentionally to expand the etching process strongly beyond usual times, in order to work out clearly all the individual steps during the etching process. The etching process was steadily controlled by determining a current passing through the foils, when applying a sinoidal test voltage of ~4 to $5\,V_{\text{peak-peak}}$ at 49.5 mHz. An alternating voltage was selected on the one hand to avoid the build-up of any polarization, and on the other hand to enable the recording of possible phase shifts and rectification (see [7]). The frequency was chosen to be quite low to avoid the build-up of excessive capacitive currents. It was intentionally slightly displaced from 50 mHz, to avoid any possible interference with eventual industrial electronic background noise. The relatively high voltage was applied to enhance even small transmitted currents up to a magnitude where they could be measured with sufficient accuracy. The roughness of all foils was measured on both sides, by means of a Mitutoyo Roughness meter SJ-201. The recorded transmitted currents as a function of etching time of all foils (for simplicity often called "etching protocols") were evaluated and their results compiled in Refs. [7, 19].

Due to the rather low etchant concentrations and temperatures applied here, etching took quite long. In the case of excessive etching times (e.g. during weekends or power failure) it eventually became necessary to interrupt the etching process and continue at a later time. There seems no general recipe to be known how to proceed in this case; suggestions from various colleagues ranged from intermediate drying of the corresponding foils up to their storage in water or some acid. The main goal in such a case is, of course, to suppress the etching process in the case of necessary interruption as fast as possible, so that it can be resumed later under controlled conditions. In the "Appendix" this question is tackled in detail. The conclusion of that work is that rapid intermediate sample drying is the only useful strategy to prevent falsification of the measuring results in an uncontrolled way by arbitrary swelling or sample contamination.

10.3 Results and Discussion

10.3.1 The Foil Side Asymmetry of Track Etching

Temperature Dependence of Asymmetric Track Etching in Fresh and Aged Foils

As a thermally driven process, etching is known to follow Arrhenius-type behaviour. This had been shown by observing the etchant breakthrough times t_C as a function of the etching temperature [21]. However, to our knowledge it has never been examined how other track etching parameters such as t_P and t_B follow this trend and especially in how far track ageing and the polymer foil side dependence might influence this behaviour. In order to find this out, we compared the track etching performance on both sides of fresh and aged samples at different etching temperatures with each other. Corresponding test measurements were performed in 2015 for both a representative fresh (∼2014) foil (foil #8 from Ref. [19]) and an aged (∼2005)[1] foil

[1] See [19] for more foil specifications.

(foil #19 from [19]). As expected, the etching times t_i (i = C, P, B) for the two sides A and B roughly follow indeed Arrhenius plots (Fig. 10.1a, b). Interestingly, the time interval Δ_P between the t_C and t_P values (or Δ_B between t_C and t_B, respectively) is broader for side B than for side A for the aged samples (Fig. 10.1b), and in both aged cases much larger than that for the fresh foil case; this is clear indication of the importance of both the foil side and the age on track etching. As there is hardly any difference seen between the Δ_P and Δ_B values for the fresh samples (Fig. 10.1a), we conclude that the side asymmetry in track etching strongly grows with sample age. In fact, the recent discussion in [19] pointed at the segregation of polymer additives (such as fillers) towards layered structures as the source of the observed side asymmetry in ion track etching, which takes years to decades time at ambient temperature.

Dividing the t_i values of both sides A and B by each other: $t_{i,A}/t_{i,B}$, one realizes that for the fresh sample these ratios are lowest at both the ambient temperature (19 ± 4) °C and at 70 °C (Fig. 10.1c). In the interval in between, the ratio has a rather constant value around 1.4 ± 0.1. For contrast, the aged sample (Fig. 10.1d) exhibits a somewhat constant $t_{i,A}/t_{i,B}$ ratio of 0.8 ± 0.1 and does not show any deviation at ambient temperature. However, above 40 °C or so, the ratio turns towards unity or beyond. Above 60 °C, the t_i values cannot be determined any longer, due to a strong overlap with the electronic signal from etchant penetration across the non-irradiated bulk (which is a typical effect of etching aged SHI-irradiated foils).

Possible Surface Contaminations

During the beginning discussion about a possible side asymmetry of etching of ion tracks in thin foils, a simple counter-argument was that such asymmetry may be just the trivial consequence of accidental surface contamination by dirt (such as oil, grease or fingerprints) during the foil production and later manipulation steps. As hitherto clear evidence against this possible source of the etching asymmetry is still missing, this question is tackled here. In the absence of any surface contamination, the onset times t_i (i = C, P, B) for the etching of the different track regimes on both foil sides should follow accurate Arrhenius correlations [18], and the same should also hold for the corresponding etching speeds v_i. However, in the case of additional surface contaminations which first have to be dissolved before track etching can start, one should observe an increase in t_i, as the dissolution of the surface contaminations would take some additional time. (Analogously, as the overall track core etching speed v_C is inversely correlated with t_C, the latter should increase for contaminated foil surfaces.) However, as the track core etching can start only after the original surface contamination is removed, both the penumbra and bulk etching speeds v_P and v_B should not be affected by that previous surface contamination. This means that, whereas t_C should become larger in the case of contaminated foils, the distances Δ_P and Δ_B should remain the same. Further, as the etchant's capability for oil/grease dissolution increases strongly with increasing etching temperature, the influence of surface contaminations on

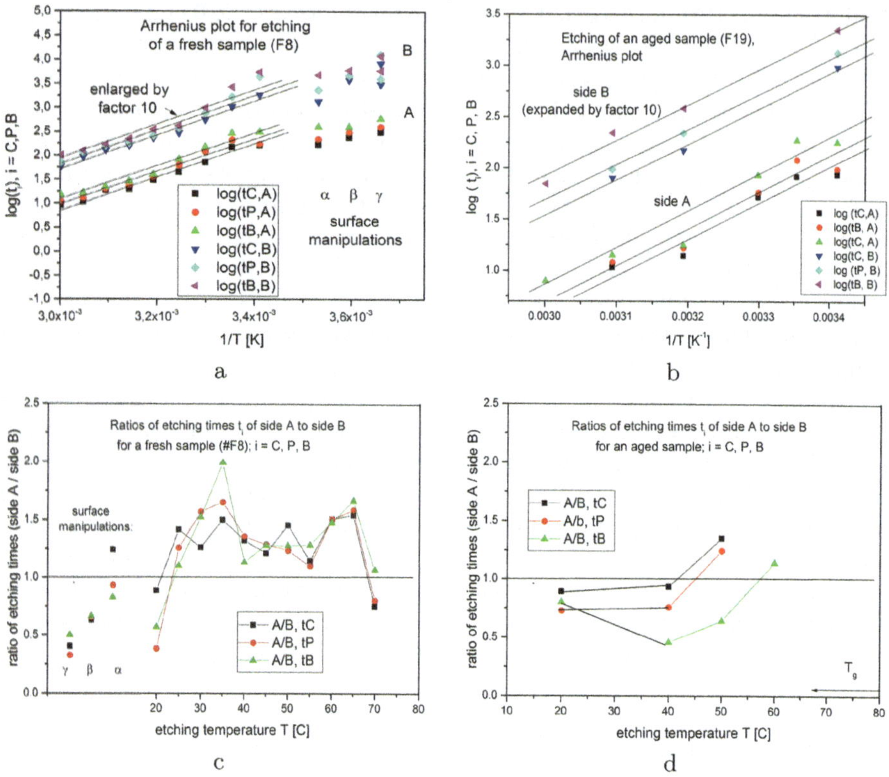

Fig. 10.1 (**a**, **b**) Temperature dependence of the etching times t_i (i = C, P, B) for sides A and B of (**a**) a fresh foil (#8) and (**b**) an aged foil (#19), depicted as Arrhenius plots. For the sake of clearness, the t_i values of sides B are enlarged by a factor 10. Due to a limited amount of available aged samples, less data points could be determined in (**b**) than in (**a**). (**c**, **d**) Temperature dependence of the ratio of the etching times t_i (i = C, P, B) for sides A and B, for (**c**) the fresh foil #8 and (**d**) the aged foil #19. The symbols α and β denote those foils which had been cleaned with household detergent or ethanol, respectively on either the side A or the side B; the symbol γ denotes those foils which had been covered intentionally with a marginal layer of oil on side A or B before etching

t_C and v_C should restrict to low temperatures only. As we saw above that the results of Fig. 10.1a (track etching in fresh foils) showed that – within our present experimental accuracy – no difference was seen between the Δ_P and Δ_B values at whatsoever temperature or foil side, this observation does not support any possible accidental surface contamination.

To give even more arguments for (or against) initial possible surface contaminations of fresh sample surfaces, we performed additional three tests: (α) cleaning the sample surface before etching with commercial household detergent on both sides (see the symbol α in Fig. 10.1); (β) treating the surface with absolute alcohol (see the symbol (β), and (γ) depositing intentionally a very thin layer of machine oil onto the sample surfaces, before performing the usual track etching tests. These studies show:

(α) The use of a commercial household detergent (used here: "Salvo" from Procter & Gamble, Mexico) before track etching yielded the same (side A) or slightly lower (side B) onset times t_i of track etching as the absence of any surface treatment before etching. Consequently, the ratio $t_{i,\mathrm{A}}/t_{i,\mathrm{B}}$ centres around unity, and is thus somewhat higher than that of uncleaned samples. This indicates that indeed some traces of surface contaminations might have existed on the surface B of fresh samples (i.e., on the foil side with the slower etching speed).

(β) Alcohol slightly increases the etching times t_i on both sides. This is understood as the consequence of the alcohol-induced polymeric swelling, resulting in an equally strong decrease of the surface-near intrinsic free volume on both sides. Consequently, the ratio $t_{i,\mathrm{A}}/t_{i,\mathrm{B}}$ roughly remains about the same as for the untreated sample.

(γ) As expected, intentional deposition of oil or grease on the sample surfaces (used here: Type 3-EN-UNO from WD-40 Company, San Diego, CA) leads to some enhancement of the times t_i on both foil sides. Apparently, the enhancement is higher on side B than on side A which indicates that the action of oil or grease as a surface diffusion barrier is the more efficient the less free volume is available on the given side. Consequently, the ratio of the etching times of both sides is smaller than in the previous cases.

Correlations Between the Track Etching Parameters and the Foil Sides

In [19] we tested various combinations of the etching parameters t_i and v_i with each other {such as $t_\mathrm{P}(t_\mathrm{C})$, $t_\mathrm{B}(t_\mathrm{P})$, $v_\mathrm{P}(v_\mathrm{B})$, $v_\mathrm{C}(t_\mathrm{C})$ and $v_\mathrm{P}(v_\mathrm{C})$}, which mostly yield clear-cut power law correlations. Though we distinguished in some of these graphs between the etching results from side A and side B, no clear-cut result emerged for the side influence from them. Therefore we plotted in Fig. 10.2 the results of the correlations of the etching parameters t_C, v_C, v_P and v_B of both sides with each other {i.e., $t_{\mathrm{C,\,B}}(t_{\mathrm{C,\,A}})$: Fig. 10.2a, $v_{\mathrm{C,B}}(v_{\mathrm{C,A}})$: Fig. 10.2b, $v_{\mathrm{P,B}}(v_{\mathrm{P,A}})$: Fig. 10.2c and $v_{\mathrm{B,B}}(v_{\mathrm{B,A}})$: Fig. 10.2d, where the first index denotes the track region (C, P or B), and the second index stands for the etched foil side (A or B)}.

First, let us consider the two graphs treating the track core etching (Fig. 10.2a, 10.2b) which reveal interesting deviations between fresh and old samples. In both cases, deviations show up insofar that the $t_{\mathrm{C,A}}$ and $v_{\mathrm{C,A}}$ track core values of aged polymer samples hardly grow any longer beyond certain values, though the corresponding $t_{\mathrm{C,A}}$ and $v_{\mathrm{C,A}}$ values of fresh samples still do. The explanation of this finding is straightforward: the presumed layer formation on the surfaces B due to long-time segregation of polymer additives (such as fillers) (on both non-irradiated and SHI-irradiated polymer foils) forms some kind of protective diffusion barrier there, which largely prevents the penetration of corrosive gases such as oxigen and moisture on that side so that the original track structure is still largely preserved.

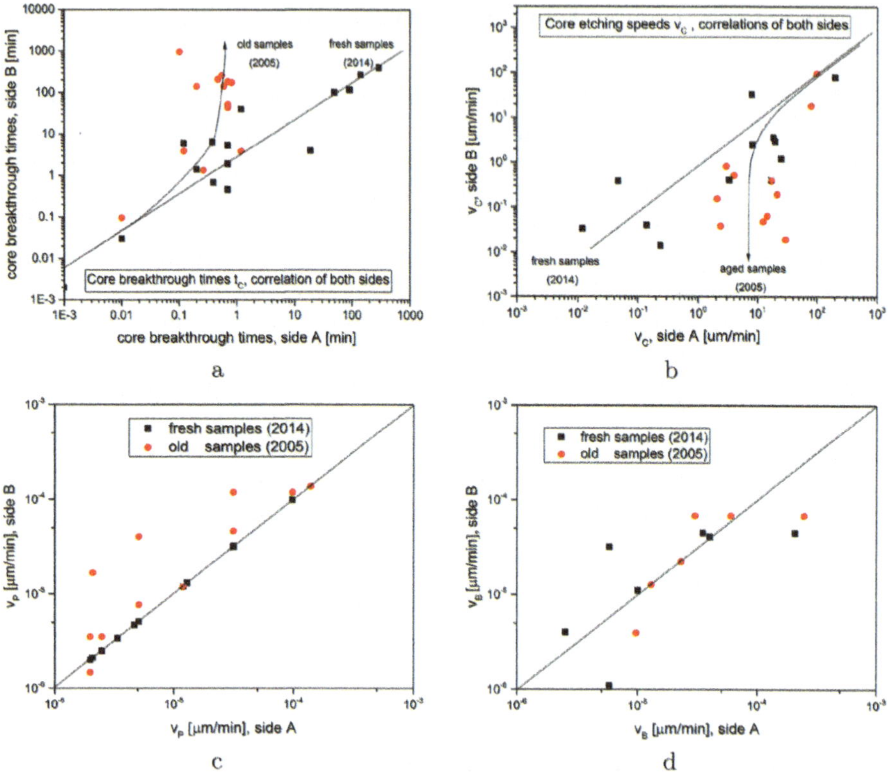

Fig. 10.2 Correlation between the track etching of both sides of all examined foils for (**a**) the onset times of core etching and (**b**) the etching speeds. Whereas most fresh samples (black dots) arrange along the diagonal lines, aged samples (red circles) usually deviate due to the very strong degradation of the sides A, leading to excessively high core etching speeds and excessively low breakthrough times. (**c**) and (**d**) show the corresponding graphs for both sides for the etching speeds of (**c**) the penumbra and (**d**) the bulk etch regions; no pronounced age dependence emerges. Lines to guide the eye

For contrast, the missing protective coverage on the opposite foil side A enables corrosive gases to readily penetrate and attack the SHI tracks so that etching will be enhanced there. Whereas foils which are just at the beginning of aging only show some slightly retarded etching on side B and some enhanced etching on side A (not shown here), our available aged foils both show such a strong etchant retardation on side B and strong enhancement on side A, so that the curves in graphs Fig. 10.2a, b virtually bend upward (towards ~ t_C values ~0.5 min for side A) and downward (towards ~ v_C values ~5 μm/min for side A), respectively, for whatever values on side B.[2] This again points at a tremendous material's destruction on foil side A due

[2]This is also the reason why – in the case of etching of aged SHI-irradiated foils at elevated temperatures – the background currents due to bulk etching start exceeding the track etching

to long-time aging, while the material on foil side B has still largely preserved its original material's properties.

Now let us turn our attention to the behaviour of the penumbra during etching: Fig. 10.2c shows that for fresh foils no deviation at all exists for the etching speeds as measured from both foil sides. However, when regarding aged samples, the penumbra etching is enhanced by roughly half an order of magnitude on side B as compared with side A. To understand this emerging difference during aging, let us recall that there exist two possible etching sequences, the most abundant one which follows the well-known 3-step sequence of track core C – penumbra P – bulk B [21], and another, less frequent one where the penumbra step is split up into two stages (the first again called penumbra P and the second called halo H): C-P-H-B [19]. The latter etching sequence occurs whenever polymeric swelling dominates. In such cases the original (usually rigid and etching-resistant) cross-linked penumbra region tends to transform to some gel (which is soft and easily etchable), with only some rigid cross-linked remnants remaining (which form the halo region), before entering the polymeric bulk. Our previous work [19] gave us some hints that on the one hand aging might promote such sol-gel transition, however degradation processes such as oxidation and hydrolysis might modify the polymer so that sol-gel transitions are hindered. If this is correct, then the etchant might easily transform the penumbra to gel in the protected region on side B, whereas etching the degenerated penumbra on side A does not allow for gel formation and hence proceeds slower. Last not least, Fig. 10.2d, illustrating the correlation between the bulk etching speeds of both sides, does not yield anything but reasonable proportionality. This just means that the polymeric bulk itself is unaffected by processes going on in the foil surfaces.

Intermediate Conclusions

From these findings we can conclude –

- that the Arrhenius plot of the etching times t_i (i = C, P, B) for the two sides of the aged foil #19 (Fig. 10.1b) looks in general similar as that one of the fresh foil #8;
- that the fresh sample #8 possibly indeed exhibits some loosely bonded surface contamination on one side which vanishes easily by either elevating the etching temperatures by at least ~5 °C or so, or by short washing with detergent;
- that the foil's glass transition temperature influences the etching behaviour of fresh sample at the highest observed temperature, 70 °C. We attribute this to the onset of structural polymeric change on one foil surface;

currents when etching from sides A above ~50 °C, so that *precise* etching parameters t_i and v_i cannot be obtained any longer above that etching temperature (however, crude estimates are still possible, see e.g., Fig. 10.1d).

- that both sides of the fresh sample (even after surface cleaning) differ in their etching behaviour by ~40%, due to intrinsic differences in either surface-near composition or density;
- that the aged sample #19 does not exhibit hints for surface contamination on any side. This indicates that any eventually earlier present surface contamination might either have been adsorbed within the polymeric bulk or equally distributed on both foil surfaces during its decade-long storage time;
- that the intrinsic differences in surface-near composition or density of both foil sides have slightly increased during the aging process by up to about 50%;
- the aged irradiated polymer foil does not show any hint for any structural PET transformation even at ~70 °C. This points at a decreasing polymer's rigidity and consistency with age, which is of course expected;
- that there exist maximum etching times (~1 min) and minimum etching speeds (~5 μm/min) for tracks on one side of aged polymers which cannot be overcome, in contrast to etching of fresh polymers.

Anyhow, the absence of such polymeric relaxation processes during hot track etching is advantageous for us as it allows us to distinguish unambiguously between the different mechanisms leading to the side dependence of track etching. Another remark: it also points at the limited value of using annealing experiments of polymer foils to simulate their long-time polymer aging, unless the polymeric relaxation times are properly taken into account.

Summarizing, we may say that one must always count with intrinsic track etching asymmetries due to local compositional or density variations[3]; apparently they get enhanced by aging. Possible slight surface contaminations may indeed play some minor role for the fresh samples at low temperatures, but no longer at elevated temperatures or for aged samples. As we concentrated in this work on the effects of track etching in *aged* polymer samples, this effect is hence of minor importance for us.[4]

[3] In this connection note that differences in surface properties on both sides of PET foils of different thickness had already been recently found on another occasion, see Ref. [22] and the explanation given there.

[4] Anyhow, the presence of surface contaminations on polymeric foils needs not always be regarded per se as disadvantageous, as – if they exhibit long-time stability – they could act as diffusion barriers to protect the polymer surfaces from in-diffusion of detrimental ambient gases such as O_2, SO_2 etc.

10.3.2 *Correlations Between Track Etching Parameters and Other External Parameters*

Correlations Between Track Etching Parameters and the Foil Thickness

When plotting the etching times t_i (i = C, P, B) of all our different samples versus the corresponding foil thicknesses, one finds a gradual increase with thickness d which – in spite of the considerably data scattering – can be described reasonably well by a square relation of the type:

$$t_i[\text{min}] = c_1 + c_2 d[\mu\text{m}]^2 \tag{10.1}$$

with $c_1 \sim 70$ min and $c_2 \sim 0.2$ min μm^{-2}, Fig. 10.3a. As in diffusive processes, the mean free diffusion paths scale with the square root of time, the observed square dependence of the foil thickness on the etching time is a strong indication that *all* the numerous processes underlying (and going on during) track etching are diffusion-controlled. Apparently this holds for all the three etched track regions. The fact that a term c_1 adds to the square relation (10.1) means that a thickness-independent incubation time (necessary for the etchant to enter the tracks, such as the initial dissolution of eventually existing diffusion barriers and sol/gel transformation of the track material before dissolution) always adds to the diffusion kinetics of track etching. In contrast to the onset times of etching of the different stages C, P and B no influence of the sample thickness could be found on neither the as-measured, nor the normalized etching speeds of C, P and B, for yet unclear reason (not shown here).

Correlations Between the Track Parameters and the SHI Fluence

The plot of the etching times t_i of the track etching stages i = C, P, B of all our samples versus the corresponding ion fluences does not yield any reasonable correlation (not shown here). However, when normalizing these as-measured etching times t_i by dividing through the bulk etching times t_B: $\tau_i = t_i/t_B$, one obtains a graph (Fig. 10.3b) which reveals a slight fluence dependence. Here, the data interval of the t_C values (black squares) increases, whereas that of the t_P values (red dots) decreases with increasing fluence.

To understand this, please take into account that effects that depend on track fluence are collective effects, i.e. are effects that deal with the interaction of tracks among each other: when a SHI enters a solid, the localized extremely high transferred energy density along the newly created latent ion track initiates within an extremely short time a shock wave that propagates radially away from the site of impact. This shock wave compresses as well the neighbouring polymer's intrinsic free volume (FV) as also the irradiation-induced FV in nearby (latent or etched) tracks. The less FV is left over in that specific polymer region, the more

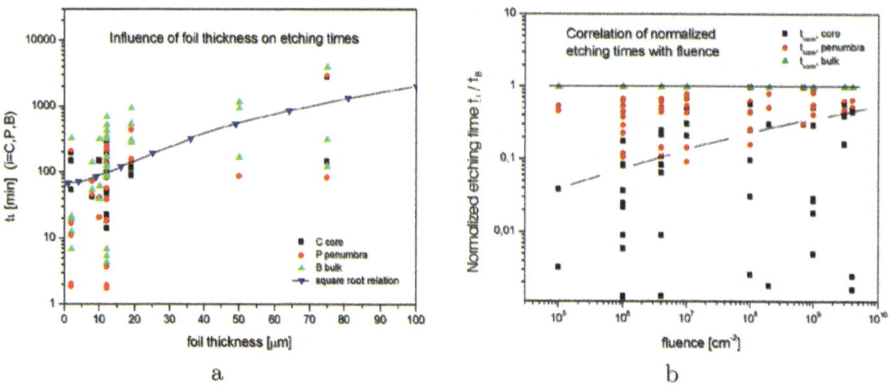

Fig. 10.3 (**a**) Dependence of all as-measured t_i results (i = C, P, B) on the foil thickness; (**b**) dependence of all normalized t_i results (i = C, P, B) on the ion fluence. (Normalization is done here by dividing through the bulk etching onset times.) (**a**) There seems to exist roughly a square root dependence with foil thickness which points at an underlying diffusional reaction kinetics (illustrated by the curve with blue standing triangles). (**b**) The increasing normalized t_C/t_B and decreasing t_P/t_B values with fluence point at the SHI-induced hammering effect as the responsible underlying mechanism. Dashed curve to illustrate the border between the t_C/t_B and t_P/t_B value ranges. There is no marked side influence found in this case. Corresponding correlations between the etching speed and foil thickness or fluence are poor and therefore omitted here

transport processes are hindered in this region. With increasing fluence the number of such compressed regions increases in the samples; consequently the overall transport along the samples (as described by diffusion) also decreases. However, due to statistical reasons there are always some regions left over which have not yet (or only little) been affected by this radiation-induced compaction [23, 24] (or "hammering" [25]) effect, so that the overall distribution of diffusion coefficients will not only shift to smaller values, but also broaden. As track etching is a diffusion-controlled process, the above-mentioned hammering effect determines directly the etchant breakthrough times t_C through the track cores: with increasing fluence the normalized core etching time interval (i.e. $\tau_C = t_C/t_B$) broadens (note: not the individual τ_C or τ_P values change, but the interval changes within which these normalized values are found). This has a direct impact on the subsequent penumbra etching, insofar as the average interval of normalized onset times of penumbra etching $\tau_P = t_P/t_B$ also decreases, see Fig. 10.3b. (Consequently, there exists also the tendency that with increasing SHI fluence, the normalized penumbra etching speed ($\nu_P = v_P/v_B$) interval decreases due to the polymer's radiation-enhanced compaction; however, this effect is not very pronounced (not shown here). It is of course trivial that in the third track etching stage, the normalized bulk etching times $t_B/t_B = 1$ are always unity.

Does there Exist Correlations Between the Track Parameters and the Foil'S Surface Roughness?

When it became clear that etching of tracks in aged polymer foils depends on the foil side, it was initially argued that this might result from the different surface roughness of both foil sides, insofar as different surface roughness might distort the impinging ion's paths somewhat by different amounts, thus resulting in different range and damage distributions. However, take into account that the very heavy SHI masses and high energies prohibit this and guarantee that the ion impact trajectory occurs independently from any surface structure and orientation. Nevertheless we will show below that there does exist an unambiguous correlation between the etchant core breakthrough times and the corresponding roughness. The discussion below tries to solve this riddle.

The surface roughness R is quantified by the deviations in the direction of the normal vector of a real surface from its ideal form. Most common for its characterization is the R_a parameter which we also use here. It is the arithmetic average of the absolute values of the vertical deviation of the measured roughness profile. Table 10.1 compiles the R_a parameters of both sides of all our SHI-irradiated foils (except for foils #13 and #18 which was too thin to exhibit the necessary stability for such a measurement, and #16 which was too brittle to enable any trustworthy measurement). Figure 10.4a shows that in most cases the roughness of both sides are well correlated with each other, which is understood by the similarity of both roll surfaces during the cold or hot rolling production process. When relating the breakthrough times t_{C} to the corresponding roughness for the sides B, we found indications for a clear-cut correlation, Fig. 10.4c: the rougher the surface, the longer is the breakthrough time. However, such a correlation is missing for the opposite sides A (Fig. 10.4b); here, t_{C} is rather independent from the corresponding roughness. We also did not obtain any clear-cut correlation between the ratios $t_{\mathrm{C,A}}/t_{\mathrm{C,B}}$ of the breakthrough times t_{C} for the etching of sides A and B, and the corresponding ratios $R_{\mathrm{A}}/R_{\mathrm{B}}$ of the surface roughness of sides A and B, respectively (not shown here). Also, no correlation exists between the etching speeds of core, penumbra and bulk on the surface roughness, for neither side A nor side B (Fig. 10.4d).

For understanding the results of Fig. 10.4b, c, let us recall what is already known about the influence of aging on the etching of the two different foil sides [19]: Side A is per definition that side which has the lower breakthrough time t_{C}, hence is characterized by a relatively high etching speed. This implies high diffusion coefficients of both the etchant and the etching products. Under an inert ambient atmosphere, this also should imply high polymeric self-diffusion, which should trigger the original surface roughness to smooth out gradually. However under usual ambient air conditions, side A also allows external impurities such as oxygen and moisture to enter easily, whereafter they considerably modify the polymer by oxidation and hydrolisation. These processes leads to degradation of the polymeric

Table 10.1 Surface roughness of both sides of all polymer foils examined in this work, as determined by a Mitutoyo SJ-201 surface roughness tester. Each value is the mean of 5 measurements. The roughness of foils #10, 13 and 18 could not be determined due to either too small available sample area or due to too thin thickness, and are therefore omitted here. Due to the brittleness of foil #16, these results are doubtful and possibly too large. Foil numbers refer to the sample list given in Ref. [19]

		Side A		Side B	
No.	Description (material, irradiation conditions, age)	Roughness [μm]	Foil sample #	Roughness [μm]	Foil sample #
0	PET, pristine, ~2008	0.30	0a	0.22	0a
1	PET, 12 μm, 170 MeV Xe, 1×10^5 cm^{-2}, 2011	0,234	240,268	0,356	290
4	Makrofol, 40 μm, 350 MeV Au, 1×10^6 cm^{-2}, 2005	1,36	292	0,54	209, 291
5	Cellulose nitrate, 10 μm, 350 MeV Au,1×10^6 cm^{-2}, 24.5.2004	0,53	295,312	0,08	293
6	PET, Melinex, 12 μm, 250 MeV Kr, 4×10^6 cm^{-2}, <2005	0,31	236,278	0,26	296
7	PET, Melinex, 12 μm, 250 MeV Kr, 4×10^6cm^{-2}, UV, ~2014	0,23	200,270, 262 –265	0.34	300,313
8	PET, Hostaphan, 19 μm, 170 MeV Xe, 1×10^7 cm^{-2}, ~2014	0,13	239,255,254	0.18	266
9	PET, 50 μm, 250 MeV Kr, 1×10^7 cm^{-2}, ~2005	0,22	246,267,275	0.17	286, 294
11	PET, Melinex, 12 μm, 170 MeV Xe, 1×10^8 cm^{-2}, 2005	0,30	238,269,289	0.32	307
12	Cellulose nitrate, 10 μm,350 MeV Au, 1×10^8 cm^{-2}, 2004,	0,29	302	0.39	301
15	Makrofol KG, 8 μm,985 MeV Au, 2×10^8 cm^{-2}, 2011	0,10	306	0.12	287,245
16	Cornstarch, 5 μm, 350 MeV Au, 7×10^8 cm^{-2}, ~2005	0,29	299	0.25	298
17	PET, Melinex, 12 μm, 160 MeV Xe, 1×10^9 cm^{-2}, ~2004	0,135	276,284	0.235	309
19	PET, Melinex, 12 μm, 750 MeV Kr, 3×10^9 cm^{-2}, ~2005	0,18	237,271,288	0.23	297
20	Makrofol KG, 10 μm,? ?, 4×10^9 cm^{-2}, <2005	0,215	277,285	0.22	308

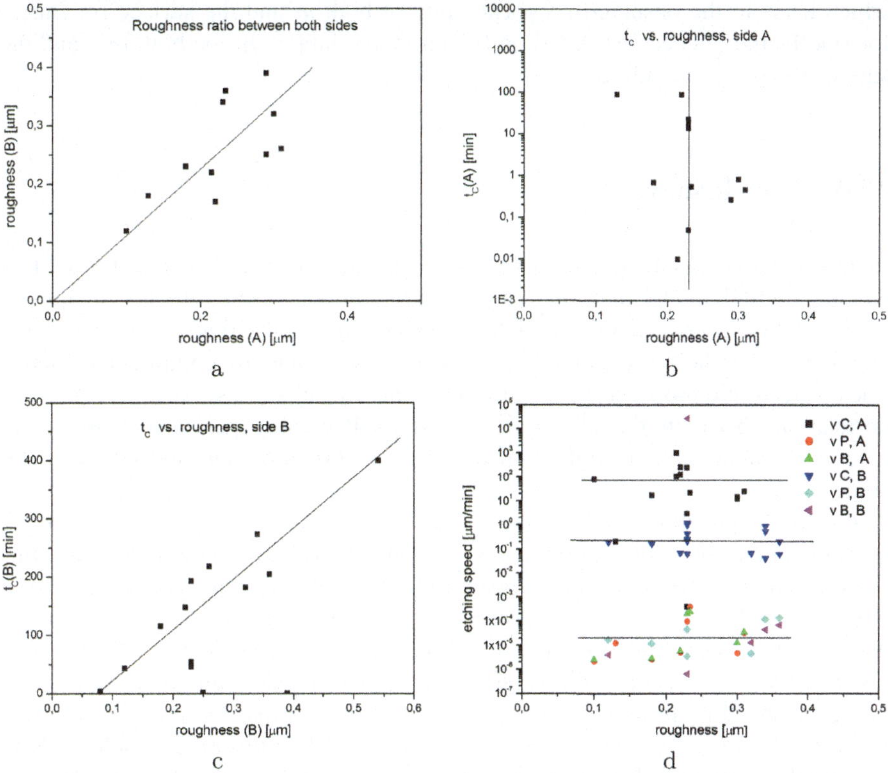

Fig. 10.4 (**a**) Correlation between the surface roughness of both sides A and B of all examined polymer foils. (**b, c**) Search for correlations between the etchant breakthrough time t_C and the surface roughness, for both foil sides A and B. Whereas for the sides B such a correlation exists, no such correlation is found for the sides A. (**d**) No correlation exists between the etching speeds of core, penumbra and bulk on the surface roughness, for neither side A nor side B

macromolecule chains, the latter largely lose their easy mobility [5] and hence leave the original surface roughness rather unaffected.

Complementarily, side B as the foil side with larger breakthrough time t_C is characterized by a relatively small etching speed due to the probable presence of a "protective" layer of segregated polymeric fillers or other additives during aging [19], which acts as a diffusion barrier against etchant attack. As this diffusion barrier also strongly retards both the oxidation and hydrolisation of the underlying polymer, the latter should still exhibit higher self-diffusion on side B than on the opposite side A. Hence gradual smoothening of the original foil's surface roughness is still possible on that side. So we think that it is the presence of this diffusion barrier

[5] The much smaller etchant molecules and the polymeric etching products will, however, be less strongly affected by the decrease of the aged macromolecule's mobility.

which leads to the observed proportionality of both t_C and the surface roughness for side B; the thicker that diffusion barrier is, the larger will be both $t_{C,B}$ and the surface roughness of side B.

10.4 Conclusions

In the first two publication of this series [7, 19], we had revisited the field of etching of SHI tracks in (preferentially) aged polymers, to establish a number of correlations between the onset times of etching of the track core, penumbra and bulk, and their corresponding etching speeds, and to compile the etched track's electronic properties. Encouraged by these results, we studied here possible correlations between the above-mentioned track etching parameters and other external parameters that might control the track formation, and indeed we found some.

One of the puzzling results in the precursor papers was that apparently the track etching parameters depend on the polymer foil side. Here, we could reconfirm this finding. The underlying reason appears to be the polymer foil's changing surface composition during aging, due to segregation and surface precipitation of polymeric additives which act as protective diffusion barriers for the underlying polymer. Usually these processes differ slightly on both foil sides, this side asymmetry being enhanced by prolonged aging. Whereas the less protected polymer foil side is strongly affected by aging due to oxidation and hydrolisation which leads to structural and chemical deterioration of the original polymer matrix, the opposite side hardly is.

There also exist correlations between the track etching parameters and the foil thickness, the SHI fluence, and – amazingly – even with the surface roughness. Whereas in the first and third cases, these correlations can be largely explained by the polymer's diffusional behaviour, in the second case the "hammering effect" (i.e. the polymer's compression by the SHI's shockwave upon impact) is thought to be responsible for this effect.

Last not least, the question had to be answered what happens during the interruption of track etching, as many of the examined foils required excessively long measuring times. The result (see the "Appendix") is unambiguous: simple sample drying leaves them unaffected from eventual track deteriorations.

This paper summarizes in how far important parameters of swift heavy-ion irradiated polymer foils are affected by aging. It establishes a first quantitative data base, from which one can estimate in how far the change of polymer and ion track properties might alter the functionality of devices made of them, such as biosensors.

Acknowledgments D.F. is grateful to the Universidad Autónoma Metropolitana-Iztapalapa, Mexico City, for the guest professorship in the frame of the Cathedra "Alonso Fenandez". The project was supported by the Grant Agency of the Czech Republic (P108-12G-108) and the Nuclear Physics Institute, Řež near Prague. D.F. is very thankful for the continuous great support and help of Prof. S. Cruz from UAM-I, Mexico City, during his stay at this university. We are further obliged very much to Dr. P. Apel from JINR Dubna, Russia for providing us with many of the examined ion-irradiated polymer foils.

Appendix

A Special Strategy for a Special Problem: Interrupted Etching

What is the Best Measuring Strategy in the Case of Interrupted Etching?

At present there does not exist any general recipe how to interrupt a continuous etching process in the case of necessity, so that etching can be resumed later under controlled conditions; present suggestions range from intermediate drying of the corresponding foils up to their storage in water or some acid. However, occasionally it was argued that even after the rapid removal of the foils from the etching vessel and short rinsing in water, the etching process might still continue for some undefined time due to the presence of some remaining etchant in the tracks, essentially in the case of thick foils. Slow sample drying could also first increase the residual etchant's concentration (and hence the etching speed) before finally stopping the etching. For contrast, if drying would occur too rapidly, internal stress or strain might build up due to the material's shrinking, which, upon resumption of etching, also could change the etching properties in an uncontrolled way. Also, immersing the etched foils transiently in water might lead to problems due to the continuing foil etching by the remaining diluted etchant, and long-time swelling by sol-gel transformation of the track region could occur in this case, which might initiate track closure. Finally, though storing the samples in acid will efficiently stop the etching process, the simultaneous formation of conducting reaction products of etchant and acid may mask the former current amplitudes fatally.

Therefore, to solve this confusion and to find out how uncontrolled sample changes during interruptions can be prevented as efficient as possible, we performed corresponding test experiments by comparing regular etchant protocols with those ones with interrupted etching with (a) intermediately drying the sample, (b) immersing the sample transiently in water, and (c) immersing the sample transiently in acid. Figure 10.5a shows that, upon transiently drying an etched sample, etching can indeed be resumed later without any major alteration in the etchant protocol shape. It appears that – at least for foils in the ~10 μm thickness range – neither excessive internal etching nor etched track closure due to the sample's swelling take place. For this sake, excessive drying (such as sample heating above 100 °C, or vacuum-drying) is not required; the sample storage within our lab conditions (~20 °C, ~60% humidity) for at least ~1/2 to 1 h is fully sufficient.

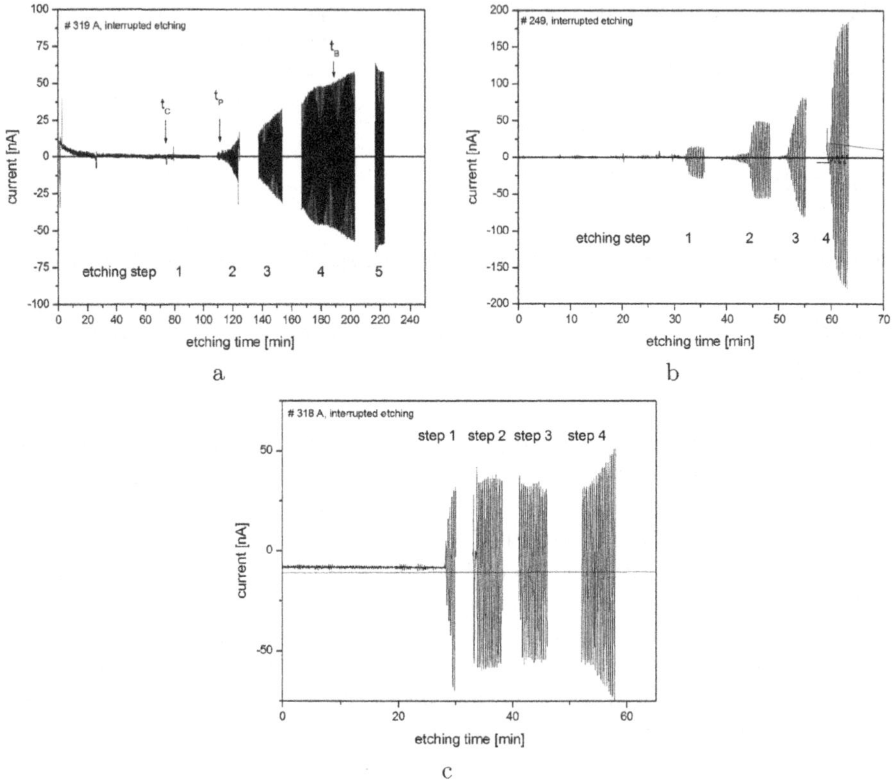

Fig. 10.5 Comparison of etchant protocols with interrupted etching with intermediate (**a**) drying, (**b**) storage in ultrapure water, and (**c**) storage in 1 M HCl acid. Used foil: #8 (PET, 19 μm thickness, fluence 1×10^7 cm^{-2}), side B

However, when submerging the samples in ultrapure water during the etching pause, the smooth continuation of the etching protocol is indeed dramatically changed by the interruption, Fig. 10.5b. In this case, the open etched tracks are not at all conserved during the interruption of the etching process. Rather, they close nearly completely during that interruption interval, so that a re-etching process has to start nearly from the beginning again. It is thought that this track closure for transmitted electric currents results from the penumbra expansion (i.e., swelling) due to gel formation. (Interestingly, it is not possible to reverse that process thereafter by simply letting the irradiated polymer foil dry. The reason for this is understood energetically, as open tracks have a much higher potential energy than closed ones.) The only possibility to re-open closed tracks is repeated etching. In such a case, however, the necessary re-etching from the beginning takes much less time – compare e.g., the incubation time of ~30 min in the first etching step with ~3 min in the second, and ~2 min in the third etching step of Fig. 10.5b. Once the etching reaches good current transmission again, it rapidly proceeds to much higher currents

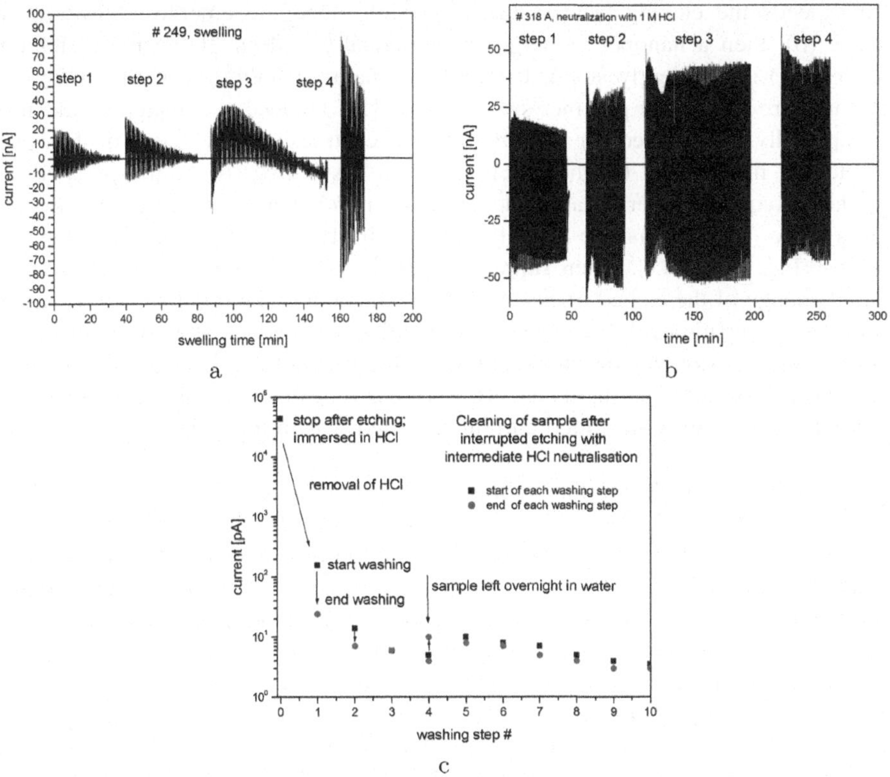

Fig. 10.6 Sample swelling (**a**) in ultrapure water and (**b**) in 1 M HCl, after the intermittent etching steps shown in (**b**) and (**c**), respectively. Finally, (**c**) illustrates how the transmitted current decreases after the last etching step and subsequent HCl treatment, by multiple washing with ultrapure water

(i.e., much larger etched track diameters) than ever before. In other words, this strategy is rather useless for interrupted etching.

One may expect that transient sample storage in acid leaves them in a similar state as in water or after drying. However, Fig. 10.5c shows that in such a case very high currents emerge. We will show below that they can be attributed to conducting contaminants (which might either stem from the conducting acid-etchant reaction products such as NaCl, or are released by the acid from the etching chamber walls). Therefore this strategy is useless for interrupted etching.

Interruptions of Etching as Disturbances of Dynamic Equilibria

During the etching of tracks in aged polymer foils, two processes counteract each other, (a) the material's removal by etching (thus widening the tracks) and (b) the polymer swelling (thus narrowing the tracks). If the swelling process

(b) exceeds the etching process (a), no visible nanopore emerges; however, if (a) < (b), then a nanopore emerges and gradually widens. If the track etching process (a) is transiently stopped, then the dynamic equilibrium shifts in favour of nanopore closure by polymer swelling (b). This tendency for nanopore closure is especially pronounced for very narrow etched tracks. If, however, in the case of etching interruption (a) the foil is dried, also swelling (b) cannot proceed as the latter requires water uptake, and therefore no change in the actual track size takes place. This is what we actually observe. In the case of intermediate sample drying (Fig. 10.5a), i.e., when suppressing both etching (a) and swelling (b), no radial change of the etched tracks occurs and the etching protocol can be resumed smoothly at any moment. If, however, the samples are intermediately stored in ultrapure water after stopping the etching (a) (Fig. 10.5b), swelling (b) steadily continues and the tracks decrease in radius [6] (Fig. 10.6a) repeatedly during each interruption[7]; the minimum transmitted current recorded was in the order of some ~1 to 3 pA.

Last not least, if during the etching interruption the samples were exposed to acid (here: 1 M HCL), the current transmission across the tracks proceeded in an uncontrolled way, Fig. 10.5c. The reason is that on the one hand, the etchant/acid reaction forms Na^+ and Cl^- ions in abundance which mask the pure current signal across the track radii, and furthermore the acid may additionally desorb adsorbed conducting contaminants from the etching vessel walls. I.e., to the previous dynamic balance between etching and swelling, contaminants are now added as a third random actor. Even if the acid is again removed before each of the etching steps continues, by replacing it with ultrapure water, the transmitted currents nevertheless do not return to the previous values, see Fig. 10.6b. To accomplish the latter, one would have to wash the chamber multiple (~10 to 20) times after each etching interruption, see Fig. 10.6c – but even after this lengthy procedure, the measurement would be falsified due to non-negligible swelling.

As a conclusion, rapid intermediate sample drying is the only recipe to resume etching measurements reliably, without encountering the danger to falsify the measuring results in an uncontrolled way by arbitrary swelling or contamination.

[6]The basic requirement for such observations is that the work must be done absolutely carefully, by cleaning both the sample and the vessel, as also the used solution as much as possible, as well mechanically as by washing. Tests show that in the beginning of cleaning, ~5 washing processes are required each to lower the measured signal in the measuring chamber by an order of magnitude, so that in total at least ~10 to 20 cleaning steps are required; see e.g., Fig. 10.6c to obtain the purity required for the shown experiment. Neglection of such precautions leads to steady increase of the transmitted currents, which does not reflect the sample's swelling, but rather the gradual desorption of excessive contaminant ions, probably stemming from the measuring chamber walls that were adsorbed there during previously examinations.

[7]A remark: The alternating sequence of track opening and closure via etching and swelling such as that shown in Fig. 10.6a somewhat resembles the biological processes of channel opening and closure in cell membranes. It is expected that especially the track closure by swelling induces strain in the polymer foil similarly as the cell channel closure induces strain in the cell membrane [26].

References

1. Fink D (2004) Ion-beam radiochemistry. In: Fink D (ed) Fundamentals of ion-irradiated polymers. Springer series in materials science vol 63, pp 251–307
2. Awasti K (2006) Transport through track etched polymeric blend membrane. Bull Mater Sci 29:261–264
3. Bethge K (ed) (1990) Ion track and microtechnology; principles and application. F. Vieweg & Sohn Verlagsgesellschaft mbH., Braunschweig
4. Sartowska B (2012) Nanopores with controlled profiles in track-etched membranes. Nucleonic 57:575–579
5. Brink LES, Elbers SJG, Robertsen T, Both P (1993) The anti-fouling action of polymers preadsorbed on ultrafiltration and microfiltration membranes. J Membr Sci 76:281–291
6. Siwy Z, Fuliski A (2002) Fabrication of a synthetic nanopore ion pump. Phys Rev Lett 89:198103
7. Fink D, Hernández GM, Cruz SA, Garcia-Arellano H, Vacik J, Hnatowicz V, Kiv A, Alfonta L (2018) Ion track etching revisited: II. Electronic properties of aged tracks in polymers. Radiat Eff Defects Solids 173:148–164
8. Fink D, Klinkovich I, Bukelman O, Marks RS, Kiv A, Fuks D, Fahrner WR, Alfonta L (2009) Glucose determination using a re-usable enzyme-modified ion track membrane. Biosens Bioelectron 24:2702–2706
9. Fink D, Hernandez GM, Alfonta L (2011) Ion track-based urea sensing. Sensors Actuators B 156:467–470
10. Arellano HG, Fink D, Hernández GM, Vacik J, Hnatowicz V, Alfonta L (2014) A nuclear track-based biosensor using the enzyme Laccase. Appl Surf Sci 310:66–76
11. Shunin Y, Fink D, Kiv A, Alfonta L, Mansharipova A, Muhamediyev R, Zhukovskii Y, Lobanova-Shunina T, Burlutskaya N, Gopeyenko V, Bellucci S (2016) Theory and modelling of real-time physical and bio-nanosensor systems. Comput Model New Technol 20:7–17
12. Fink D, Hernández GM, Arellano HG, Vacík J, Havranek V, Kiv A, Alfonta L (2016) Nuclear track-based biosensing: an overview. Radiat Eff Defects Solids 171:173–185
13. Fink D, Vacik J, Hnatowicz V, Hernandez GM, Garcia-Arrellano H, Alfonta L, Kiv A (2017) Diffusion kinetics of the glucose/glucose oxidase system in swift heavy ion track-based biosensors. Nucl Instrum Methods Phys Res Sect B Beam Interactions Mater Atoms 398:21–26
14. Shunin Y, Fink D, Kiv A, Mansharipova A, Muhamediyev R, Zhukovskii Y, Lobanova-Shunina T, Burlutskaya N, Gopeyenko V, Bellucci S (2017) Modelling of real-time physical and bio-nanosensors for medical applications and ecological monitoring. In: Borisenko VE, Gopeyenko SV, Gurin VS, Kam CH (eds) Physics, chemistry and application of nanostructures. World Scientific, Singapore, pp 220–223
15. García-Arellano H, Muñoz H G, Fink D, Vacik J, Hnatowicz V, Alfonta L, Kiv A (2018) Dependence of yield of nuclear track-biosensors on track radius and analyte concentration. Nucl Inst Methdos B 420:69–75
16. Ermakov V, Kruchinin S, Fujiwara A (2008) Electronic nanosensors based on nanotransistor with bistability behaviour. In: Bonca J, Kruchinin S (eds) Proceedings NATO ARW "Electron transport in nanosystems". Springer, pp 341–349
17. Ermakov V, Kruchinin S, Hori H, Fujiwara A (2007) Phenomena in the resonant tunneling through degenarate energy state with electron correlation. Int J Mod Phys B 11:827–835
18. Repetsky P, Vyshyvana IG, Nakazawa Y, Kruchinin SP, Bellucci S (2019) Electron transport in carbon nanotubes with adsorbed chromium impurities. Materials 12:524
19. Fink D, Muñoz H. G, García A. H, Vacik J, Hnatowicz V, Kiv A, Alfonta L (2018) Ion track etching revisited: I. Correlations between track parameters in aged polymers. Nucl Inst Method B 420:57–68
20. Daub M, Enculescu I, Neumann R, Spohr R (2005) Ni nanowires electrodeposited in single ion track templates. J Optoelectr Adv Mat 7:865–870

21. Apel PY, Fink D (2004) Ion-track etching. In: Fink D (ed) Transport processes in ion-irradiated polymers. Springer series in materials science, vol 65, pp 147–202
22. Kolska Z, Reznickova A, Hnatowicz V, Svorcik V (2012) Surface properties of poly(ethylene terephthalate) foils of different thickness. J Mater Sci 47:6429–6435
23. Fink D (2004) High-fluence effects, pp 189–190, and: Fink D, Hnatowicz V, Apel PY Microstructural changes: an overview, pp 309–312, and: Hnatowicz V, Fink D Bulk changes, pp 349–350, In: Fink D (ed) Fundamentals of ion-irradiated polymers. Springer series in materials science, vol 63
24. Fink D, Hnatowicz V, Apel PY (2004) Fluence dependence. In: Fink D (ed) Transport processes in ion-irradiated polymers. Springer series in materials science, vol 65, pp 110–114
25. Hedler A, Klaumünzer SL, Wesch W (2004) Amorphous silicon exhibits a glass transition. Nat Mater 3:804
26. Ranade SS, Syeda R, Patapoutian A (2015) Mechanically activated ion channels. Neuron 87:1162–1179

Chapter 11
Plasmon Metal Nanostructures Formation in Piezocomposite Material Controllable in Micrometric Level for Detection and Sensing Cell-Biological Particles

T. Janusas, S. Urbaite, and G. Janusas

Abstract Plasmon metal nanostructures are fundamental components integrated in sensing system of MEMS applications, mostly medicine and pharmacy. This paper cover main design aspects of microperiodic structures formed in different type nanocomposite films: metal/polymer with SPR effect and metal/polymer piezocomposite elements for detection and sensing purposes. Mathematical model of the system, using designed piezocomposite material with controllable properties by ultrahigh frequency excitation, was developed. This model allows to analyze the mechanical parameters of the system, which enables to control motion of cell-biological mechanisms in bio suspension and to force them to be attracted by metal nanoparticles, deposited on designed microperiodic structure in nanocomposite layer.

Keywords Nanostructures · SPR · Piezocomposite · Silver nanoparticles · Microperiodic structure · Diffraction efficiency · Bioparticle · Detection · Sensing

11.1 Introduction

The molecular interactions and cell-biological mechanisms behind the adhesive behavior of a cell are not properly understood till now [1]. In the last decades, the development of nanocomposite materials and the hope for elaboration of successful bacteria/virus or other particle detection accelerated the search for appropriate materials and conditions, which may help to keep particles functioning and detect

T. Janusas (✉) · S. Urbaite · G. Janusas
Kaunas University of Technology, Kaunas, Lithuania
e-mail: giedrius.janusas@ktu.lt

J. Bonča, S. Kruchinin (eds.), *Advanced Nanomaterials for Detection of CBRN*,
NATO Science for Peace and Security Series A: Chemistry and Biology,
https://doi.org/10.1007/978-94-024-2030-2_11

them on artificial surfaces [2]. One of the solutions to detect those particles in biologically active medium is insertion of metal nanoparticles in a polymer matrix [3]. Designed metal/polymer based nanocomposites with microperiodic structures are like small sensing elements based on surface plasmon resonance (SPR) and works like a litmus paper – nanocomposite surface with metal nanoparticles is covered with investigated biologically active medium, where antigen and antibody interaction appears [4]. Bacteria/virus or other particles, existing in observed medium, are automatically attracted by metal nanoparticles to the surface of nanocomposite [5]. Nanocomposite materials having piezoelectric properties are a field of investigations, because of their scientific significance and technological implications [6–9]. From both scientific and technological points of view, it is highly desirable to pattern piezoelectric nanocomposite films with feature size down to micro- or nanometer scale. Moreover, designing an SPR element with piezoelectric properties may lead to a new research area of particle motion in the fluids when in contact with nanocomposite surface [10].

When sensing element (microperiodic structure formed in nanocomposite layer) is in contact with biologically active medium (liquid (i.e. blood) or gas (i.e. CO2)), it is essential that particles in that medium would be attracted to the surface of nanocomposite in order to detect them. The microperiodic structure couples the light of a He-Ne laser at a given resonance angle into the waveguide layer [11]. This incoupling angle is very sensitive to the presence of adsorbed particles and to any change of refractive index of the medium covering the surface of the nanocomposite. By precise measurement of the incoupling angle, the adsorbed particles can be determined with ultrahigh sensitivity. However, certain limitations appear – too high concentration of investigated medium or too big size of particles in that medium. In this case, metal nanoparticles are not able automatically to attract them to nanocomposite surface. For this purpose, piezoelectric materials are incorporated in nanocomposite, i.e. allowing to tune the microstructure and to control the motion of particles in the investigated medium.

Because of the properties of piezocomposite material controllable in micrometric level by ultrahigh frequency excitation, the mathematical model of the system is developed. This model allows analyzing the mechanical parameters of the system, which enable to control motion of cell-biological mechanisms in bio suspension and to force them to be attracted by metal nanoparticles, deposited on designed microperiodic structure in nanocomposite layer in order to be prepared for analyzing them using single molecule analytical tools.

The optical, electrical and mechanical characteristics of created piezocomposite material with controllable properties in micrometric level, and the design and functionality of the microperiodic structure on nanocomposite layer to be used for a single molecule analysis, methodology are presented in this paper.

11.2 Theoretical Evaluation of SPR in PMMA Thin Film

In the design process of sensing element having plasmonic properties, it is essential to select biologically compatible materials for synthesis of nanocomposite thin films in order to obtained SPR effect. For these investigations, metal – silver (Ag) nanoparticles were used because of their unique optical, thermal, and electrical properties. When affecting Ag nanoparticles with the light at a specific wavelength, strong interaction forms between them. This phenomenon occurs because conduction electrons on the surface of metal nanoparticles sustain a collective oscillation. Various metal nanoparticles may be integrated into products ranging from photovoltaic to biological and chemical sensors.

In this research paper, silver (Ag) nanoparticles and polymethyl methacrylate (PMMA) polymer were selected for SPR investigation. The refractive index of PMMA-Ag thin film depends from the size of Ag nanoparticles. Before experimental investigation of SPR in PMMA thin film, a theoretical evaluation of extinction versus wavelength for the various sizes of Ag nanoparticles (Fig. 11.1) was done with MiePLOT software.

According to theoretical optical spectra (Fig. 11.1), spherical silver nanoparticles, embedded in PMMA dielectric media, were simulated in terms of the Mie electromagnetic theory. This theory allows evaluating the extinction cross section Q_{ext} and elastic scattering Q_{sca} for a wave incident on a part. In this case, the extinction-cross-section of silver nanoparticles decreased with spherical silver nanoparticles size. Thus, based on these theoretical calculations, Ag nanoparticles of 60 nm diameter were selected for the experimental investigation. Ag nanoparticles

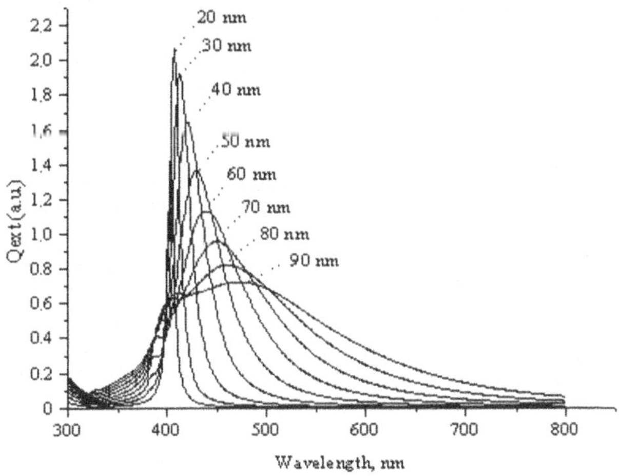

Fig. 11.1 Theoretical evaluation of extinction versus wavelength for the various size of Ag nanoparticles

(60 nm diameter) ensure the good SPR for the wide wavelength interval, because of sharp and wide peak.

Moreover, the refractive index of PMMA-Ag thin film depends from the size of Ag nanoparticles and is controlled by geometry, surface morphology, mechanical and optical properties, and by type of synthesized materials.

11.3 Synthesis and Formation of Nanocomposites

Basic principle for design of SPR nanostructures is incorporation of selected metal (Ag) nanoparticles in polymer PMMA matrix. Thin films of PMMA and PMMA-Ag were formed on glass substrates (cleaned for 10 min in ultrasonic bath of acetone) using a spin-coating technology. PZT-PMMA-Ag is formed on copper foil with Thin films of blended solutions were formed by centrifuge Dynapert Precima dried in the electric furnace of 100 °C for 10 min and then radiated by UV light (Hg Philips lamp of 1.2 kV) for 60 s. Thus, incorporation of silver nanoparticles in polymer matrix gives strong interactions with light, i.e. SPR phenomena occurs. For piezocomposite formation Lead Zirconate Titanate (PZT) was added to PMMA-Ag solution (see Table 11.1).

PZT composite material with PMMA presented in paper [12] demonstrates the best piezoelectric and hydrophobic properties. Silver nanoparticles synthesized with polymeric material may contain bioactive properties that can help to suppress bacteria's growth.

Periodical microstructures were fabricated onto thin films by lowcost and high throughput replication technology – a hot embossing process. This process allows simultaneous and single step formation of gratings on various microstructures with high accuracy. Thin film layer and master grating were placed together and fixed in the metallic mandrel between plates and putted in the furnace of 110 °C for 12 min (see Table 11.1).

Table 11.1 Materials and synthesis of nanocomposites

Materials	Preparation of blends
PMMA	5% PMMA solution (obtained dissolving PMMA (average M_W = 15,000) in dimethylformamide) obtained from Sigma Aldrich
PMMA-Ag	PMMA (average M_W = 15,000) was dissolved in dioxane and dimethylformamide (5 wt. % polymer in solvent). 100 μl of aqueous (500 mg of $AgNO_3$ were dissolved in 1 ml water) solution of silver nitrate ($AgNO_3$) were mixed with 3 ml of PMMA solution. All obtained from Sigma Aldrich. Millipore–Q deionized water used for synthesis. Different UV exposure time was used
PZT-PMMA-Ag	In PMMA-Ag solution an oxalic acid/water based for synthesis nano powders of Lead Zirconate-Titanate ($Pb(Zr_x, Ti_{1-x})O_3$) with $x = 0.52$, also known as PZT (52/48) was added. The other reagents used were oxalic acid, deionized water, acetic acid, ammonia solution

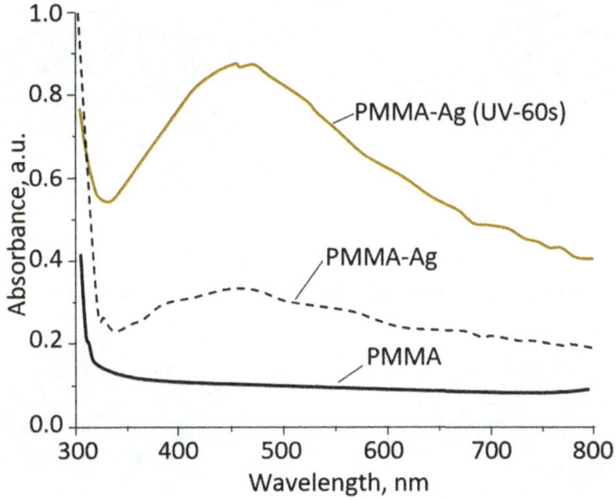

Fig. 11.2 Absorbance spectra of PMMA (M_W = 15,000) and Ag-PMMA before and after UV irradiation

The hot embossing procedure in polymers was discussed in previous researches [13].

11.4 Investigation of Nanocomposite Thin Films Properties

As it was stated in introduction, PMMA can be used for formation of sensing elements for detection of cells or other bio particles. For this purpose, further investigations were concentrated on designing nanostructures formed from PMMA-Ag nanocomposites. Properties of PMMA-Ag nanocomposites were compared with pure PMMA. PZT was introduced in PMMA-Ag composite for piezoelectric properties enabling to tune designed microperiodical structure.

11.4.1 Nanocomposites with SPR Effect

Optical properties of PMMA-Ag nanocomposite and pure PMMA were investigated using UV/VIS/NIR Avantes spectrometer.

The shifting and narrowing of SPR peak indicates the silver nanoparticles agglomeration. UV radiation induces the formation of photoinitiator radicals. The latter can create new cross-linked nanocomposites with PMMA. These radicals reduces Ag ions in nanocomposites and promotes Ag growth in coloidal solutions. Thus, SPR peak for the polymer PMMA is not typical, polymer is clear in all visible light diapason (Fig. 11.2, black line). When nanoparticles of Ag are introduced to

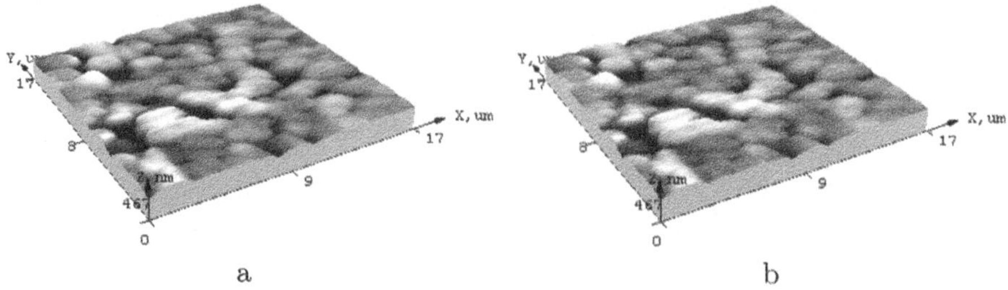

Fig. 11.3 Surface morphology of PMMA (**a**) and Ag-PMMA nanocomposite (**b**)

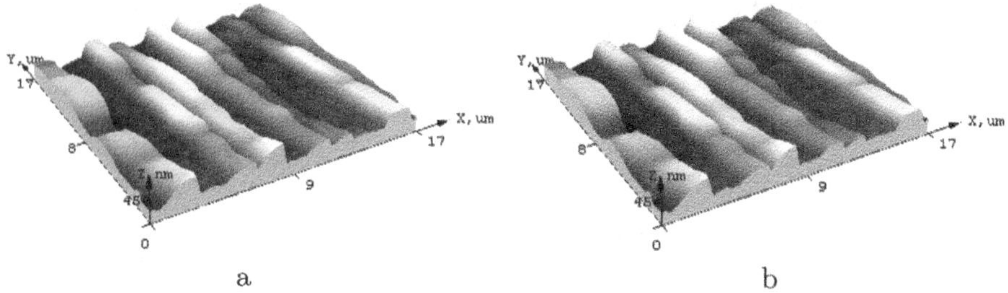

Fig. 11.4 Diffraction grating imprinted on (**a**) PMMA layer and (**b**) Ag-PMMA nanocomposite layer

polymer PMMA matrix, peak is observed at 450 nm with absorbance of 0.34 a.u. (Fig. 11.2, dotted line). The shift of 9 nm in wavelength and increase in absorbance from 0.34 a.u. to 0.89 a.u., when Ag-PMMA thin film is irradiate with UV for 60 s, is observed (Fig. 11.2, brown line).

The evaluation of surface morphology of PMMA and PMMA-Ag nanocomposite (Fig. 11.3) was done using Atomic Force Microscope NT-206. Surface roughness of formed PMMA-Ag thin film (Rq = 4.1 nm, thickness – 694 nm) is about seven times smaller with only some nano-size irregularities compared with PMMA thin film (Rq = 29.9 nm, thickness – 724 nm). Thus, UV exposure makes significant improvement in surface morphology of nanocomposite thin film.

In the next step periodic microstructures were formed in pure PMMA and PMMA-Ag nanocompsite using Nickel master matrix (Table 11.2) and hot embossing technique described in paper [13]. Surface morphology of the formed microstructures were analyzed using Atomic force Microscopy (AFM) (Fig. 11.4).

Table 11.2 Technological data for formation of periodical microstructures

Master grating periodicity	Depth	700 nm
	Periodicity	4 μm
Master grating dimensions	Length	2 mm
	Width	16 mm
Grating lines	Parallel to the short edge	
Pressure in metallic mandrel	12 MPa	

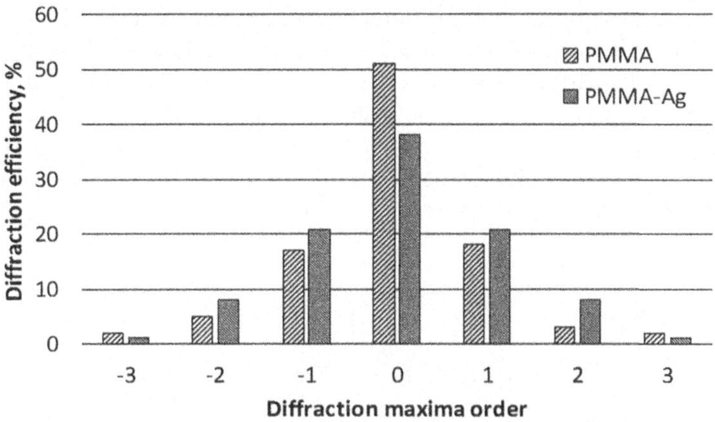

Fig. 11.5 Diffraction efficiency of grating imprinted on PMMA and Ag-PMMA nanocomposite layers

According to AFM measurements, microstructure formed on PMMA-Ag nanocomposite layer, has rather smooth surface (roughness $Rq = 64.6$ nm) with periodic microstructure properties of average grating depth 0.54 μm, and period 4 μm compared to periodical microstructure embossed on pure PMMA layer with surface roughness $Rq = 91.0$ nm, average depth 0.21 μm, and period 4 μm. The best results similar to master periodic microstructure were obtained for using PMMA-Ag nanocomposite material.

Distribution of the transmittance diffraction efficiencies (Fig. 11.5) of the formed periodic microstructures were measured using laser ($\lambda = 632$ nm) diffractometer. Distribution of diffraction efficiencies was observed in 0, ±1, ±2, ±3 diffraction maxima, because these maxima are the most important when evaluating optical properties of formed microstructures.

Figure 11.5 shows, that 51% of diffraction efficiency is concentrated in zero order maxima and 17% in ±1 maxima for periodic microstructure formed in PMMA. Distribution of diffraction efficiency of periodic microstructure formed on PMMA-Ag is better, i.e. efficiency of zero order maxima is smaller by 13%, while fist order maxima have better efficiency by 6%.

11.4.2 *Nanocomposites with Combination of SPR and Piezoelectric Effect*

Investigations showed, PMMA-Ag nanocomposite material could be used for design and fabrication of periodic microstructures. When this element is in contact with biologically active medium, it is essential that particles in that medium would be attracted to the surface of nanocomposite in order to sort, filtrate and detect them. But, because of too high or too low concentration, density variation of the investigated medium, metal nanoparticles are not able automatically to attract them to nanocomposite surface. Therefore, Ag nanoparticles were integrated into PZT-PMMA piezoelectric nanocomposite material using techniques described in paper [12]. Surface of the PZT-PMMA-Ag nanocomposite is presented in Fig. 11.6.

Optical properties of PZT-PMMA and PZT-PMMA-Ag were investigated by UV/VIS/NIR Avantes spectrometer (Fig. 11.7).

Fig. 11.6 SEM view of PZT-PMMA-Ag nanocomposite

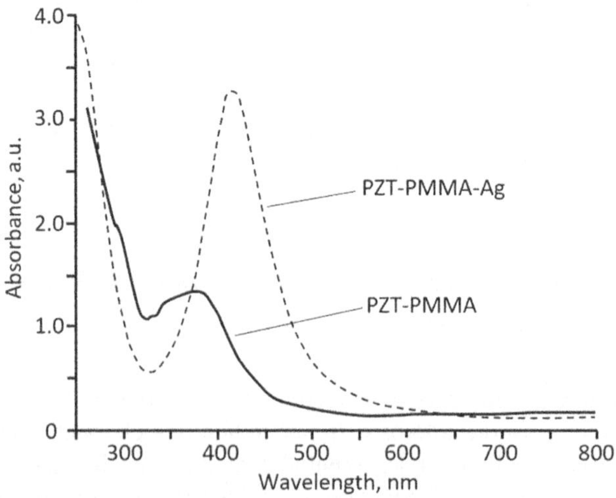

Fig. 11.7 Absorbance spectra of piezoelectric nanocomposites

Synthesizing PMMA with nanoparticles of PZT leads to two dominating peak at wavelength 376 nm (Fig. 11.7 black line). Significant changes were observed when PZT-PMMA piezoelectric nanocomposite was synthesized with silver nanoparticles, i.e. sharp dominating SPR peak was investigated at 424 nm (Fig. 11.7 dotted line). Incorporation of Ag nanoparticles in piezoelectric nanocomposites, shifts a dominating peak into region of visible light spectrum.

PMMA here is used as material improving thermo – plasticity of PZT composite material. This is essential when forming periodical microstructures imprinted in formed piezocomposites. Using hot embossing process described in paper [13], periodical microstructures were imprinted on PZT-PMMA and PZT-PMMA-Ag nanocomposites (Fig. 11.8).

Imprinted structures on nanocomposite materials have the same geometrical properties, i.e. it could be stated, Ag nanoparticles has no influence on the replication procedure of microstructure.

Moreover, diffraction efficiency measurements (Fig. 11.9) shows that for a periodical microstructure imprinted on PZT–PMMA, 13% of the diffracted energy was concentrated in its zero order and about 10% in its first order maxima. Incorporation of Ag nanoparticles in PZT nanocomposite increases the diffraction

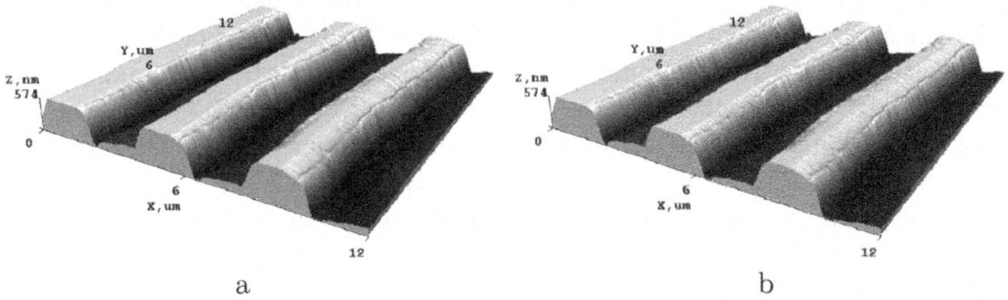

Fig. 11.8 AFM view of periodic microstructure formed on (**a**) PZT-PMMA and (**b**) PZT-PMMA-Ag nanocomposites

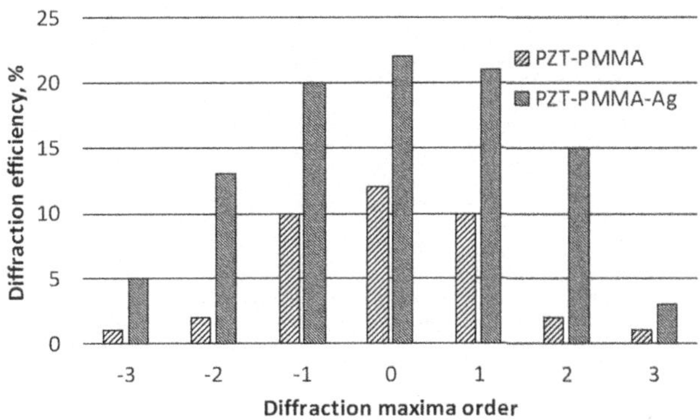

Fig. 11.9 Diffraction efficiencies of gratings imprinted on piezoelectric nanocomposite thin films

efficiency in its zero order by 8% and in its first orders of maxima by 10%. Thus, results imply that silver nanoparticles significantly improve optical response of designed piezocomposite microstructure.

11.5 Simulation of Bulk Acoustic Waves in Periodic Microstructure Formed in PZT-PMMA-Ag Composite Material

PZT-PMMA-Ag nanocomposite material demonstrates piezoelectric properties in micrometric level. Therefore, it could be used for selection, sorting and manipulation of bioparticles in periodic microstructure using ultrahigh frequency excitation. Firstly, numerical model of one micro channel (period) formed in PZT-PMMA-Ag composite material was created for the investigation of particle manipulation using bulk acoustic waves.

Geometry and mesh of the microstructure is presented in Fig. 11.10. It was used to analyze dynamics of particles with density of 500 kg/m^3 and diameter 3 μm excited by normal acceleration of 1.5×10^6 m/s^2.

In the first step, a distribution of the acoustic pressure in the cross-section of the microchannel was calculated at different excitation frequencies. For the dynamical investigation of particle motion the first eigenmode of acoustic pressure observed at 0.43 MHz was selected, because it has horizontal nodal line of zero pressure (Fig. 11.11). It means, that particles during the manipulation process will be focused along this line.

Particle motion was investigated using particle tracing method for fluid flow. Position of particles after 20 ms of excitation is presented in Fig. 11.12. The color of particle represents the length of particles path. The particles focusing process is very fast and this technique could be used for detection or investigation of bioparticles in the dynamic systems, when fluid has laminar flow. Moreover, if the process takes

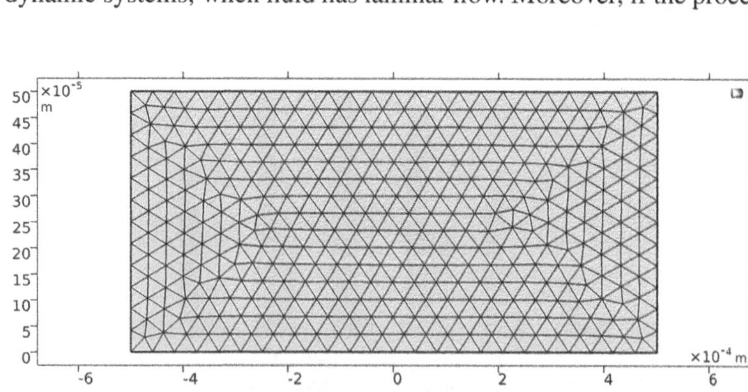

Fig. 11.10 Geometry and mesh of the microstructure

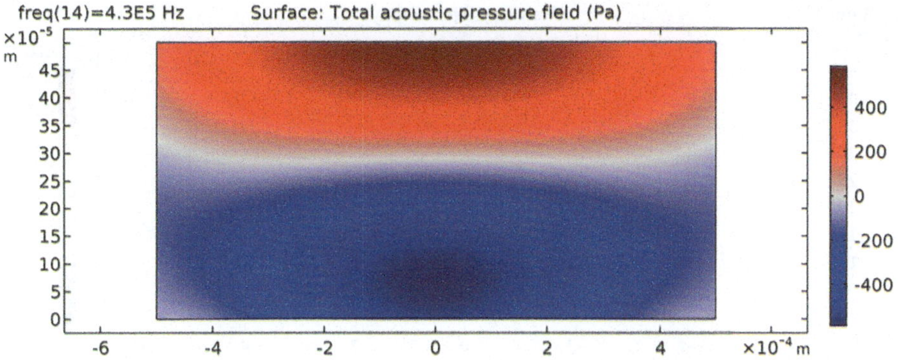

Fig. 11.11 Distribution of acoustic pressure in microchanel excited at 10,000 Hz

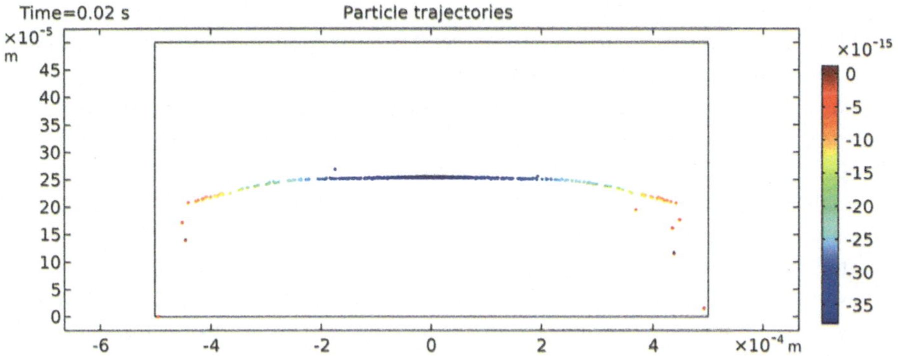

Fig. 11.12 Trajectories of particles after 0.02 s

longer time, i.e. at least 0.1 s microparticles will be focus into one spot (Fig. 11.13). This phenomena allows to do particles separation from the medium.

The numerical model of the system allows to analyse the dynamical parameters of the system which enable to control motion of cell-biological mechanisms in bio suspension and to force them to be attracted by designed microperiodic structure on nanocomposite layer in order to be prepared for analysing them using single molecule analytical tools.

11.6 Conclusions

Investigation results showed, that SPR effect is not typical phenomenon in piezocomposite materials. Implementation of metal (Ag) nanoparticles increases the sensitivity of the designed element, i.e. Ag nanoparticles in piezoelectric nanocomposites shift a dominating peak into region of visible light spectrum. For sensitive

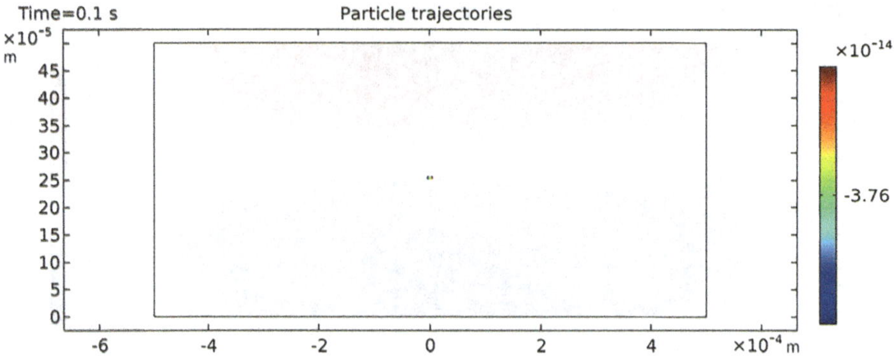

Fig. 11.13 Trajectories of particles after 0.1 s

and functional system piezoelectric and SPR effect based elements should work in the wavelength range from 372 to 424 nm.

Thus, PZT-PMMA-Ag nanocomposite material demonstrated good piezoelectric properties in micrometric level and were used for periodical microstructure formation in it. Designed sensing element may be used for in-situ monitoring of analytes and particles in it. Therefore, based on obtained results and using ultrahigh frequency excitation, a mathematical model of the described system was created. This numerical model of one micro channel (period) was created for the investigation of particle manipulation using bulk acoustic waves. It allowed to analyse the dynamical parameters of the system and to control motion of cell-biological mechanisms in bio suspension and to force them to be attracted to the surface of designed microperiodic structure.

Acknowledgments This research was funded by a grant (No. S-MIP-19-43) from the Research Council of Lithuania.

References

1. Bryaskova R, Georgieva N, Andreeva T, Tzoneva R (2013) Cell adhesive behavior of PVA-based hybrid materials with silver nanoparticles. Surf Coat Technol 235:186–191
2. Mestek O, Loula M, Kaòa A, Vosmanská M (2020) Can ultrafast single-particle analysis using ICP-MS affect the detection limit? Case study: silver nanoparticles. Talanta 210:1–4
3. Mansour A, Poncin-Epaillard F, Debarnot D (2020) Distribution of metal nanoparticles in a plasma polymer matrix according to the structure of the polymer and the nature of the metal. Thin Solid Films 699:1–9
4. Zeng L, Chen M, Yan W, Li Z, Yang F (2020) Si-grating-assisted SPR sensor with high figure of merit based on Fabry–Pérot cavity, Opt Commun 457:1–7
5. Li X, Xu H, Chen ZS, Chen G (2011) Biosynthesis of nanoparticles by microorganisms and their applications. J Nanomater 2011:1–16

6. Orlova T, Chernozem RV, Ivanova AA, Bartasyte A, Mathur S, Surmeneva MA (2019) Hybrid lead-free polymer-based nanocomposites with improved piezoelectric response for biomedical energy-harvesting applications: a review. Nano Energy 62:475–506
7. Ermakov V, Kruchinin S, Fujiwara A (2008) Electronic nanosensors based on nanotransistor with bistability behaviour. In: Bonca J, Kruchinin S (eds) Proceedings NATO ARW "Electron transport in nanosystems". Springer, Berlin, pp 341–349
8. Filikhin I, Peterson T, Vlahovic B, Kruchinin SP, Kuzmichev YB, Mitic V (2019) Electron transfer from the barrier in InAs/GaAs quantum dot-well structure. Phys E Low-dimensional Syst Nanostruct 114, Article 113629
9. Ermakov V, Kruchinin S, Pruschke T, Freericks J (2015) Thermoelectricity in tunneling nanostructures. Phys Rev B 92:115531
10. Fernández-Benavides DA, Cervera-Chiner L, Jiménez Y, Arias de Fuentes O, Muñoz-Saldaña J (2019) A novel bismuth-based lead-free piezoelectric transducer immunosensor for carbaryl quantification. Sens Actuators B Chem 285:423–430
11. Yaremchuk I, Tamulevičius T, Fitio V, Gražulevičiūte I, Bobitski Y, Tamulevičius S (2013) Guide-mode resonance characteristics of periodic structure on base of diamond-like carbon film. Opt Commun 301–302:1–6
12. Cekas E, Janusas G, Guobiene A, Palevicius A, Vilkauskas A, Ponelyte Urbaite S (2018) Design of controllable novel piezoelectric components for microfluidic applications. Sensors (Basel) 18(11):1–23
13. Sodah A, Gaidys R, Narijauskaite B, Sakalys R, Janusas G, Palevicius A, Palevicius P (2019) Analysis of microstructure replication using vibratory assisted thermal imprint process. Microsyst Technol 25:477–486

Chapter 12
Detection of CBRN Agents at the Interface with P(VDF-TRFE) Film by Scanning Third Harmonic Generation

S. G. Ilchenko, R. A. Lymarenko, V. B. Y. Taranenko, V. V. Multian, V. Ya. Gayvoronsky, S. A. Pullano, D. C. Critello, and A. S. Fiorillo

Abstract This work is aimed at developing nonlinear optical method of non-destructive and surface sensitive, real time (in situ) detection of CBRN agents. Ferroelectric polymer polyvinylidene fluoride PVDF and its copolymer P(VDF-TrFe) film is used as chemical-stabile and sensing element. Results showed that combination of the interface scanning technique at third harmonic, the second harmonic generation readout, and spatial profile analysis is promising for nondestructive remote sensing applications.

Keywords Polyvinylidene fluoride (PVDF) · Third harmonic generation (THG) · Second harmonic generation (SHG) · Laser beam self-action · Optical damage · Interface scanning technique

12.1 Introduction

Polyvinylidene fluoride (PVDF) (-CH_2–CF_2-$)_n$ is a semi-crystalline, high purity thermoplastic fluoropolymer, which has crystalline phases named β, α, γ, and δ [1, 2]. The C–F bonds are polar, and the highest dipole moment is obtained (from the α-phase after stretching and poling) with the alignment of all dipoles of the polymer in the same direction, corresponding to the β-phase of the PVDF, which has ferroelectric properties.

S. G. Ilchenko (✉) · R. A. Lymarenko · V. B. Y. Taranenko
International Center "Institute of Applied Optics" National Academy of Sciences of Ukraine, Kyiv, Ukraine
e-mail: Svitlana-ilchenko@ukr.net

V. V. Multian · V. Y. Gayvoronsky
Institute of Physics National Academy of Sciences of Ukraine, Kyiv, Ukraine

S. A. Pullano · D. C. Critello · A. S. Fiorillo
Department of Health Sciences, University Magna Græcia, Catanzaro, Italy

J. Bonča, S. Kruchinin (eds.), *Advanced Nanomaterials for Detection of CBRN*, NATO Science for Peace and Security Series A: Chemistry and Biology,
https://doi.org/10.1007/978-94-024-2030-2_12

The β-phase can be also obtained by adding a co-polymer; in such a case only electrical poling is required, and the stretching is no longer necessary. A common PVDF copolymer is VDF with trifluoroethylene (TrFe) [3–6]. The structure of such a copolymer is very similar to that of PVDF, with the insertion of TrFe (-CF_2–CFH-)$_n$ units among the fundamental monomers of PVDF molecular chains.

The most relevant use of ferroelectric polymers is in the manufacturing of sensors in robotics and medicine applications. PVDF-based impact sensors are widely used for sensing, as artificial sensitive skin, acoustic detector, pressure sensor integrated on catheter surface, and blood pressure detection [7, 8]. The term "impact" refers to the ability of the sensor to detect any mechanical vibration or shock induced on its surfaces. Basically, an impact sensor made of the PVDF-based film is a two-plate capacitor with the piezoelectric polymer as the dielectric material where the applied force induces a charge across the capacitor itself [9]. When a pressure is applied, polarization of molecules decreases because the magnitude of polarization is proportional to the distance between hydrogen and fluorine atoms. In terms of detected electric signal, the output voltage is negative as response to the compression. If the applied force is removed, the distance between atoms changes and positive voltage output is generated.

Piezoelectric and pyroelectric polymer-based sensors conjugated with a front-end MOSFET transistor were also developed [10]. Compared to piezoceramics, such integrated transducers have improved signal to noise ratio, better acoustic impedance match to body constituents (tissue and water) and high sensitivity. Besides the piezoelectric polymer film, in the form of POSFET (Piezoelectric Oxide Semiconductor Field Effect Transistor) device, it has been also used in sensing applications with the aim of mimicking the human touch, which consists of various receptive structures arranged within the layers of the skin [10]. The role of the thin polymer film is to manifest a charge as a response of an applied force, modulating the drain current of the transistor. A matrix structure of POSFET elements is desirable to mimic a real human touch [11]. Electronic skin (E-skin) development require to identify the functional material to enable certain sensing capabilities. Piezoelectric polymer films of PVDF have been chosen as meeting the target requirements of mechanical flexibility, high sensitivity, detectability of dynamic touch and robustness [12–16].

Recently, an ultrasonic sensor inspired by the shape of a bat's cochlea, which is part of one of the most sophisticated echolocation systems to be found among mammals, has been developed [17, 18]. In this case, PVDF was used for the manufacturing of the sensor, which is assembled by folding a sheet of the piezopolymer according to a logarithmic spiral geometry.

PVDF and its copolymers exhibit also a spontaneous electric polarization as a result of temperature variation, an effect known as pyroelectricity. In this context, thin film piezopolymer materials are characterized by low dielectric constants and dielectric loss, low cost, ease of handling and manufacturing, and stable performance. Pyroelectric materials are significant in the development of temperature-sensing devices as they are capable to respond to changes in detectable radiation intensity, for instance the infrared (IR) energy. The pyroelectric vidicon tube represents a worthy example of thermal detector: the IR radiation as a source of stimulation hits a disk-like pyroelectric target, and the resulted electric charges are read out by an electron beam scanning [19].

Conventional approach to study the PVDF nonlinear optical properties is to measure second-harmonic generation efficiency [20]. In this paper we are looking for new field of polymer films application via interface scanning technique. We report nonlinear optical properties of ferroelectric polymer polyvinylidene fluoride PVDF and its copolymer P(VDF-TrFe) film deposited on glass cover slip. Combination of the quadratic and cubic nonlinear optical responses is promising for nondestructive remote sensing at the functionalized interfaces of the studied material.

Nonlinear optical (NLO) methods are powerful and versatile tools for surface and interface studies, providing information and contrast even on a molecular level. Due to the NLO methods are non-destructive, contact-free and often surface sensitive can be used for real time (in situ) monitoring of physical and chemical properties change of investigated media. Furthermore, the chemical stable PVDF polymer can be used as promising material for the opportunity to investigate of the specimens in the biomedical field or for non-destructive testing via NLO methods.

We investigated NLO properties of PVDF polymer via optical methods based on the second and third order of polarization and evaluate their potential for applications in sensorics. Primarily, experimental configurations in which optical methods are presented for characterization of the material providing complementary information of polymer, especially photoinduced variations and saturation effects.

12.2 Experimental Setup

Experimental setup of the interface scanning technique presented in Fig. 12.1a. The excitation is performed by the pulsed YAG:Nd laser (42 ps FWHM, repetition rate 40 Hz, $\lambda = 1064$ nm) with the Gaussian spatial profile. The peak laser intensity of incident light is varied with the gradient neutral attenuator (A). Laser beam is focused by the focusing lens L with focal distance 3.25 cm on the sample S which that positioned on the high precision translation stage. The input laser power is registered by photodiode PD1. SH and TH signals are collected by lens and measured with a photomultiplier PMT (Hamamatsu H10721-210). A for conventional interface scanning technique a set of filters were used to remove the excitation beam and to extract the SH and TH responses. To scan spectral response in range 300–600 nm monochromator were placed before PMT.

Studying of the self-action effect was done with experimental setup presented in Fig. 12.1b, c. In general, this setup was based on the previous one described in Fig. 12.1a that was modified: after lens L1 a set of photodiodes were placed in order to analyze the photoinduced refractive index and optical absorption coefficient variations (see Fig. 12.1b, c).

The experimental setup allows to measure both the total (Fig. 12.1b) and on axis (Fig. 12.1c) transmittance variations versus the incident peak laser intensity. For the

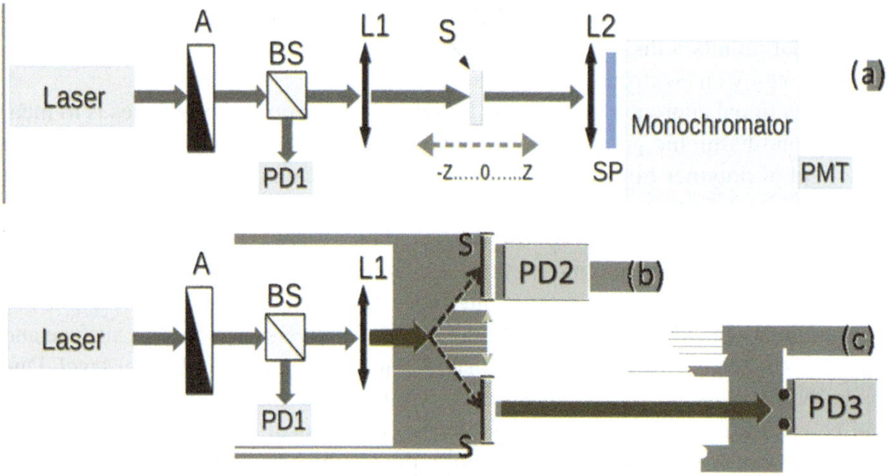

Fig. 12.1 Scheme of experimental setups for (**a**) SHG and THG measure; (**b**) photoinduced variation of total transmittance; (**c**) laser beam self-action technique for analysis of photoinduced variation of on-axis transmittance: *A* attenuator, *S* sample, *BS* beam splitter, *L* focusing lenses, *D* finite diaphragm ($d = 2$ mm), *PD1, PD2, PD3* photodiodes, *SP* shortpass filter, *PMT* photomultiplier tube

setup (b) case the sample S is positioned directly at the face of the wide aperture (1 cm) photodiode in order to avoid the scattering impact. The total transmittance of the studied polymer is then determined by the signals of the photodiodes PD3 corrected to apparatus function. In case (c) the transmittance signal of polymer through a finite aperture in the far field as a function of the intensity were measured. To interpret the on-axis transmittance in the far field the signal of were registered by photodiode PD2. The laser beam intensity increasing leads to self-lensing in the sample. It is sensitive to convergence /divergence of the laser beam which is caused by self-focusing/defocusing phenomena in the media.

12.3 Sample Preparation

For analysis of nonlinear optical properties ferroelectric polymer polyvinylidene fluoride based films two types of materials were used: perpendicularly oriented PVDF and its copolymer P(VDF-TrFE) without initial orientation. Orienting of dipoles in PVDF polymer was realized by typical approach that provides perpendicular orientation of the dipoles in PVDF. It consists of two steps: (i) stretching and (ii) high voltage poling.

To prepare polymer pellet forms of P(VDF-TrFE) were preferably dissolved in methyl ethyl ketone (MEK) with a molar ratio of 70/30. The coverslip was then annealed on a hot plate at 80 °C for about 10 min in order to allow the

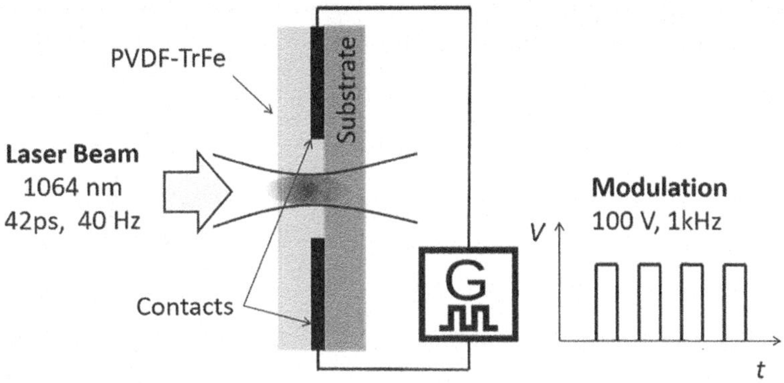

Fig. 12.2 Schematic representation of P(VDF-TrFe) sample for analysis of the poling influence on NLO response, G – waveform generator

solvent evaporation. For thicker films, a recursive procedure of the above-mentioned steps could be performed. However, in case of multilayer structure an adjunctive annealing step at 170 °C for one hour, and then a cooling to room temperature was required to increase the crystallinity of the film. The unit cell of the lattice of P(VDF-TrFe) consists of two lateral chains whose antiparallel dipole components thus neutralizing each other, which becomes regularly antiparallel upon heat treatment at high temperatures. It should be noted that the presence of TrFe in PVDF leads to partial dipole orientation. It makes it more efficient for tuning response by applying of electrical field.

Two types of samples were investigated. Type 1 – free PVDF 40 μm film with perpendicular to the surface poling. Type 2 – P(VDF-TrFe) copolymer 10 μm film with electrodes without initial poling. To prepare the sample, first, the mask were developed to deposite Ag metal stripes with thickness about 100 nm as electrodes. After that the polymer layer were deposited on top (see Fig. 12.2).

12.4 Results and Discussions

12.4.1 Interface Scanning Technique

Interface scanning technique at third harmonic (ISTH) is very sensitive approach for analyzing cubic nonlinear susceptibility of wide range of materials. In this technique readout should be realized by tight focused laser beam at the interface between of two media with different nonlinear susceptibilities [21, 22]. In general, the interface scanning technique allows analyze one parameter – amplitude of the THG signal at the interface between the reference material and studied media [23–25]. This value depends on the ratio between cubic susceptibility of the media and reference [22]. Sensitivity of such approach defined by two factors: the THG signal registration

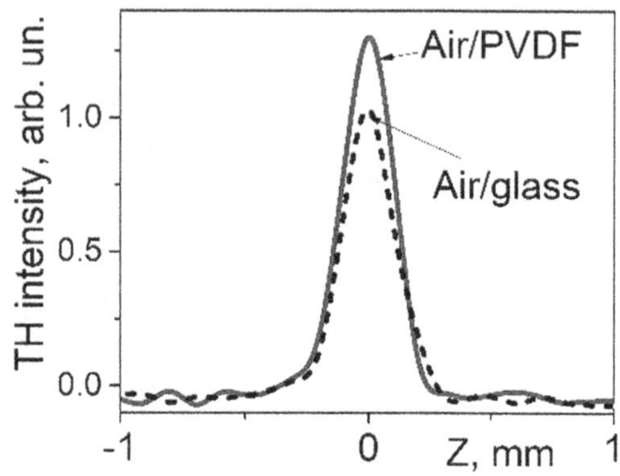

Fig. 12.3 Comparison of the TH signals generated at the interfaces air/PVDF and air/glass

system efficiency and the value of cubic susceptibility of the reference material. On practice the sensitivity of registration system is limited by the optical components and sensor available on the market and budget of designed solution. In this case it is more efficient to vary the reference material or apply materials with smart nonlinear optical response [26].

The second step of improving the interface scanning technique is to add more nonlinear effects that can be analyzed from the same area such as SH and self-action response.

This work was focused on studying of PVDF based polymers with different chemical structure modification and orienting that is promising for nonlinear optical applications. It is known for such kind of material that it is possible to vary the properties of the film by applying of electrical field. An objective of the research is to study the properties of PVDF based films as reference material for interface scanning technique and sensing applications based on it.

For this porpoise the free perpendicularly poled PVDF film with thickness 40 μm was studied by IS@TH with picosecond laser excitation at 1064 nm and signal registration at 355 nm (see Fig. 12.3). Comparison was done with the TH signal from the glass substrate with thickness 1 mm at the interface air/glass. It should be noted that measurements were done with relatively long beam waist (about 500 μm) in comparison with the film thickness. In this case for the 40 μm film the THG signal can be represented as effective signal from the interfaces air/PVDF and PVDF/air. Applied approach demonstrated up to 20% the higher efficiency of THG in PVDF at and allowed to estimate the effective cubic susceptibility of the PVDF at third harmonic about $\chi^{(3)}(3\omega) \sim 1.8 \times 10^{-14}$ esu.

Specifics of PVDF based on possibility of polymer orienting makes it promising for SHG applications as it was described in [27–29]. Supplying of the interface scanning technique by SH signal readout (ISSH) is promising for sensing, because SHG is highly sensitive to the orientation, birefringence and local symmetry properties of a medium while THG is sensitive to the interface and combining of these techniques allows estimate which kind of material is under investigation.

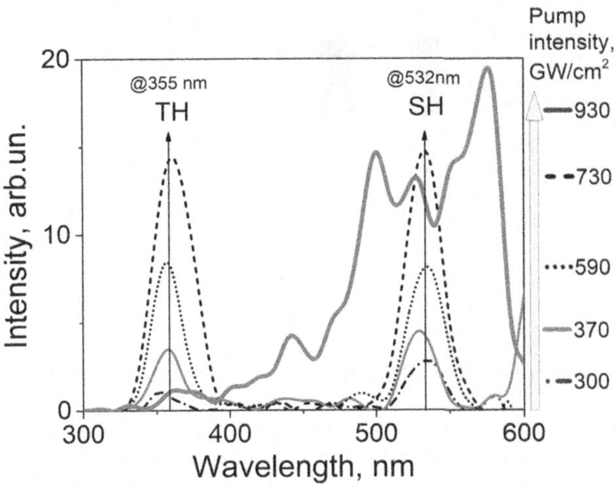

Fig. 12.4 Synchronous growing of TH (355 nm) and SH (532 nm) peaks in spectral response of the perpendicularly oriented PVDF film under picosecond laser excitation at 1064 nm at different intensities up to 1 TW/cm^2

For measurements the interface scanning technique was used with modification for spectral analysis realized by monochromator and PMT. Obtained results for perpendicularly poled PVDF films presented on Fig. 12.4. It should be noted that in order to increase the sensitivity of the setup the input and output slits of the monochromator were optimized on maximal signal that affected on the spectral resolution. However, this approach allows to easily identify the peaks of SH and TH in wide pump intensities range.

Processes of the second and the third harmonics generation require the high laser intensities. In order to obtain robust data for the laser radiation frequency conversion in the PVDF films we should work with laser intensity at least 100 GW/cm^2.

In general, for nonlinear bulk crystals the intensity of SHG $I_{2\omega}$ should demonstrate the quadratic growth with the intensity I_ω of the laser pump like $I_{2\omega} \sim (I_\omega)^p$, where $p = 2$. At the same time the intensity of TH signal $I_{3\omega}$ can be described by cubic dependence ($p = 3$) on the excitation intensity. In our case we observe a synchronous growth of TH and SH up to 800 GW/cm^2. It can be explained by competition between these two nonlinear optical processes that depends on excitation intensity [30]. While SHG is dominant at low excitation intensities, the nonlinear response will eventually be governed by THG at high excitation intensities. In order to observe such a competition, the damage threshold for polymer needs to be large enough so that the intensity of THG can exceed that of SHG. Obtained data for SH and TH signals demonstrate factor $p = 1.8$ in range up to 800 GW/cm^2 and $p = 2.9$ in range up to 600 GW/cm^2. At pump intensities $I > 800$ GW/cm^2 we have observed broad supercontinuum generation with consequent optical damage at higher excitation levels. Obtained results demonstrate the perspectives of the oriented PVDF films as reference material for analysis of the

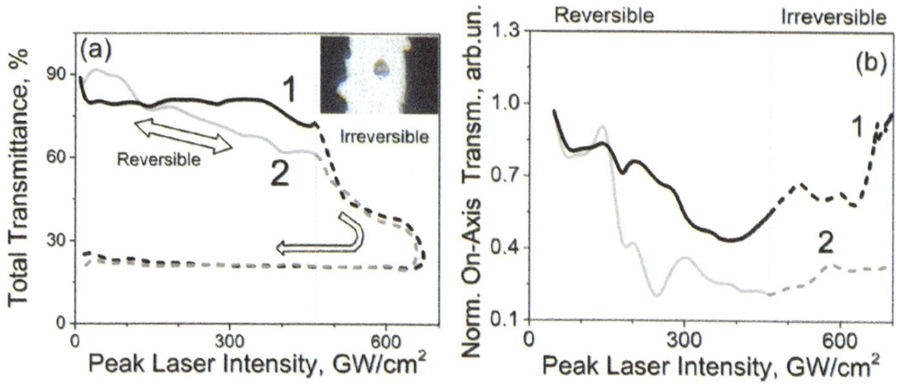

Fig. 12.5 The photoinduced variations of the (**a**) total and (**b**) normalized on-axis transmittances versus the peak laser intensity of picosecond laser pulses at 1064 nm for the structure based on P(VDF-TrFe) polymer (see Fig. 12.2) in initial state **1** and with modulated electrical field **2**. Solid line corresponds to reversible response, dashed – irreversible

cubic (THG) and quadratic (SHG) responses of studied materials. This information can be used as two-parameter sensing technique that increase the sensitivity and provides more information about studied media.

12.4.2 Laser Beam Self-Action Effects

Taking to account the possibility of tuning of the properties of the PVDF-based materials by orienting of polymers with electrical field we designed the special cell (see Fig. 12.2). In our case this cell allows to simultaneously orient the polymer by applying the electrical field with different waveforms and to study the nonlinear optical response.

In order to study NLO properties of polymer self-action technique was applied. This technique allows analyze the absorptive and refractive cubic NLO response and it is optimal for previous analysis of materials due to the high sensitivity and robustness.

Figure 12.5a, b shows the photoinduced total and on-axis transmittance variations versus the peak laser intensity in correspondingly for the studied P(VDF-TrFe) samples in initial state **1** and with modulated electrical field **2**.

Total (Fig. 12.5a) and on-axis transmittance in the far field (Fig. 12.5b) of the sample were normalized by glass substrate response in order to compensate the impact of the apparatus function on refractive NLO response. Initial sample demonstrate sharp decrease of the transmittance in range up to 30 GW/cm^2. The efficiency of this process can be described by estimated imaginary part of cubic susceptibility $\mathrm{Im}(\chi^{(3)}) = 3.0\times 10^{-15}$ esu. At higher intensities this process saturates and stays constant up to 400 GW/cm^2. Further increase of intensity leads to decrease of transmittance with $\mathrm{Im}(\chi^{(3)}) = 7.7 \times 10^{-15}$ esu that finished by optical damage at about 800 nm. It should be noted that self-action measurements were done with the longpass filters on the photodiodes in order to cut the signals of the SH, TH and photoluminescence.

In case of the applied modulated electrical field the response of the sample is significantly changing. The initial photodarkening in range up to 30 GW/cm^2 for initial sample without poling turns to photobleaching with efficiency $\mathrm{Im}(\chi^{(3)}) = -8.4 \times 10^{-15}$ esu. Further increase of intensity leads to photodarkening of the sample ($\mathrm{Im}(\chi^{(3)}) = 2.8 \times 10^{-15}$ esu) up to 400 GW/cm^2. At higher intensities the response is the same as for initial sample.

It was shown that the initial samples demonstrate the self-defocusing effect at low intensities (<40 GW/cm^2) that corresponds to the positive photoinduced variation of the refractive index with efficiency $\mathrm{Re}(\chi^{(3)}) = -5.8 \times 10^{-11}$ esu. Increasing of the laser intensity leads to the growing of self-defocusing effect efficiency with $\mathrm{Re}(\chi^{(3)}) = -2.0 \times 10^{-11}$ esu that saturates at about 400 GW/cm^2. At higher intensity range sample demonstrate self-focusing that can correspond to the activation of additional nonlinear effects.

Similar refractive response at low intensities (<40 GW/cm^2) was observed for sample with applied electrical field. Increase of intensity leads to more efficient self-defocusing manifestation with $\mathrm{Re}(\chi^{(3)}) = -4.7 \times 10^{-11}$ esu that saturates at about 300 GW/cm^2.

Obtained results demonstrate (i) the sensitivity of the spatial profile analysis technique for readout the cubic nonlinear response of proposed reference films of P(VDF-TrFe) and (ii) possibility of cubic nonlinear response tuning by electrical field.

Comparison of the photoinduced response of the film with previous spectral measurements for perpendicularly oriented PVDF demonstrated the high connection between SHG, THG and self-action effects. Photoinduced response of P(VDF-TrFe) significantly changing in intensity range up to 400 GW/cm^2 that correspond to the variation of rapid growing of THG in comparison of SHG. At a stage of supercontinuum generation at higher excitation intensities (see Fig. 12.4) the photoinduced absorptive and refractive NLO responses of P(VDF-TrFe) become irreversible. Joint SHG, THG, and self-action effects manifestation data processing provides amplification of the interface scanning technique sensitivity due multiple kinds of the NLO (quadratic, cubic degenerate/nondegenerate) responses readout.

To investigate laser irradiation effect on P(VDF-TrFe) with irreversible nonlinear response manifestation at the high laser intensity the UV-VIS optical transmission spectra were measured. For this purpose, the area about 2 × 2 mm was irradiated at intensity about 500 GW/cm^2 in step-by-step mode. In this mode all area firstly was divided by the grid with sector size 50 × 50 μm that corresponds to the size of the waist of the focused laser beam and secondly irradiated sector-by-sector with exposition time 2 s. The studied sample was covered by the diaphragm in order to study response from the irradiated area only.

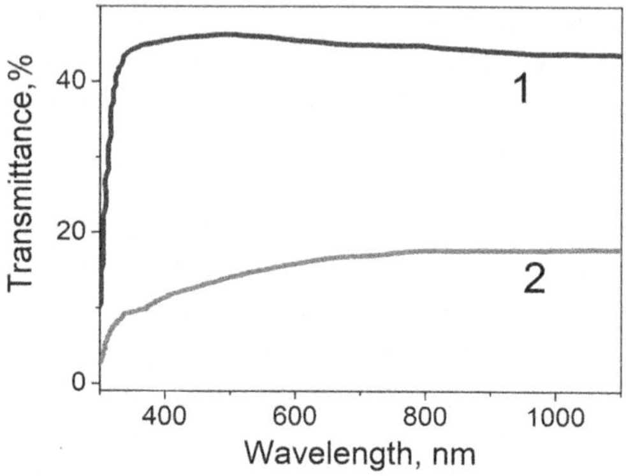

Fig. 12.6 The optical transmittance spectra of the PVDF thin films: **1** – initial (reversible) and **2** – after high dose laser irradiation at the beginning of the intensity range with irreversible response (see Fig. 12.5)

The UV-VIS transmittance spectra in the region 190–1100 nm for initial and irradiated P(VDF-TrFe) polymer films are presented in Fig. 12.6. Obtained transmission spectra for initial film demonstrate flat spectral response of the polymer in UV, VIS up to near IR with slight decrease of transmittance by few percent. The optical transmission spectra of the laser irradiated P(VDF-TrFe) film is in two times lower in comparison with initial sample with lower transmittance in UV range. The reason for this behavior is that the high intensity laser beam can leads to (i) damage or modification of the polymer structure or (ii) light induced reorientation of the polymer dipoles.

12.5 Application

To develop sensor properties the polymer surface should be activate by high sensing response, biocompatible, cost-effectivity, long-term stability, and simple fabrication material. For this purpose, the P(VDF-TrFe) can be used as reference material for interface scanning technique – sensitive approach for studying of nonlinear properties of materials. For sensing applications this approach can be improved by covering the surface of P(VDF-TrFe) film, for example, ZnO nanoparticles [30]. ZnO have some advantages over other metal-oxide based gas sensors. NPs like ZnO is eco-friendly, low cost and non-toxicity chemoresistive sensor. The change in resistance of ZnO highly depended on the presence of chemisorbed oxygen ions. Presented in atmospheric air the oxygen molecules are adsorbed on the surface of ZnO. The extraction of electrons from the conduction band of ZnO which increases the resistance of ZnO. The gas sensing mechanism of ZnO gas sensors relies on the alteration of a depletion layer on the surface of ZnO nanostructures [31]. That can be impact on the SH and TH signal and we expect that further investigation leads to improving highly sensitive, selective, and cost-effective ZnO nanoparticles-based gas sensors.

12.6 Conclusions

Nonlinear optical properties of PVDF and its copolymer P(VDF-TrFe) film were investigated. Due to their anisotropic structure polymers demonstrate high nonlinear response efficiency.

Within the conventional study of the second harmonic generation (SHG) efficiency in PVDF based materials we have investigated two kinds of the cubic nonlinear optical response: (i) nondegenerate – third harmonic generation (THG) effect and (ii) degenerate – photoinduced variation of refractive index and optical absorption under picosecond laser excitation with pump wavelength at 1064 nm. Combination of the quadratic and cubic nonlinear optical responses is promising for nondestructive remote sensing at the functionalized interfaces of the studied material.

Obtained results demonstrate the perspectives of the oriented PVDF films as reference material for analysis of the cubic (THG) and quadratic (SHG) responses of studied materials. This information can be used as two-parameter sensing technique that increase the sensitivity and provides more information about studied media.

Measurements of the second and the third harmonic generation efficiencies showed the synchronous growing of the SHG and THG signals in wide range of pump intensity up to $\sim$900 GW/cm^2. This efficient response of PVDF film can be used for remote analysis of analyte content and concentration variations at the interface of PVDF film and designing of a highly efficient sensing devices for wide range of applications.

References

1. Lovinger AJ (1983) Ferroelectric polymers. Science 220(4602):1115–1121. https://doi.org/10.1126/science.220.4602.1115
2. Lando JB, Doll WW (1968) The polymorphism of poly(vinylidene fluoride). I. The effect of head-to-head structure. J Macromol Sci Part B 2(2):205–218. https://doi.org/10.1080/00222346808212449
3. Jung S-W, Yoon S-M, Kang SY, Yu B-G (2008) Properties of ferroelectric P(VDF-TRFE) 70/30 copolymer films as a gate dielectric. Integr Ferroelectr 100(1):198–205. https://doi.org/10.1080/10584580802541106
4. Seo J, Son JY, Kim W-H (2019) Structural and ferroelectric properties of P(VDF-TrFE) thin films depending on the annealing temperature. Mater Lett 238:294–297. https://doi.org/10.1016/j.matlet.2018.11.156
5. Legrand JF (1989) Structure and ferroelectric properties of P(VDF-TrFE) copolymers. Ferroelectrics 91(1):303–317. https://doi.org/10.1080/00150198908015747
6. Septiyani Arifin DE, Ruan JJ (2018) Study on the curie transition of P(VDF-TrFE) copolymer. IOP Conf Ser Mater Sci Eng 299:012056. https://doi.org/10.1088/1757-899X/299/1/012056
7. Marsili R (2000) Measurement of the dynamic normal pressure between tire and ground using PVDF piezoelectric films. IEEE Trans Instrum Meas 49(4):736–740. https://doi.org/10.1109/TIM.2000.863916

8. Sharma T, Aroom K, Naik S, Gill B, Zhang JXJ (2013) Flexible thin-film PVDF-TrFE based pressure sensor for smart catheter applications. Ann Biomed Eng 41(4):744–751. https://doi.org/10.1007/s10439-012-0708-z
9. Wang T, Farajollahi M, Choi YS et al (2016) Electroactive polymers for sensing. Interface Focus 6(4):20160026. https://doi.org/10.1098/rsfs.2016.0026
10. Van der Spiegel J, Fiorillo AS (1993) U.S. Patent 5,254,504, 19 Oct 1993
11. Dahiya RS, Adami A, Collini C, Valle M, Lorenzelli L (2013) POSFET tactile sensing chips using CMOS technology. In: 2013 IEEE SENSORS, Baltimore. IEEE, pp 1–4. https://doi.org/10.1109/ICSENS.2013.6688149
12. Seminara L, Pinna L, Ibrahim A et al (2014) Electronic skin: achievements, issues and trends. Procedia Technol 15:549–558. https://doi.org/10.1016/j.protcy.2014.09.015
13. Kappassov Z, Corrales J-A, Perdereau V (2015) Tactile sensing in dexterous robot hands – review. Robot Auton Syst 74:195–220. https://doi.org/10.1016/j.robot.2015.07.015
14. Hosoda K, Tada Y, Asada M (2006) Anthropomorphic robotic soft fingertip with randomly distributed receptors. Robot Auton Syst 54(2):104–109. https://doi.org/10.1016/j.robot.2005.09.019
15. Ermakov V, Kruchinin S, Fujiwara A (2008) Electronic nanosensors based on nanotransistor with bistability behaviour. In: Bonca J, Kruchinin S (eds) Proceedings NATO ARW "Electron transport in nanosystems". Springer, pp 341–349
16. Ermakov V, Kruchinin S, Hori H, Fujiwara A (2007) Phenomena in the resonant tunneling through degenarate energy state with electron correlation. Int J Mod Phys B 11:827–835
17. Fiorillo AS, Pullano SA, Bianco MG, Critello CD (2019) Bioinspired US sensor for broadband applications. Sens Actuators Phys 294:148–153. https://doi.org/10.1016/j.sna.2019.05.019
18. Fiorillo AS, Pullano SA, Critello CD (2019) Spiral–shaped biologically–inspired ultrasonic sensor. IEEE Trans Ultrason Ferroelectr Freq Control 1. https://doi.org/10.1109/TUFFC.2019.2948817
19. Yamaka E (1984) Pyroelectric IR sensor using vinylidene fluoride-trifluoroethylene copolymer film. Ferroelectrics 57(1):337–342. https://doi.org/10.1080/00150198408012772
20. Bergman JG, McFee JH, Crane GR (1971) Pyroelectricity and optical second harmonic generation in polyvinylidene fluoride FILMS. Appl Phys Lett 18(5):203–205. https://doi.org/10.1063/1.1653624
21. Barbano EC, Harrington K, Zilio SC, Misoguti L (2016) Third-harmonic generation at the interfaces of a cuvette filled with selected organic solvents. Appl Opt 55(3):595–602. https://doi.org/10.1364/AO.55.000595
22. Multian VV, Riporto J, Urbain M et al (2018) Averaged third-order susceptibility of ZnO nanocrystals from third harmonic generation and third harmonic scattering. Opt Mater 84:579–585. https://doi.org/10.1016/j.optmat.2018.07.032
23. Débarre D, Supatto W, Beaurepaire E (2005) Structure sensitivity in third-harmonic generation microscopy. Opt Lett 30(16):2134. https://doi.org/10.1364/OL.30.002134
24. Barbano EC, Harrington K, Zilio SC, Misoguti L (2016) Third-harmonic generation at the interfaces of a cuvette filled with selected organic solvents. Appl Opt 55(3):595. https://doi.org/10.1364/AO.55.000595
25. Débarre D, Beaurepaire E (2007) Quantitative characterization of biological liquids for third-harmonic generation microscopy. Biophys J 92(2):603–612. https://doi.org/10.1529/biophysj.106.094946
26. Gayvoronsky V, Galas A, Shepelyavyy E et al (2005). Giant nonlinear optical response of nanoporous anatase layers. Appl Phys B 80(1):97–100. https://doi.org/10.1007/s00340-004-1676-2
27. Jones J, Zhu L, Tolk N, Mu R (2013) Investigation of ferroelectric properties and structural relaxation dynamics of polyvinylidene fluoride thin film via second harmonic generation. Appl Phys Lett 103(7):072901. https://doi.org/10.1063/1.4817519
28. Lim J, Moon J, Yu K, Lim E, Park G, Lee S-D (1999) Birefringence and second harmonic generation in PVDF/PMMA blends. Ferroelectrics 230(1):257–261. https://doi.org/10.1080/00150199908214928

29. Kubono A, Kitoh T, Kajikawa K et al (1992) Second-harmonic generation in poly(vinylidene fluoride) films prepared by vapor deposition under an electric field. Jpn J Appl Phys 31(Part 2, No. 8B):L1195–L1197. https://doi.org/10.1143/JJAP.31.L1195
30. Dai J, Yuan M-H, Zeng J-H et al (2014) Controllable color display induced by excitation-intensity-dependent competition between second and third harmonic generation in ZnO nanorods. Appl Opt 53(2):189. https://doi.org/10.1364/AO.53.000189
31. Bhati VS, Hojamberdiev M, Kumar M (2019) Enhanced sensing performance of ZnO nanostructures-based gas sensors: a review. Energy Rep S2352484719304160. https://doi.org/10.1016/j.egyr.2019.08.070

Chapter 13
Nanoscale-Specific Analytics: How to Push the Analytic Excellence in Express Analysis of CBRN

B. A. Snopok and O. B. Snopok

Abstract Nanomaterials (NM) are rapidly finding a wider application in modern technological products due to their unique properties associated with their small dimension. Accurate detection of nano-scale materials by fast and precise analytical techniques is essential in order to control their safety and security. However, analyzing nanoscale analytes remains extremely challenging due to their limitless diversity, complex and dynamic nature. For this reason, traditional analytical techniques, hyphenated systems or "univariate" arrays thereon are being pushed to their limits; new innovative approaches to face state of the art problems are required. Comprehensive analysis is crucial for the detection of the limitless list of NMs: functionality and reliability of new approaches depend on the ability to simultaneously control the key descriptors of NMs (at the very least, their geometric features, composition and interfacial functionality). An overview of the features of the relationship between analytical chemistry and nanoworld is presented here. We discuss some traditional methods and the emergence of powerful new technologies in nanophotonics. One of the most powerful analytical techniques for the express identification of NMs is plasmonic scattering interferometry. It is based on an NM-induced interferometric "amplification" of the disturbance caused by a nano-object and its transformation into a micron-range image (NM's "fingerprint"), which is easy to visualize using standard technical means of ordinary microscopy. The NM's "fingerprint" is its virtual marker similar to regular QR- or bar-codes. In the end, a brief discussion of multivariate approaches in the development of new analytical methodologies for identifying and quantifying nanomaterials, along with a short outlook on the key takeaways summarizing the most important points in nanoscale analytics is also included.

Keywords Nanoscale analytics · Nanomaterials · Engineered nanoparticles

B. A. Snopok (✉) · O. B. Snopok
V.E. Lashkaryov Institute of Semiconductor Physics, National Academy of Sciences of Ukraine, Kyiv, Ukraine
e-mail: snopok@isp.kiev.ua

J. Bonča, S. Kruchinin (eds.), *Advanced Nanomaterials for Detection of CBRN*,
NATO Science for Peace and Security Series A: Chemistry and Biology,
https://doi.org/10.1007/978-94-024-2030-2_13

13.1 Introduction

The methods for generating, storing and processing information specified by the structure of material objects are largely similar to those used in conventional digital technologies [1]. Any complex system (e.g. DNA biopolymers), containing as a whole a large amount of such "analog" information, is built "from the bottom up" and has in its basis a hierarchical design with several levels of organization. At each of these levels, the "behavior" of the acting forces is "regulated" by a certain set of "rules" (stereochemical complementarity of nucleotides, physical laws etc.), unchanged in a given range of scales and mutually consistent with the lower and higher levels of the corresponding boundary conditions [2, 3].

The "lowest", basic level of organization is usually represented by unstructured and indistinguishable elements – unique "bricks" of the entire structure (be it "0" or "1" in a binary code, "the most elementary" particles etc.). The next level is formed by a given set of primitives (letters, atoms etc.), which themselves do not carry the load of the functional identification of a material object. And only at the next level – "words", "molecules", "nucleotides", bytes etc. – does *the primitive structure contain information that somehow identifies an object or its action*; this information can be read through special selection-interpretation rules implemented at this level of organization. In the generally accepted classification of physical, chemical or biological systems, *this boundary is what defines the lower range of spatial dimensions, on which the simplest functions are realized* – functions that somehow manifest themselves at the macro level (color, chemical reactivity etc.). This is why the prefix "nano" is so widely used in publications in various fields of science and technology and, in particular, distinguishes the task of detecting nano-objects from the entire class of analytical chemistry problems. The variety of nanoobjects due to the strong dependence of the properties of nanobodies on their size, shape and composition, is what determines the complexity of analytical analysis. Only a slight change in shape or size on the scale of space under consideration can radically change the functions of a nano-object. All of this leads to the presence of "hidden" from "ordinary" analytical analysis procedures properties of nano-objects, and, accordingly, to the need to rethink the ideology of the analytical analysis of nanobodies.

Humans have always been exposed to tiny particles *via* various natural processes (dust storms etc.), and our bodily systems are well adapted to protect us from these potentially harmful intruders. Particles originating from human activities have existed for millennia as well – for example, smoke from combustion and lint from drapery, but recent industrial development has profoundly increased the particulate pollution. What's important is that the technological advancement has also changed the character of the particulate pollution, increasing the proportion of nanometer-sized particles ("nanoparticles") and expanding the variety of chemical compositions. Engineered nanoparticles (ENP) are being developed and incorporated in a number of commercial products, raising the potential of human exposure during their manufacture, use and disposal. The manipulation of matter specific for the nanoscale ("nanotechnology") is creating many new materials with characteristics which are not always easily predicted from current knowledge. Within the nearly limitless diversity of these materials, some happen to be toxic to biological systems; others are relatively benign, while others yet confer health benefits. Such uncertainty surrounding the potential associated risks for both human population and environment makes designing well-founded and robust technologies to ensure safe development and use of nano-enabled products particularly challenging. But the need for these technologies is urgent: recent studies have shown a strong correlation between particulate levels, respiratory and cardiovascular diseases, various cancers, and mortality – all of them finally resulting in more and more exacting regulation requirements for industrial stakeholders [4, 5]. The key to understanding the toxicity of nanoparticles is that their unpredictable reactivity and minute size – smaller than cells and cellular organelles – allows them to penetrate these basic biological structures, disrupting their normal function [6–12].

Delivering analytical and characterization excellence in nanomaterials assessment tools requires a high level of preparation, understanding and achievement in the following aspects:

- **motivations and challenges**: the social demand for safety and security is the driving force that stimulates the activity in the area of nanoscale analytics;
- **enabling technologies**: starting with simple tools for assistance in data acquisition and analyte tracking, and up to real-time data manipulation and analytical conclusions;
- **analytical process**: analytical methods need to account for the enormous diversity of materials and the dynamics of nanomaterials;
- **analytical thinking**: starting with a simple description of facts, proceeding to context explanation and the interpretation of the mechanisms in order to estimate what might happen in the future; moreover, all information must be presented in a form that can be processed by a human user, since the evaluation progress depends on the results derived from the analysis and the interpretation of collected data.

Finally, to recognize the analytical excellence at the nanoscale, we must go beyond traditional analytical chemistry. Analytics today is not just recording and processing a wider variety of information; the obtained information is used to make the most important decisions in technology, environmental safety and security of humans. Analytics is more and more integrated with artificial intelligence and tries to find the answer not only to "To which extent does the data generate value?", but also to "How to reveal to a specific decision-maker the underlying significance of selected target information?"

13.2 Nanoscale Analytes: Complexity and Diversity – the New Normal

The nanosized particles turned out to be a fundamentally new object for various areas of analytical science and safety control in ecology and medicine. For example, noteworthy are some of their unusual properties – in particular, their toxicity, which does not correlate with the properties of the material in the atomic-molecular state or in a solid. This is well illustrated by the example of the effect of nano-objects on viruses and bacteria, for which a large amount of experimental material has been collected. For example, it has been shown that the antiviral effect of spherical gold nanoparticles in a citrate shell against the H1N1 influenza virus is fully determined by the size: the inhibitory action of nanoparticles with a diameter of 7÷10 nm is stronger than the action of nanoparticles with a diameter of 20 nm and is observed for a lower concentration, whereas no meaningful effect is observed for nanoparticles with a diameter greater than 30 nm [13]. Summarizing the results obtained for various nanoparticles, viruses and bacteria, it's been established that:

1. the antiviral and antimicrobial effect of nanoparticles shows certain "versatility": nanoparticles of different metals, oxides and polymeric nanoparticles essentially have the same effect on different viruses and bacteria;
2. the effectiveness of the antiviral and antimicrobial effect of nanoparticles depends primarily on their size [14–17]: nanoparticles exhibit better antiviral activity that is higher compared to ions of the same substance [18, 19] or microparticles;
3. the effectiveness of the antiviral and antimicrobial effect of nanoparticles also depends on their concentration (dome-shaped dependence with a maximum, [13, 17, 19–22]), chemical composition, shape (triangular pyramids are better antimicrobial agents compared to spherical ones [23, 24]) and the virus/microbe type.

The patterns noted above demonstrate one of the most important differences between nanomaterials and typical objects of analytical chemistry: *the antimicrobial and antiviral effects of nanoparticles are due to some physical phenomenon*, rather than a chemical interaction of one type or another. Nanoscale analytes induce the change of the toxicity paradigm: physical effects come to the forefront, not the features of the chemical structure. This means that special approaches are needed – the ones that take into account not only the chemical composition of the object, but also its geometric characteristics, such as size and shape. It is necessary to stress that the size is not important in itself – what's important is new functional properties that can be realized only in a limited range of sizes of matter. Therefore, the size of the object indicates the need to take into account new properties, depending on what function the given nanobody realizes (for example, absorption of plasmon particles of noble metals [25]).

Summarizing all of the above, we can conclude that due to the limitless diversity of nanomaterials, they should no longer be viewed simply as "small objects" and the current knowledge of the properties of "bulk" materials may not suffice in reliably predicting their characteristics. It is fundamentally impossible to predict which characteristics of a nano-object will be optimal for identification, especially in a situation when it comes to screening raw multicomponent samples in an industrial or natural environment. The need to select at least a few main interpretative descriptors for nanomaterials in terms of their "personal identity" is a very pressing challenge.

The practical implementation of the concept of analytical analysis involves a reasonable choice of key identification descriptors and, accordingly, analytical methods intended to identify the set of specific features that can segregate identifiable objects (e.g. nanoparticles in a solution) into separate classes (e.g. gold or silver nanoparticles with the size of 20 or 30 nm). It is a common feeling that the smallest possible group of descriptors includes geometry features, chemical composition and dominated interfacial functionality.

13.3 Generalized Workflow of a Measurement Process in Analytical Chemistry

Classical analytical chemistry assumes that the information content of a measurement increases with the increase of the concentration of the analyte itself (Fig. 13.1). Usually this is due to external causes (changes in measurement conditions, equipment errors etc. [26]) which do not allow to uniquely formalize the detector response when the analyte concentration is low. The situation is fundamentally different in the case of nanoscale analytes – registration and even tracking of individual particles in many cases is not that difficult; the problem arises at the classification stage (selectivity), for the implementation of which it is necessary to determine a

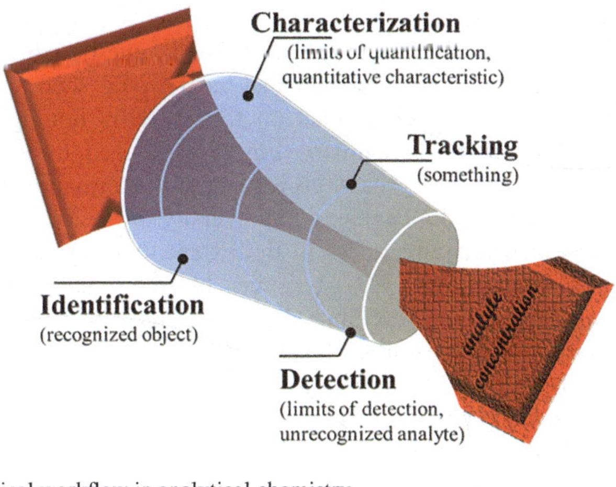

Fig. 13.1 A typical workflow in analytical chemistry

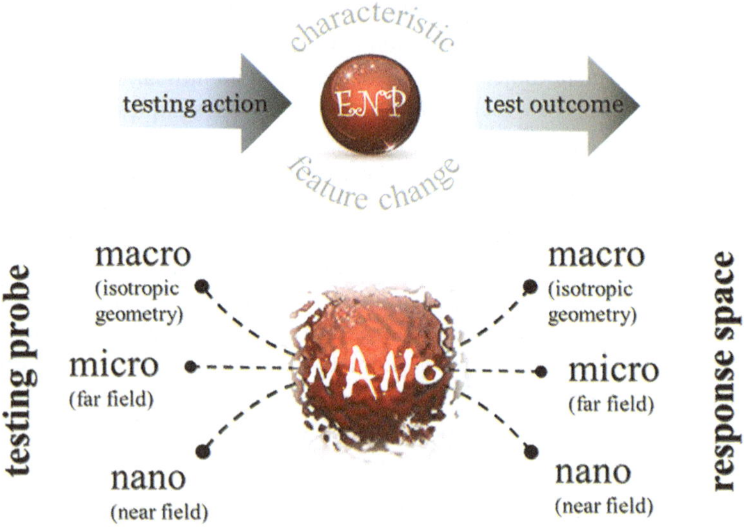

Fig. 13.2 A typical concept of the analytical process of "active" analysis methods

certain set of characteristics of ENPs being determined. This makes it necessary for the registration system to generate a certain space of heterogeneous responses, which are in one way or another connected with the characteristic properties of the nanobody. As noted above, to achieve a minimum acceptable result and to reduce the level of uncertainty when classifying an object, it is necessary to know at least the size, shape and material of the nanoparticle. Realizing that the process of detecting nanoscale analytes must be of multivariate nature, one can reformulate the priority goal of the analytical analysis of nano-objects: what we really need is to in-situ *detect and identify nanoscale analytes* embedded in complex matrices in a rapid, accurate and reproducible way.

Important from the practical point of view is the question of how to obtain such information. This is due to the fact that as it would seem, the size of the analyte should impose some restrictions on the spatially homogeneous testing probes customary for analytical chemistry (Fig. 13.2). Indeed, the standard method of averaging identical elementary reactions leading to the formation of a response is not applicable in the case of ENPs due to the diversity of the latter. The concept of selectivity (as a graded measure refers to the extent to which other components interfere with particular analyte [27]) becomes blurred, since close in terms of size, but differing in shape or something else ENPs can have completely different functional properties. Such features of nano-objects necessitate a more detailed analysis of methods for their identification.

Figure 13.2 shows a generalized diagram of typical "active" analysis methods. As usual, the object is subjected to some external influence of a known magnitude, while the detector registers the influence of the object on this test influence. For example, the absorption capacity of a material is determined from the decrease of the intensity of the light beam initially falling on the object. We are primarily interested in the following question: how should the spatial dimensions of the testing probe and the detector and the size of a single nanoparticle correlate in order to obtain an analytically correct result for determining the characteristics of a given object?

13.4 Enabling Technologies: Is It Possible to Save Money by Using Traditional Analytics?

Table 13.1 summarizes the conventional methods typically used in nanoscale analytics, with the classification criterion being the typical key descriptors of nanosystems – geometric parameters, surface composition and its properties. The vast majority of the methods presented in the Table are variations of standard research tools and are not intended to solve analytical problems outside of stationary laboratories with highly qualified personnel and appropriate infrastructure [28, 29].

The key to understanding the problems related to the identification and characterization of nano-objects is their minute size – smaller than typical probes usually used in analytical science. The most promising approaches for analytical express analysis are the "fastest" analytical techniques that calculate nanoscale-specific characteristics from macroscopic measurements using a mathematical transformation in line with some "validated" algorithms. Most of these techniques are based on the analysis of the Brownian motion of nanoparticles, including under the influence of an external pulling field (z-potential). The most common technique utilizing Dynamic Light Scattering (DSL) does not allow to analyze individual NPs and instead analyzes the behavior of an ensemble of nanoparticles. The more advanced Nanoparticle Tracking Analysis (NTA) automatically locates and follows the center of each and every particle and measures the average distance moved per frame; however, to calculate the characteristics of the nanoobject, the same macroscopic diffusion equation is used. Therefore, the "fastest" commercially available instrumentation utilizing DSL or NTA methods are inherently "integral analyzers" [30]. Methods with the highest level of spatial resolution like TEM or high-speed-AFM are too slow for real-time analysis and cannot recognize the nanomaterials within a complex environment [31]. The situation is similar with other properties of nanomaterials – shape, chemical composition, interfacial activity etc.: each method has both its advantages and disadvantages, but there is still no single "universal" technique that fully meets various stakeholders' needs. Moreover, due to the limitless diversity of nanomaterials, the development of the equipment for each individual class of nano-objects with specific functionality is economically inexpedient, and fundamentally impossible for real-time analysis. In order to obtain more information about a nanoscale analyte, it is appropriate to organize the methods in an integrated screening scheme, which allows for a rational physical-chemical materials characterization.

Table 13.1 A summary of techniques acceptable for the characterization of nanomaterials

Key descriptor	Group of methods	Limitations
Geometry features	Electron (TEM, SEM etc.) and scanning probe microscopy (AFM, STM etc.)	Time consuming, cannot detect NPs in-situ in a native environment, very expensive
	Light-based static or dynamic scattering (DLS), particle tracking analysis (NTA)	Nonlinear size dependence limiting the size range, strong interference with the matrix
	X-ray diffraction methods (XRD, etc.)	Time consuming, very expensive
	Advanced mass-spectroscopy in combination with various techniques (TG-IR GC/MS, SP-ICP-ToF-MS, HDC-RC-MS, SIMS etc.)	Time consuming, sample decomposition, expensive instrumentation
Chemical composition	Optical techniques including surface and tip-enhanced spectroscopies (UV-VIS, Raman, SERS, TERS, FTIR etc.)	Typically expensive with low spatial resolution, sometimes with strong interference with the matrix; however, in theory can be fast enough
	Specific high-energy spectroscopes (XPS, AES, ToF-SIMS, EDX etc.)	Requires difficult pretreatment of the sample, very specific and expensive instrumentation
	Conventional chromatography (HPLC)	Time consuming, sample decomposition, expensive instrumentation
Dominated interfacial functionality	Separation and fractionation in gradient fields and dynamics flows (disc centrifuge CLS/CPS, AF4, centrifugal FFF etc.)	Advanced versions can be efficiently used for multiple acquisition under the reduction of separation time
	Various selective surface-based sensors (QCM, QCM-D, SPR, SAW, WGLS etc.), reactivity assay on a chip, electrochemical methods	Cannot be applied in complex media due to nonselective adsorption/reactions

A typical solution used to improve the performance of each of a number of single methods is "hyphenation" – an integration of two or more techniques in a single platform. This concept is based on "univariate" arrays with completely independent channels but fusing all the data into an ENP virtual marker; in other words, TEM gives information concerning size and shape, UV-VIS – absorption, EDX – composition etc. The main advantages of this concept are as following: high precision analysis of preselected properties, the usage of conventional time-proven techniques and typically direct measurements of nanostructures. However, a characteristic feature of all these platforms is combining approaches that are heterogeneous in their structure for the most complete characterization of the ENPs. As a result, it is quite difficult to coordinate such different analytical techniques within a single setup. Common problems with these systems are typically related to

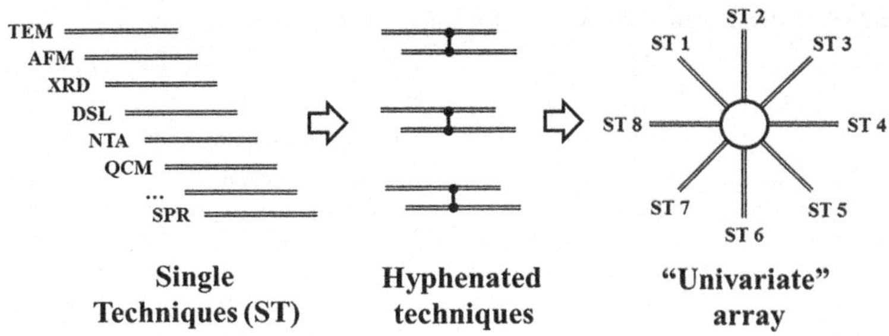

Fig. 13.3 Evolution of analytical concepts based on a hardcoded properties list

low throughput of the analysis, short analyte range and expensive instrumentation. Moreover, due to the properties list being hardcoded, the number of objects that can be classified and, accordingly, identified by such a system directly depends on the number of "hyphenated" methods (the dimension of the space of "orthogonal" responses) and their resolution (Fig. 13.3). Restrictions on the number of types of objects to be identified, as well as the possibility of classification only by given measured properties, significantly reduce the practical value of this approach. The most promising area of application for such systems is high precision characterization of specific properties and secondary standards validation. Therefore, investment in an analytical platform for real-time analysis of CBRN based on hyphenated technologies "for the sake of more data" in many cases is suboptimal. Such approaches to evaluate the presence of nanomaterials are prohibitive in terms of time and expense.

This short overview of the state-of-the-art in nanomaterials characterization allows us to conclude that the application of traditional methods based on the broadly used "univariate" concept incorporating the "arithmetic addition" of different nanoscale-specific instrumentation is not sufficiently reliable for the identification of unexpected situations, which are the most dangerous in an industrial environment or in the field; *a more sophisticated analytical concept must be developed, practically realized and validated.*

In order to find the best ways for express identification of nanomaterials, it is necessary to analyze the potential possibilities and typical limitations of various instrumentation methodologies suitable for the analysis of nanomaterials. Table 13.2 summarizes the main methodological approaches specific to optical detection methods, which allow obtaining information about nano-objects.

Systematic evaluation of current analytical methodologies and key nanoscale-specific methods (see Tables 13.1 and 13.2) unambiguously identifies the hotspots in nano safety assessment tools: no analytical strategy or instrumental platform exist for real-time identification of a priori unknown nanomaterials in complex environments. With regard to a general approach specific for safety and security, the need to develop new interpretative descriptors for nanomaterials in terms of their "personal identity" is a very pressing challenge. Approaches using different

Table 13.2 Analytical methodologies used for nanoscale analytes

Spatial scale				
Testing probe	Response space	Analytical methodology	Group of methods	Comments
Macro (non-local)	Macro (far field)	Macroscopic sampling and measurements; statistical averaging of local processes	Classical optical methods like UV-VIR-IR spectroscopies, DLS in colloidal solutions, elastic scattering etc.	Conventional approaches using the analysis of ensembles of nanoparticles
Nano (near field)	Nano-macro (near-far fields)	Local excitations based on delivering and shaping light to the nanoscale area	Methods based on the local concentration of light in a given nano-sized area: SNOM, TERS etc.	Expensive and very sensitive equipment
Micro-macro	Micro-macro	Nanoparticle-induced effect with or without local amplification	Typically light scattering techniques with or without subsequent interferometrical amplification	Tracking systems (NTA) or microscopy with interferometrical detection (WF-SPRM)

variants of light scattering seem to be the most promising, especially in those cases when the effect is local in nature and is associated with some amplification of the response due to the organization of the registration system.

13.5 Multivariate Approach: The Effectiveness of a Method Is Determined by Its Results

Identification of unknown or well-known, but very different in their properties objects, is only possible with the greatest efficiency when using the concept of multifactor fingerprinting, when the image of an object (the "fingerprint") in the response space depends on different characteristics of the classified object in a non-linear manner (Fig. 13.4). All systems of "quick response" of living organisms (smell, taste, sight) are built according to this very principle [32, 33]: we can easily recognize the coffee flavor even though it contains thousands of different components. The situation with nanomaterials is the same: multifunctional nano-objects can be described not only by the sum of their individual characteristics (size, shape, mass etc.) but also by some abstract representation – a virtual fingerprint, whose form is specific to the visualization system (like human fingerprints or iris scanners) [33, 34]. These virtual fingerprints can be used for further classification (a multivariate "calibration" procedure) and identification (e.g. by screening during the

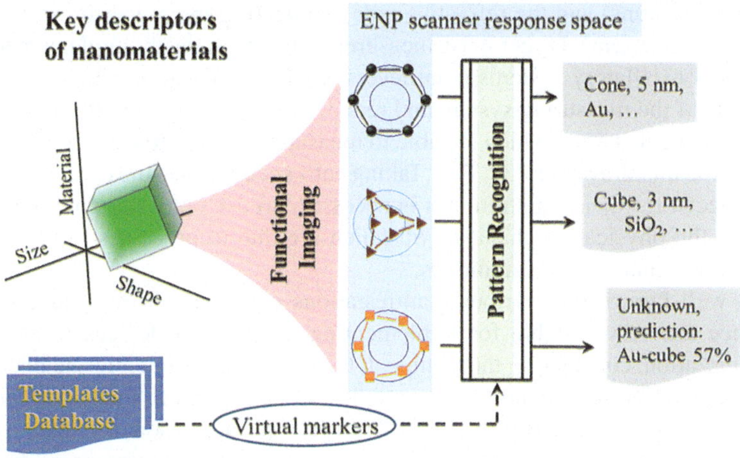

Fig. 13.4 Fingerprint technology for nanosystems without single "markers"

analysis) based on image recognition techniques. The multivariate concept offers some advantages over "univariate" arrays, such as: sensitivity to a wider range of ENMs, improved selectivity and the ability to recognize both "simple" and "complex" ENMs. The relationship between an ENM and its virtual fingerprint goes through the stages of resolution, classification and calibration – typical for any system utilizing data recognition procedures as a basic operational algorithm [33].

To elaborate the concept of "multivariate" arrays, there is a need for a device encoding the "personality" of a nano-object with multifunctional imaging and, accordingly, correlating the object with a specific fingerprint – its "virtual marker" (with pattern recognition methods using databases of validated reference standards). The most promising from a practical point of view would be a device that on the hardware level forms some ENP representation or graphic image, since this technology is widely used for product labeling etc. using QR or bar codes. It is necessary to build a scanner that can create and then read these codes, which would solve the problem of combining the stages of detection and identification (Fig. 13.1) within the framework of the modern paradigm of analytical chemistry. To avoid uncertainties when performing mathematical conversions typical for DLS, forward looking strategies must be strongly focused on actual particle counting techniques.

13.6 Imaging Technologies: Seeing the Invisible

At the heart of a successful approach for reliable screening and characterization of nanomaterials must be the "universal sensitivity", which takes into account the extreme diversity of ENM composition. It is known that the most "universal" characteristics of matter available for measurement are the mass (roughly speaking, the sum of protons) and the refractive index (reflecting the properties of the entire electronic subsystem). Direct mass measurement is most easily implemented using mechanical oscillatory systems (cantilevers, QCM, SAW etc.), which complicates the design of the measuring system and reduces its reliability [35, 36]. On the other hand, optical converters, which are able to measure the refractive index of materials, do not have this drawback [37, 38]. Taking into account the sub-wave character of nano-objects, for their registration it is necessary to use optoelectronic registration methods, the physical mechanisms of which are due to processes in the range of units-tens to hundreds of nanometers.

It is well known that various manifestations of the effect of surface plasmon resonance are very suitable for implementing in practical devices for nanoscale characterization, considering the technical, technological, economic and operational requirements [39, 40]. Numerous examples of its practical implementation are available on the market in the form of integral surface resonance spectrometers produced by many companies [41]. Various versions of SPR microscopy have also been developed, however, for obtaining objects on the surface they turned out to be not promising due to the "very large" propagation length of plasmon-polaritons on the surface reaching tens and hundreds of micrometers [42, 43].

A principal feature of the plasmonic scattering interferometry is that this technique can visualize nano-objects, whereas the resolution of conventional optical microscopes is about half the wavelength (i.e. 200–300 nm) due to the diffraction limit [44]. The physical mechanism that causes these unprecedented capabilities is as follows. The propagation of a surface wave (SW) excited in an SPR transducer can be roughly described by classical physical optics ([45], a more accurate theoretical description can be found in [46, 47] for Wide-Field Surface Plasmon Resonance Microscopy and in [48, 49] for interferometric plasmonic imaging). When an object with a size that is much smaller than the SW wavelength appears on the propagation path of a surface wave, a part of this wave is reflected back, while another part is scattered – depending on the size, shape and material of the object. Since all the waves in the plane of the surface are coherent, this leads to the interference of the reflected, scattered and incident waves on the object. The resulting interference pattern is of microscopic size (since all waves "participating" in this process move at a speed only slightly lower than the speed of light) and can be easily detected by the technical means of conventional optical microscopy. This can be illustrated by imagining rain droplets on the water surface: even if one is not able to see single droplets, the secondary circular waves are easily visible. An essentially analogous "trigger" effect is used in the Wilson chamber, biosensors [50] and the like, when the energy and the organization of the surrounding medium are used for local "amplification" and an object of microscopic size is formed, which is easy to visualize using standard technical means of ordinary microscopy.

WF-SPRM allows to detect and characterize single nanoparticles on a macroscopic sensor surface of $\sim$1.3 mm^2 which is extendable to $\sim$3–5 mm^2. A large field of view of the developed device is fundamental for analytical applications since it enables real-time detection of single nanoparticles at very low concentrations in the suspension. It also allows to achieve a high dynamic range (from a single NP up to a million of NPs). The developed prototype is able to detect many types of nanoparticles not only in aqueous suspensions but also in such complex media as wine or juice [51]. Detections of single nanoparticles at concentrations lower than ~2 pg/mL (mass) and 10^6/mL (number) were routinely achieved [52]. This enables real-time monitoring of the process of the interaction of the nano-object with the surface, and the image recorded during this process carries information about the size, shape and material of the object on the surface.

The response space of WF-SPRM (a fingerprint in the form of a fringe pattern [46, 47, 51, 52]) represents the result of the interaction between the ENM and the "testing probe" – propagating surface plasmon-polariton waves (Fig. 13.5). Evaluating the effect of the ENM on the overall changes in surface-plasmon-polariton propagation triggered by the ENM seems significantly more informative and universal compared to a set of single push-type markers. All of this allows us to create a simple and highly efficient detector of individual nanoparticles using the multivariate detection approach.

The most acceptable version of plasmonic scattering interferometry for practical use today is the modification based on WF-SPRM [44]. At the same time, the full 3D indicatrix of the radiation scattering into the upper hemisphere above the surface of the sample can give much more information about the object. As shown in [39, 53–56], radiation scattering of plasmons contains a lot of specific information about the scattering of objects in the nanometer range.

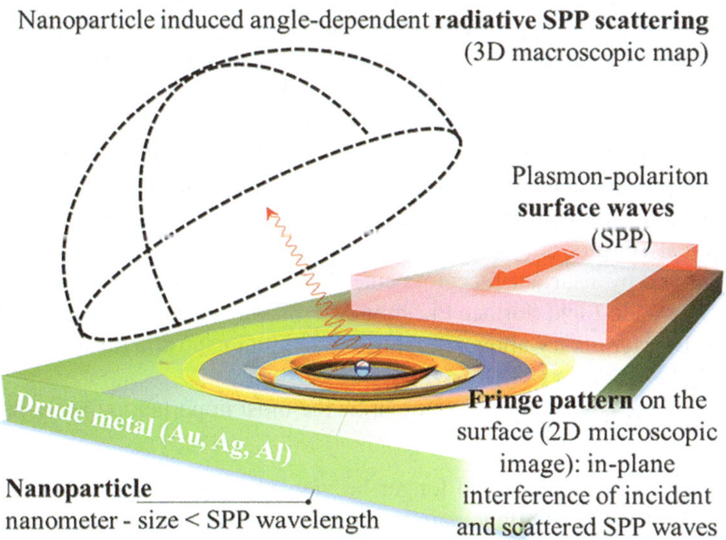

Fig. 13.5 Simplified schematic presentation of the amplification mechanism specific for plasmonic scattering interferometry: ENM (i) disturbs the SPP resonance condition (angle dependent intensity of the reflected light), (ii) induce the ENM specific 2D fringe pattern ("fingerprint") formation and (iii) give rise to 3D unique map of SPP radiative scattering in far field

The main advantages of this new type of microscopy are fast and high throughput identification using multivariate pattern recognition, "universal" sensitivity in respect to ENP materials and limitless list of analytes. Issues that require further development include the formation of chemical images of analytes, the formation of databases of the reference images and the development of high-performance computing, which enables online identification of objects.

Typical analytical applications of the interferometric plasmon spectroscopy are *real-time analysis, fast identification of unknown components and alarm-like systems.*

13.7 Concluding Remarks: Transforming the Way We Think

It is obvious that it is the real-time systems that will need to be further developed, standardized and validated in order to provide useful screening data on the presence of nanomaterials. It is also necessary to develop new interpretative descriptors for nanosystems. Due to the high variability of sizes, shapes and chemical compositions of nanomaterials, integrated global profiling is required to fully determine unknown nanomaterials in complex environments.

Analytical and characterization excellence at the nanoscale can be achieved by switching *from the "orthogonal" characterization of single compounds with a hardcoded properties list to the multivariate imaging of complex analytes and the corresponding pattern recognition* in nanoscale analytics. Plasmonic scattering interferometry may be the best solution here, bridging the nano and macro worlds and providing an opportunity for the express analysis of nanoscale analytes.

The development logic of SPR devices well demonstrates this vector of development of analytical methods. Indeed, historically first SPR instruments have been developed for bulk refractometry – to monitor the index of refraction in a homogeneous infinite space above the sensor [57]. Later came SPR-based (bio)chemical sensors and sensor arrays for the detection of minute changes of matter captured by a surface receptor center [37, 38, 58, 59]. And today, approaches based on the Wide-Field Surface Plasmon Microscopy utilizing in-plane Plasmonic Scattering Interferometry allow to identify individual nanostructures on the surface of a physical transducer [44, 46, 47, 51, 52].

The following key takeaways summarize the most important points in nanoscale analytics:

- nanoscale analytes should no longer be viewed simply as "small objects"; the need to select at least a few main interpretative descriptors for nanomaterials in terms of their "personal identity" is a very pressing challenge;
- the smallest possible list of key descriptors is as following: geometry features (size, shape, distribution), chemical composition and dominated interfacial functionality;

- nanoscale analytes induce the change of the toxicity paradigm: physical effects come to the forefront, not features of the chemical structure;
- it can be argued that an adequate analytical platform must be based on direct nanoparticle counting with real-time local characterization of each single particle separately;
- the identification of ENP's "personal identity" can be achieved by using the multivariate approach, coding and identifying ENPs with "virtual markers" like QR or bar-codes;
- one of the most promising techniques for the practical realization of ENP "multivariate" coding-disclosure procedures is WF-SPRM with an ENP-induced interferometric "amplification" of the disturbance caused by a nano-object and its transformation into a micron-range image;
- the expansion of data volume and its organization requires their processing in-situ;
- the need for efficient online data processing and the complexity of this process unambiguously show that artificial intelligence is a strategic imperative in nanoscale analytics.

Modern analytics has become more of a philosophy for decision-making, rather than a suite of technologies. This trend clearly indicates the increasing integration of individual measurements at the hardware level and the rising involvement of artificial intelligence technologies in solving the urgent problems of mankind.

Acknowledgments Authors thank Tetiana Snopok for help in manuscript preparation.

References

1. Frenkel SV, Tsygelny IM, Kolupaev BS (1990) Molecular cybernetics. Svit, Lviv
2. Snopok BA, Snopok OB (2018) Information processing in chemical sensing: unified evolution coding by stretched exponential. In: Nanostructured materials for the detection of CBRN. Springer, New York
3. Snopok BA (2014) Nonexponential kinetics of surface chemical reactions (review). Theor Exp Chem 50(2):67–95
4. Statement on emerging health and environmental issues (2018) Emerging issues and the role of the SCHEER, Position Paper (2018), EC scientific committee on health, environmental and emerging risks SCHEER, 14 Jan 2019. https://doi.org/10.2875/33037. ISBN: 978-92-76-00232-1
5. Valcárcel M, Simonet BM, Cárdenas S (2008) Analytical nanoscience and nanotechnology today and tomorrow. Anal Bioanal Chem 391:1881–1887
6. Revel M, Châtel A, Mouneyrac C (2018) Micro(nano)plastics: a threat to human health? Curr Opin Environ Sci Health 1:17–23
7. Dolez PI (ed) (2015) Nanoengineering: global approaches to health and safety issues. Elsevier, Amsterdam
8. Oliveira M, Almeida M, Miguel I (2019) A micro(nano)plastic boomerang tale: a never ending story? TrAC Trends Anal Chem 112:196–200

9. Ermakov V, Kruchinin S, Fujiwara A (2008) Electronic nanosensors based on nanotransistor with bistability behaviour. In: Bonca J, Kruchinin S (eds) Proceedings NATO ARW "Electron transport in nanosystems". Springer, Berlin, pp 341–349
10. Ermakov V, Kruchinin S, Hori H, Fujiwara A (2007) Phenomena in the resonant tunneling through degenarate energy state with electron correlation. Int J Mod Phys B 11:827–835
11. Kruchinin S, Pruschke T (2014) Thermopower for a molecule with vibrational degrees of freedom. Phys Lett A 378:157–161
12. Ermakov V, Kruchinin S, Pruschke T, Freericks J (2015) Thermoelectricity in tunneling nanostructures. Phys Rev B 92:115531
13. Lozovski V, Lysenko V, Piatnitsia V, Scherbakov O, Zholobak N, Spivak M (2012) Physical point of view for antiviral effect caused by the interaction between the viruses and nanoparticles. Bionanosci J 6:109
14. Lysenko V, Lozovski V, Lokshyn M, Gomeniuk YV, Dorovskih A, Rusinchuk N, Pankivska Y, Povnitsa O, Zagorodnya S, Tertykh V, Bolbukh Y (2018) Nanoparticles as antiviral agents against adenoviruses. Adv Nat Sci Nanosci Nanotechnol 9:025021
15. Azam A, Ahmed AS, Oves M, Khan MS, Memic A (2012) Size-dependent antimicrobial properties of CuO nanoparticles against Gram-positive and -negative bacterial strains. Int J Nanomed 7:3527
16. Khylko O, Rusinchuk N, Shydlovska O, Lokshyn M, Lozovski V, Lysenko V, Marynin A, Shcherbakov A, Spivak M, Zholobak N (2016) Influence of the virus-nanoparticles system illumination on the virus infectivity. J Bionanoscience 10(5):1–7
17. Raghupathi KR, Koodali RT, Manna AC (2011) Size-dependent bacterial growth inhibition and mechanism of antibacterial activity of zinc oxide nanoparticles. Langmuir 27:4020
18. Lu L, Sun RW, Chen R, Hui CK, Ho CM, Luk JM, Lau GK, Che CM (2008) Silver nanoparticles inhibit hepatitis B virus replication. Antivir Ther 13:253
19. Shionoiri N, Sato T, Fujimori Y, Nakayama T, Nemoto M, Matsunaga T, Tanaka T (2012) Investigation of the antiviral properties of copper iodide nanoparticles against feline calicivirus. J Biosci Bioeng 113:580
20. Siddiqi KS, ur Rahman A, Tajuddin, Husen A (2018) Properties of zinc oxide nanoparticles and their activity against microbes. Nanoscale Res Lett 13:141
21. Lysenko V, Lozovski V, Spivak M (2018) Nanophysics and antiviral therapy. Ukr J Phys 58:77
22. Morones JR, Elechiguerra JL, Camacho A, Holt K, Kouri JB, Ramirez JT, Yacaman MJ (2005) The bactericidal effect of silver nanoparticles. Nanotechnology 16:2346
23. Pal S, Tak YK, Song JM (2007) Does the antibacterial activity of silver nanoparticles depend on the shape of the nanoparticle? A study of the Gram-negative bacterium Escherichia coli. Appl Environ Microbiol 73:1712
24. Dong PV, Ha CH, Binh LT, Kasbohm J (2012) Chemical synthesis and antibacterial activity of novel-shaped silver nanoparticles. Int Nano Lett 2:9
25. Klimov V (2014) Nanoplasmonics. Jenny Stanford Publishing, Singapore
26. Boltovets PM, Snopok BA (2009) Measurement uncertainty in analytical studies based on surface plasmon resonance. Talanta 80:466–472
27. Vessman J, Stefan R, van Staden J, Danzer K, Lindner W, Burns D, Fajgelj A, Muller H (2001) Selectivity in analytical chemistry (IUPAC Recommendations 2001). Pure Appl Chem 73(8):1381
28. Sattler KD (ed) (2020) 21st century nanoscience: a handbook (10-volume set). CRC Press, Boca Raton
29. Soriano ML, Zougagh M, Valcárcel M, Ríos Á (2018) Analytical nanoscience and nanotechnology: where we are and where we are heading. Talanta 177:104–121
30. Berne BJ, Pecora R (2000) Dynamic light scattering: with applications to chemistry, biology, and physics (Dover Books on Physics). Dover Publications, Mineola
31. Sattler KD (ed) (2020) 21st century nanoscience – a handbook: advanced analytic methods and instrumentation 3. CRC Press, Boca Raton
32. Nakamoto T (ed) (2016) Essentials of machine olfaction and taste. Wiley, Singapore

33. Snopok BA, Kruglenko IV (2002) Multisensor systems for chemical analysis: state-of-the-art in electronic nose technology and new trends in machine olfaction. Thin Solid Films 418(1):21–41
34. Burlachenko J, Kruglenko I, Snopok B, Persaud K (2016) Sample handling for electronic nose technology: state of the art and future trends. Trends Anal Chem 82:222–236
35. Kravchenko S, Snopok B (2020) "Vanishing mass" in the Sauerbrey world: quartz crystal microbalance study of self-assembled monolayers based on tripod-branched structure with tuneable molecular flexibility. Analyst 145:656–666
36. Johannsmann D (2015) The quartz crystal microbalance in soft matter research: fundamentals and modeling. Springer International Publishing, Heidelberg
37. Snopok B (2020) Biosensing under surface plasmon resonance conditions 19. In: Sattler KD (ed) 21st century nanoscience – a handbook: nanophotonics, nanoelectronics, and nanoplasmonics, vol 6. CRC Press, Boca Raton
38. Snopok BA (2012) Theory and practical use of surface plasmon resonance for analytical purposes (rev.). Theor Exp Chem 48(5):265–284
39. Raether H (1985) Surface plasmons on smooth and rough surfaces and on gratings. Springer, Berlin
40. Maier SA (2007) Plasmonics: fundamentals and applications. Springer Science + Business Media LLC, Berlin
41. Lindon J, Tranter GE, Koppenaal D (eds) (2016) Encyclopedia of spectroscopy and spectrometry. Academic, San Diego
42. Schasfoort RBM, Tudos AJ (eds) (2008) Handbook of surface plasmon resonance. The Royal Society of Chemistry, Cambridge
43. Knoll W (1997) Thin films for optical coating 13. In: Hummel RE, Guenther KH (eds) Handbook of optical properties, vol 1. CRC Press, Boca Raton
44. Nizamov S, Mirsky VM (2018) Wide-field surface plasmon resonance microscopy for in-situ characterization of nanoparticle suspensions. In: Kumar CSSR (ed) In-situ characterization techniques for nanomaterials. Springer, Berlin/Heidelberg
45. Nikitin PI, Grigorenko AN, Beloglazov AA, Valeiko MV, Savchuk AI, Savchuk OA, Steiner G, Kuhne C, Huebner A, Salzer R (2000) Surface plasmon resonance interferometry for micro-array biosensing. Sensors Actuators A Phys 85(1–3):189–193
46. Nizamov S, Scherbahn V, Mirsky VM (2017) Wide-field surface plasmon microscopy of nano- and microparticles: features, benchmarking, limitations, and bioanalytical applications. SPIE Opt Sens 10231:1023108
47. Nizamov S, Kasian O, Mirsky VM (2016) Individual detection and electrochemically assisted identification of adsorbed nanoparticles by using surface plasmon microscopy. Angew Chemie Int Ed 55(25):7247–7251
48. Qian C, Wu G, Jiang D, Zhao X, Chen HÂ, Yang Y, Liu XW (2019) Identification of nanoparticles via plasmonic scattering interferometry. Angew Chem 58(13):4217–4220
49. Yang Y, Shen G, Wang H, Li H, Zhang T, Tao N, Ding X, Yu H (2018) Interferometric plasmonic imaging and detection of single exosomes. Proc Natl Acad Sci 115(41):10275–10280
50. Snopok B, Kruglenko I (2015) Analyte induced water adsorbability in gas phase biosensors: the influence of ethinylestradiol on the water binding protein capacity. Analyst 140:3225–3232
51. Nizamov S, Scherbahn V, Mirsky VM (2016) Detection and quantification of single engineered nanoparticles in complex samples using template matching in wide-field surface plasmon microscopy. Anal Chem 88(20):10206–10214
52. Scherbahn V, Nizamov S, Mirsky VM (2016) Plasmonic detection and visualization of directed adsorption of charged single nanoparticles to patterned surfaces. Microchim Acta 183(11):2837–2845
53. Lysenko SI, Snopok BA, Sterligov VA, Kostyukevich EV, Shirshov YM (2001) Light scattering by molecular-organized films on the surface of polycrystalline gold. Opt Spectrosc 90(4):606–616
54. Lysenko SI, Snopok BA, Sterligov VA (2010) Scattering of surface plasmons and normal waves by thin gold films. Opt Spectrosc 188(4):618–628

55. Savchenko A, Kashuba E, Kashuba V, Snopok B (2007) A novel imaging technique for the screening of protein–protein interactions using scattered light under surface plasmon resonance conditions. Anal Chem 79:1349–1355
56. Savchenko A, Kashuba E, Kashuba V, Snopok B (2008) Imaging of plasmid DNA microarrays by scattering light under SPR conditions. Sens Lett 6:705–713
57. Manilo M, Boltovets PM, Snopok B, Barany S, Lebovka NI (2017) Anomalous interfacial architecture in laponite aqueous suspensions on a gold surface. Colloids Surf A Physicochem Eng Asp 520:883–891
58. Boltovets P, Shinkaruk S, Vellutini L, Snopok B (2017) Self-tuning interfacial architecture for Estradiol detection by surface plasmon resonance biosensor. Biosens Bioelectron 90:91–95
59. Nguyen HH, Park J, Kang S, Kim M (2015) Surface plasmon resonance: a versatile technique for biosensor applications. Sensors (Basel) 15(5):10481–10510

Chapter 14
Model of Interaction Between TiO_2 Nanostructures and Bovine Leucosis Proteins in Photoluminescence Based Immunosensor

A. Tereshchenko, V. Smyntyna, and A. Ramanavicius

Abstract A model of interaction between photoluminescent TiO_2 nanoparticles and Bovine Leucosis proteins *gp*51 during the formation of optical immunosensor for the determination of *gp*51 antibodies has been proposed. The main reason of the changes in the photoluminescence (PL) spectra (i.e. PL maxima shifts and PL intensity variations) as a result of Bovine Leucosis proteins adsorption on the surface of TiO_2 thin film is an electrostatic interaction between the TiO_2 surface charge and partial uncompensated charges of *gp*51 proteins.

Keywords TiO_2 nanoparticles · Photoluminescence · Bovine Leucosis · Immunosensor

14.1 Introduction

Nanostructured Titanium dioxide (TiO_2) is a popular material for biosensors application due to its good bio-compatibility and high chemical stability. Being a wide bandgap semiconductor that has an intense photoluminescence (PL) at room temperature, TiO_2 is broadly applied in optical bio- and immune sensors [1, 2]. Immunosensors belong to the class of biosensors that based on the reaction between antibody and antigen by formation of an immune complex [3], where

A. Tereshchenko (✉) · V. Smyntyna
Department of Experimental Physics, Faculty of Mathematics, Physics and Information Technologies, Odesa National I.I. Mechnikov University, Odesa, Ukraine
e-mail: alla_teresc@onu.edu.ua

A. Ramanavicius
Department of Physical Chemistry, Faculty of Chemistry and Geosciences, Vilnius University, Vilnius, Lithuania

J. Bonča, S. Kruchinin (eds.), *Advanced Nanomaterials for Detection of CBRN*,
NATO Science for Peace and Security Series A: Chemistry and Biology,
https://doi.org/10.1007/978-94-024-2030-2_14

interaction between antigen-antibody couple is highly specific and selective one. Recently, immunosensors based on optical transducers that use photoluminescence, absorbance, reflectance or fluorescence signal are of great interest because they demonstrate simple, fast and accurate determination of the target analytes [4]. The main advantage of the optical systems is that optical signal can detect the bio-molecular interaction contactless, i.e. without contamination or significant damage of the bio-samples [1, 3, 4]. Besides, no additional labels of the target analytes (such as dyes or quantum dots) and no contacts for the electrical measurements are required.

In this research an optical immunosensor based on TiO_2 thin film which consisted of TiO_2 nanoparticles (anatase crystal phase) for the determination of Bovine Leucosis antibodies has been developed. In our case, the changes in the photoluminescent properties of TiO_2 nanoparticles as a result of the adsorption of Bovine leucosis proteins have been determined. Biosensors based on the changes in the photoluminescence spectra from nanostructured semiconductors like the shift of PL-maximum and variation of PL-signal intensity have been developed in the range of works [4–10]. However, the interaction of proteins and semiconductors and the reasons of the changes in the photoluminescence spectra were poorly discussed. Although the mechanism of the interaction between semiconductor and proteins is the key in solving many of problems, which are still arising during the development of TiO_2-based immunosensors, such as an improvement of sensitivity and selectivity [4]. This research is aiming to explain the origin of the changes in the photoluminescence spectra of TiO_2 resulted after the protein adsorption on its surface during the formation of biosensitive layer, and after its interaction with the target analyte.

14.2 Experimental

The TiO_2 thin film consisted of TiO_2 nanoparticles (purchased from Sigma Aldrich) was formed by sol-gel synthesis on the glass substrates. The details of the deposition procedure and structural characterization of TiO_2 thin film are described in some previous authors' works [5, 11]. The PL spectrum of TiO_2 nanoparticles, shown in Fig. 14.1a, is characterized by broad non-symmetric maximum around 500 nm, which can be splinted (using Origin 8.0 Pro) in two peaks, related to self-trapped exciton (STE) emission and luminescence caused by oxygen vacancies ($V_{[O]}$) [12].

Bovine leucosis antigens *gp*51 were adsorbed on the surface of a nanostructured TiO_2 thin film by direct adsorption similarly using the process described in [5, 11]. The scheme of such optical immunosensor is shown in Fig. 14.2. It was found that immobilization of *gp*51 leukemia antigens on the surface of TiO_2 is accompanied by an increase of photoluminescence signal of the sample as well as the shift of the photoluminescence peak from 517 to 499 nm was observed after modification of the TiO_2 by adsorbed *gp*51 antigens (Fig. 14.1b). Further interaction of immobilized *gp*51 antigens with *gp*51 antibodies resulted in reversed changes in

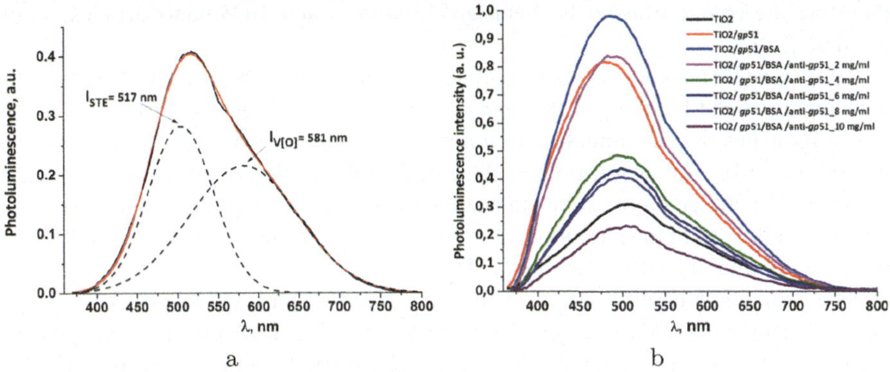

Fig. 14.1 PL spectrum of TiO_2 thin film on glass substrate (**a**); PL spectra of TiO_2/*gp*51 immunosensor after its interaction with anti-*gp*51 antibodies (**b**)

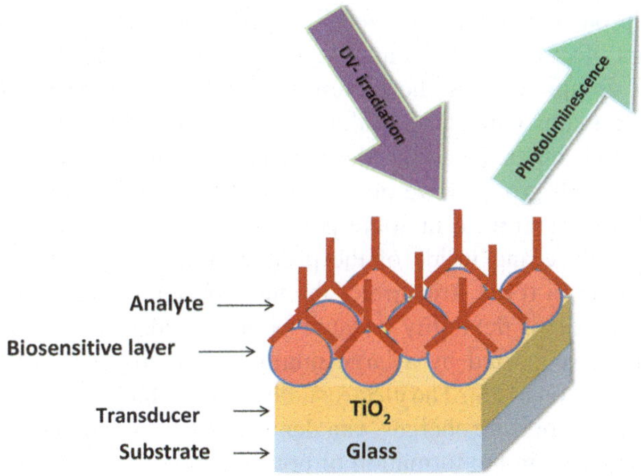

Fig. 14.2 Photoluminescence based immunosensor scheme

TiO_2 photoluminescence spectra, i.e. a decrease in PL intensity and the backward PL peak shift from 499 nm to 516 nm. The sensitivity of the obtained immunosensor was in the range of 2–8 mg/ml [5, 11].

14.3 Results and Discussion

Considering the interaction of proteins and semiconductor nanostructures, a few main possible mechanisms of interaction can take place: charge transfer, electrostatic interaction, resonance energy transfer, etc. Bovine Leucosis protein *gp*51 is not a redox protein, i.e. it cannot be involved in reduction-oxidation reactions therefore the charge transfer between *gp*51 antigens and TiO_2 nanoparticles is not possible [13].

TiO_2 (anatase) is known as a semiconductor of n-type conductivity, usually with an 'upward' band bending of the energy levels when closing the surface of TiO_2, which indicates the accumulation of a negative charge (bound at surface levels) on its surface [14]. The adsorption of the most of molecules is known to introduce an additional charge on the solid state surface and it can change the existing surface energy levels or form the additional ones that are involved in the exchange of charges with the volume of a solid material [15].

The proteins consist of amino acids that might contain positively and/or negatively charged radicals that are determining the charge of the different protein domains [16]. A large quantity of negatively charged groups such as aldehyde (–CHO), hydroxyl (–OH), carboxyl (–COOH) and primary amine (–NH_2) and some other groups, which are involved into the structure of amino acids, are responsible for the partial ($\delta+$ and $\delta-$) charges of particular protein domains. Therefore the proteins are characterized by electrostatic properties, and sometimes even significant electrostatic 'asymmetry of protein molecule' because the atoms and functional groups forming the protein molecules are charged differently both in their sign and in absolute charge value. Naturally, the charges at least partly are compensating each other, but since the ternary structure of proteins is relatively rigid and the charged groups have only limited degree of freedom to move within the protein globule, therefore in some parts of the protein some uncompensated charge on the surface and inside of the protein still remains. It should be taken into account that even if the structure of the most proteins is at some extent 'rigid' there is some degree of flexibility because both secondary and tertiary structures of the protein are supported by a large number of hydrogen bonds but many of them are not very strong [13]. The electrostatic bonds, which are based on Coulomb forces, between the opposite charges, van der Waals forces and disulfide bonds also play an important role in the formation of both secondary and tertiary structures of protein.

14.3.1 Interaction Between TiO_2 and gp51 Proteins

A *gp*51 protein molecule has a molecular mass of 51 KDa and its characteristic geometric size is about 6 nm in diameter. The authors [17], which have published a research on the formation of *gp*51 virus based capsid of BLV, have constructed an image of *gp*51 virus structure from the X-ray crystallography data and they have reported that this protein is extra-flexible, which provides very high functionality and the ability to associate and/or dissociate of BLV capsid from the membrane of BLV infected cell. Therefore, it is expected that on the surface of TiO_2 *gp*51 forms a well-ordered monolayer. The formation of such layer was confirmed in other our researches by spectroscopic ellipsometry [18–20]. Although the *gp*51 protein is not a redox-protein, however, like many others, it contains a number of partially charged groups and domains, represented as partial charges "$\delta-$" and "$\delta+$", which mostly are lower in value than the total electron charge (1.6×10^{-19} coulombs) per charged atom or group. The presence of these partial charges suggests that the electrostatic influence on the surface charge of TiO_2 from the side of partially uncompensated charges in those parts of the *gp*51 protein that located on the surface of TiO_2 is responsible for the adsorption of this protein on the TiO_2 surface. The Coulomb interaction takes place between charged groups in the *gp*51 protein and the negatively charged surface of the TiO_2 because such electrostatic interactions are very strong at a distances ranging from several Angstroms to few nanometers. Therefore, among the others interactions such as hydrogen bonds, disulfide bonds, Van der Waals interaction, etc., which also have significant role during the adsorption of proteins, the electrostatic interaction plays one of the most important role during the adsorption of proteins to electrically charged surfaces, such as TiO_2. In addition, the local electric fields of charged domains of adsorbed proteins are affecting the PL-centers of TiO_2 and it causes the shift in the photoluminescence spectra of TiO_2 nanoparticles. Therefore, the photoluminescence maximum caused by STE shifts from 517 to 499 nm (i.e., to 18 nm), which corresponds to ~0.086 eV that is less than 0.1 eV, and it is one of the proofs of electrostatic interaction based physical adsorption of *gp*51 [15, 21].

The splitting of the photoluminescence spectra into Gaussian curves at each stage of the experiment shows that after the adsorption of *gp*51 protein molecules on the TiO_2 surface the energy value of excitation levels, which are responsible for the luminescence and are associated with oxygen vacancies $I_{V[O]}$, almost does not change remaining at a value of 605 ± 2 nm. At the same time, the photoluminescence maximum caused by recombination of self-trapped excitons (STE) shifts to short wavelengths, changing its position from 517 ($STE_1 = 2.39$ eV) nm to 499 ($STE_2 = 2.48$ eV) nm (Fig. 14.3). Since the involvement of the STE level in the process of radiative recombination is regulated by the surface, this indicates that STE level is located either on the surface plane or not very deeply within the surface layer of the TiO_2. The displacement of the light emitting recombination peak indicates that the energy level of STE is complex and has its basic and excited states. The appearance of luminescence in the region of 499 nm indicates a radiative transition

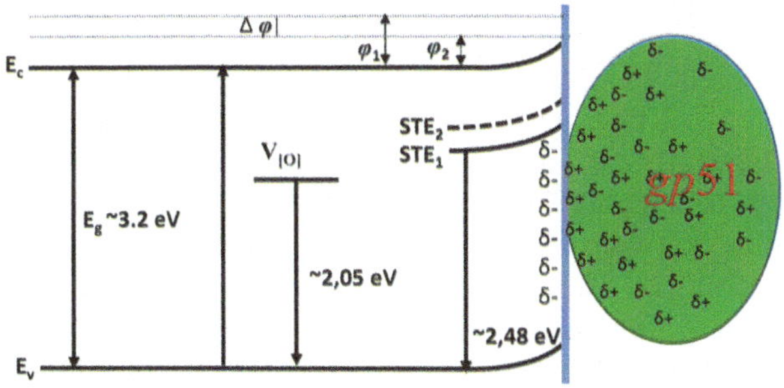

Fig. 14.3 Energetic levels of TiO_2/*gp*51

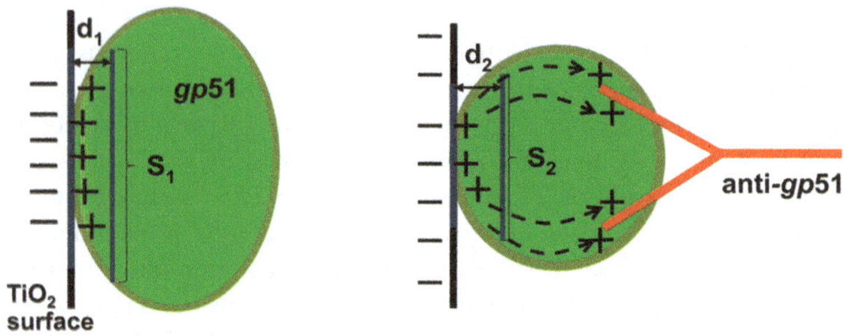

Fig. 14.4 Flat capacitor based model of the charges interaction between TiO_2 surface and *gp*51 proteins: (**a**) electrostatic interaction of uncompensated charges of immobilized protein *gp*51 with charges located on the surface of TiO_2; (**b**) model of interaction that takes into account the electrostatic interaction of charges within *gp*51 antigens and anti-*gp*51 antibodies

from the excited STE level. The blue-shift of the photoluminescence maximum by 18 nm as a result of adsorption of the *gp*51 protein, which corresponds to $\Delta E_{STE} = STE_2 - STE_1 = 0.086\,eV$, also indicates that the initial value of the potential barrier ϕ_1 on the TiO_2 surface has decreased by a value of 0.086 eV ($\Delta\varphi$) (Fig. 14.3). Variation of the potential barrier means that the value of negative charge localized on the TiO_2 surface has also changed, due to the charge-charge-based interaction with adsorbed protein *gp*51. Positively charged atoms and groups, which are provided by the *gp*51 protein, partially compensates the surface charge of TiO_2 and therefore reduces the energy of electrons localized at the surface levels, which are the most responsible for the generation of PL-signal (Fig. 14.4a). Taking into account the fact that the total negative charge predominates on the TiO_2 surface, the positively charged parts of the *gp*51 protein electrostatically interact with the negatively charged TiO_2 surface. As a result, a partial decrease of the surface charge reduces the electric field in the TiO_2 surface region (Fig. 14.4a).

14.3.2 Interaction of TiO_2/gp51 Immunosensing Structure with Anti-gp51 Proteins

Further interaction of TiO_2/*gp*51 structure with anti-*gp*51, which is also a protein, leads to the inverse changes in the photoluminescence spectra, i.e., to UV-shift of the spectrum (Fig. 14.1b) and decrease the photoluminescence intensity to the value that corresponds to the pure TiO_2. The latter effect is based on the formation an immune complex between immobilized antigens *gp*51 and anti-*gp*51 antibodies, which were present in aliquot. Formation of this immune complex, besides of the van der Waals interaction and other interactions, at a very high extent is based on the interaction between oppositely charged domains, functional groups and atoms in *gp*51 and anti-*gp*51 molecules (including the formation of number of hydrogen bounds, which can be estimated as specific kind of electrostatic interaction). It can be assumed that uncompensated charges ($\delta+$ and $\delta-$) of both proteins are involved in electrostatic interactions during the formation of immune complex. As a result, some of the charged groups that were originally involved in the interaction between *gp*51 and TiO_2 are at least partially compensated by the opposite charge of the anti-*gp*51 protein groups, thereby reducing the direct electrostatic effect from immobilized *gp*51 proteins to the charged surface of TiO_2 and to light emitting centers (Fig. 14.4b). The effects described above have an effect on the shift of PL-maximum and on the decrease in the potential barrier on TiO_2/*gp*51 interface due to the charge-charge interaction between TiO_2 and *gp*51. The potential barrier at the interface between TiO_2 and *gp*51 has greater value in TiO_2/*gp*51 structure in comparison with that in TiO_2/*gp*51/anti-*gp*51 due to partial compensation (decrease in value) and/or delocalization of charges, which were initially involved into interaction between TiO_2 and *gp*51 after formation of TiO_2/*gp*51 structure [21].

The distribution of charges in TiO_2/*gp*51 structure can also be interpreted as a model based on an 'imaginary flat capacitor' (Fig. 14.4), formed as a result of the electrostatic interaction between oppositely charged protein *gp*51 layer and the TiO_2 surface [21]. The capacitor is formed as a result of *gp*51 protein adsorption on TiO_2 surface, after which the charges are distributed in the most favorable way (energetically), partially compensating each other. Thus, the positive 'imaginary capacitor plate' is based on the positive charges, which are predominant in the protein *gp*51 area that after adsorption appears in close proximity to TiO_2/*gp*51 interface and/or due to the negative electrostatic effect of TiO_2 are induced/attracted closer to negatively charged surface. These charged atoms/groups/domains of *gp*51, localized in the close proximity to the TiO_2 surface, electrostatically affect the TiO_2 emission centers and the energy value of the surface potential barrier. Therefore, the position of the energy levels of the TiO_2 emission maximum depends on TiO_2 surface modification stage (TiO_2 or TiO_2/*gp*51) and shifts from/backwards the initial position of the demarcation level. Figure 14.4a represents an imaginary flat capacitor consisting of a negatively charged plate on the surface of TiO_2 and an 'imaginary positively charged plate' formed in *gp*51 protein in close proximity to TiO_2/*gp*51 interphase. Hence, the interaction of TiO_2/*gp*51 with anti-*gp*51 antibodies and the formation of *gp*51/anti-*gp*51-based immune complex leads to a 'deformation' and the reduction of charge 'stored' on 'the positive imaginary capacitor plate' (Fig. 14.4b). This is mainly due to the redistribution and partial compensation of charges during the formation of the *gp*51/anti-*gp*51 immune complex, which in turn reduces the charge of 'the imaginary capacitor plate' based on *gp*51 ($q_2 < q_1$). Due to this reduced charge it can be interpreted as the reduction of the area of the same plate (S_2) and/or the increase of the distance (d_2) between the two imaginary capacitor plates based on *gp*51 and TiO_2 which leads to the decrease of capacitance according to equation (14.1):

$$C = \frac{\varepsilon\varepsilon_0 S}{d}. \tag{14.1}$$

This effect is observed because some of the *gp*51 protein charges move from the TiO_2/*gp*51 interface towards interacting anti-*gp*51 protein and are partially compensated by the charge present in anti-*gp*51, whereby an imaginary positive *gp*51-based capacitor plate of the capacitor is reduced in imaginary surface area and/or correspondingly moving apart from the negative TiO_2 plate. This effect leads to a decrease in the capacitance of this imaginary capacitor and the electric field induced by *gp*51 becomes reduced. Therefore, after the interaction of TiO_2/*gp*51 with anti-*gp*51 antibodies and the formation of *gp*51/anti-*gp*51 complex, which is involved into TiO_2/*gp*51/anti-*gp*51 structure, the electrostatic effect of *gp*51 initially adsorbed on TiO_2 towards the TiO_2 surface significantly decreases. The PL shifts are attributed to the variations in the self-trapped exciton energy level, which were induced by the changes of electrostatic interaction between positively charged atoms and groups, provided by the adsorbed *gp*51 protein and negatively charged surface of TiO_2.

14.4 Conclusions

A model of interaction mechanism between nanostructured TiO_2 layer and Bovine Leukemia virus proteins *gp*51, during the formation of PL-based immunosensor, have been developed. The main reason of changes in the photoluminescence spectra of TiO_2 as a result of adsorption of *gp*51 antigens is the electrostatic interaction between the TiO_2 surface charge and the partial uncompensated charges of *gp*51 proteins. Subsequent interaction of the immunosensing structure of TiO_2/*gp*51 with the target analyte anti-*gp*51 leads to inverse changes in the photoluminescence spectra due to the charge distribution as a result of the formation of an immune complex. The charge–charge-based interaction in the double charged layers *gp*51/TiO_2 can also be interpreted as a model based on 'imaginary capacitor', formed as a result of the electrostatic interaction between oppositely charged protein *gp*51 layer and the TiO_2 surface. The proposed interaction mechanism provides the general understanding of the interaction between TiO_2 and proteins, what is a key in the development of new PL-immunosensors and solving of many issues related to an improvement of performance of PL-based immunosensors, first of all, related to the sensors' sensitivity and selectivity.

Acknowledgments This work was supported by joint Ukrainian-Lithuanian research project "Application of hybrid nanostructures which are based on TiO_2 or ZnO and modified by biomolecules, in optoelectronic sensors", funded by Lithuanian Research Council (project No. P-LU-18-53).

References

1. Tereshchenko A, Bechelany M, Viter R, Khranovskyy V, Smyntyna V, Starodub N, Yakimova R (2016) Optical biosensors based on ZnO nanostructures: advantages and perspectives. A review. Sensors Actuators B Chem 229:664
2. Preclíková J, Galáø P, Trojánek F, Daniš S, Rezek B, Gregora I, Nìmcová Y, Malý P (2010) Nanocrystalline titanium dioxide films: influuence of ambient conditions on surface- and volume-related photoluminescence. J Appl Phys 108:113502
3. Tereshchenko A, Fedorenko V, Smyntyna V, Konup I, Konup A, Eriksson M, Yakimova R, Ramanavicius A, Balme S, Bechelany M (2017) ZnO films formed by atomic layer deposition as an optical biosensor platform for the detection of Grapevine virus A-type proteins. Biosens Bioelectron 92:763
4. Tereshchenko A, Smyntyna V, Konup I, Geveliuk SA, Starodub MF (2016) Metal oxide based biosensors for the detection of dangerous biological compounds. In: NATO science for peace and security series a: chemistry and biology "nanomaterials for security" 2016. https://doi.org/10.1007/978-94-017-7593-9_22
5. Viter R, Tereshchenko A, Smyntyna V, Ogorodniichuk J, Starodub N, Yakimova R, Khranovskyy V, Ramanavicius A (2017) Toward development of optical biosensors based on photoluminescence of TiO_2 nanoparticles for the detection of Salmonella. Sensors Actuators B Chem 252:95
6. Tereshchenko A, Viter R, Konup I, Ivanitsa V, Geveliuk SA, Ishkov Y, Smyntyna V (2013) Proceedings of SPIE, the international society for optical engineering 9032, Nov 2013. https://doi.org/10.1117/12.2044464
7. Viter R, Savchuk M, Iatsunskyi I, Pietralik Z, Starodub N, Shpyrka N, Ramanaviciene A, Ramanavicius A (2018) Analytical, thermodynamical and kinetic characteristics of photoluminescence immunosensor for the determination of Ochratoxin A. Biosens Bioelectron 99:237–243
8. Ermakov V, Kruchinin S, Fujiwara A (2008) Electronic nanosensors based on nanotransistor with bistability behaviour. In: Bonca J, Kruchinin S (eds) Proceedings NATO ARW "Electron transport in nanosystems". Springer, Dordrecht, pp 341–349
9. Ermakov V, Kruchinin S, Hori H, Fujiwara A (2007) Phenomena in the resonant tunneling through degenarate energy state with electron correlation. Int J Mod Phys B 11:827–835
10. Kruchinin S, Pruschke T (2014) Thermopower for a molecule with vibrational degrees of freedom. Phys Lett A 378:157–161
11. Viter R, Smyntyna V, Starodub N, Tereshchenko A, Kusevitch A, Doycho I, Geveluk S, Slishik N, Buk J, Duchoslav J, Lubchuk J, Konup I, Ubelis A, Spigulis J (2012) Novel immune TiO_2 photoluminescence biosensors for leucosis detection. Procedia Eng 47:338–341
12. Sildos I, Suisalu A, Kiisk V, Schuisky M, Mändar H, Uustare T, Aarik J (2000) Effect of structure development on self-trapped exciton emission of TiO_2 thin films. Proc SPIE 4086:427–430
13. Ogawa T (ed) (2013) Protein engineering – technology and application. IntechOpen. ISBN: 978-953-51-1138-2
14. Gupta SM, Tripathi M (2011) A review of TiO_2 nanoparticles. Chin Sci Bull 56:1639
15. Smyntyna V (2013) Electron and molecular phenomena on the surface of semiconductors. Nova Publishers, New York
16. D.Nelson L, Cox MM, Lehninger AL (2000) Principles of biochemistry. Worth Publishtrs Inc., New York
17. Obal G, Trajtenberg F, Carriön F, Tomé L, Larrieux N, Zhang X, Pritsch O, Buschiazzo A (2015) Conformational plasticity of a native retroviral capsid revealed by X-ray crystallography. Science 5182:1–7
18. Balevicius Z, Baleviciute I, Tumenas S, Tamosaitis L, Stirke A, Makaraviciute A, Ramanaviciene A, Ramanavicius A (2014) In situ study of ligand-receptor interaction by total internal reflection ellipsometry. Thin Solid Films 571:744

19. Balevicius Z, Makaraviciute A, Babonas GJ, Tumenas S, Bukauskas V, Ramanaviciene A, Ramanavicius A (2013) Study of optical anisotropy in thin molecular layers by total internal reflection ellipsometry. Sensors Actuators B Chem 181:119–124
20. Baleviciute I, Balevicius Z, Makaraviciute A, Ramanaviciene A, Ramanavicius A (2013) Study of antibody/antigen binding kinetics by total internal refection ellipsometry. Biosens Bioelectron 39:170
21. Tereshchenko A, Smyntyna V, Ramanavicius A (2018) Interaction mechanism between TiO_2 nanostructures and bovine leukemia virus proteins in photoluminescence-based immunosensors. RSC Adv 8:37740–37748

Chapter 15
Superconducting Gravimeters Based on Advanced Nanomaterials and Quantum Neural Network

V. Yatsenko, D. Kruchinin, B. Vlahovic, I. Morozova, and S. Kruchinin

Abstract This paper focuses on a new methodology, superconducting sensor for the estimation of gravitation perturbations, optical interferometer, and applied software that are capable to realize the method under the noise influence. This paper is oriented on the development of a cryogenic sensitive element with an accuracy of about 10^{-10} g. This sensor can be used for the analysis of geodynamical processes and for the accumulation of experimental data under various conditions, which is of importance in order to develop future methods.

Keywords Cryogenic sensitive element · Magnetic levitation · Nanostructured materials · Quantum computing · Quantum neural network

15.1 Introduction

Today, the remote sensing is one of the fastest growing technologies in the world. It is a multibillion-dollar industry, and remote thematic images are routinely used in an increasing number of the implementation fields. The solution of many important practical problems depends on a large-scale usage of the measurement systems and underlying physical principles. These problems include the monitoring of natural

V. Yatsenko
Space Research Institute of NASU-NSAU, Kyiv, Ukraine

D. Kruchinin
Taras Shevchenko Kyiv National University, Kyiv, Ukraine

B. Vlahovic
North Carolina Central University, Durham, NC, USA

I. Morozova
National Aviation University, Kyiv, Ukraine

S. Kruchinin (✉)
Bogolyubov Institute for Theoretical Physics, NASU, Kyiv, Ukraine

J. Bonča, S. Kruchinin (eds.), *Advanced Nanomaterials for Detection of CBRN*, NATO Science for Peace and Security Series A: Chemistry and Biology,
https://doi.org/10.1007/978-94-024-2030-2_15

resources based on the analysis of gravity anomalies, studying of global geodynamic processes and evolution of the Earth gravity field, analysis of the movement of Earth's poles, etc. Detailed information on Earth's gravity field is necessary for many branches of applied science (geology, navigation, Earth-shape science, etc.). A gravimeter is a fine instrument intended for measuring the gravity (g). The accuracy of the best land-surface gravimeters is about 10^{-8} g. The best results of measurements on the board of moving vehicles do not exceed 10^{-7} g for sea gravimeters and 10^{-6} g for air gravimeters. The development of the superconductive magnetic levitation in Ukraine known as Magnetic Potential Well (MPW) started in the late 1962 as a natural consequence of the development of applied low-temperature superconductivity, the electromechanical energy transformation theory, and the control theory methods. This branch of levitation is usually defined as the stable equilibrium of a superconductive ring or another loop of zero electrical resistance without contact with other bodies on the base of the use of magnetic forces.

15.2 Magnetic Potential Well

The Magnetic Potential Well (MPW) phenomenon discovered in 1975 means that the magnetic suspension between two spaced magnets can be changed into the magnetic repulsion only owing to a decrease of the spacing between these magnets [1]. Before this discovery, the magnetic attractive force between two spaced magnets was considered as a force, which was increasing with the spacing decrease. In other words, the MPW means the minimum of the magnetic potential energy as a function of the distance, though, before the MPW discovery, the magnetic interaction was considered as a monotonic one without the magnetic potential energy minimum anywhere with the exception of boundary points.

In order to understand better the MPW and MPW-levitation, it is desirable to dwell on a simple comparison between a magnetic force and an elastic one. The well-known property of a mechanical spring (Fig. 15.2) is the increase in the attractive force P, if the spacing x becomes greater, as Fig. 15.1 shows. It is also known from the experience that some position x_0 of a body of weight G balanced by such spring is stable in the gravity field.

In contrast, the attractive force of two magnets decreases, if the spacing increases, as shown in Fig. 15.3. Thus, the equilibrium position balanced by the magnetic attraction is unstable. Naturally, in the latter case in order that the stability to be achieved, the P force law in Fig. 15.3 must be transformed into the P law of Fig. 15.2.

As is shown in Fig. 15.3, this can be accomplished with an electromagnet obeying the unstable force laws at different values of currents I_1, I_2, I_3 by the feedback control of the I value depending on the spacing, as depicted by the dashed line in Fig. 15.3. This is the solution with the electromagnetic controlled levitation. Finally, Fig. 15.4 demonstrates the MPW-force behavior. At a small spacing, the magnetic

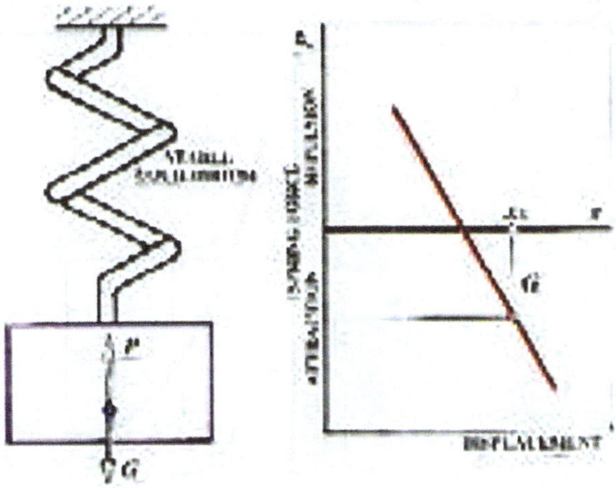

Fig. 15.1 Spring-suspension

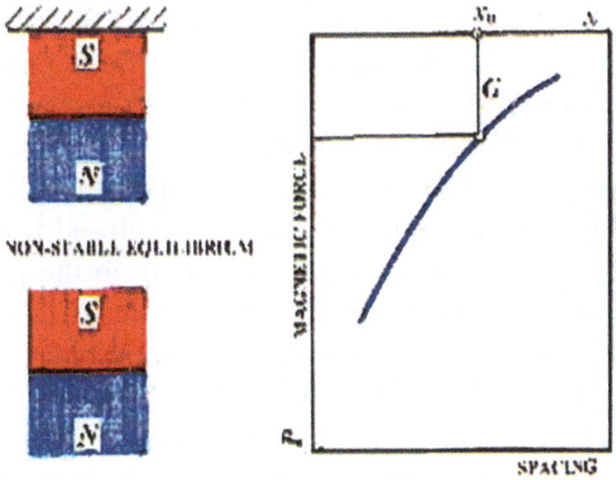

Fig. 15.2 Two-magnet non-stable suspension

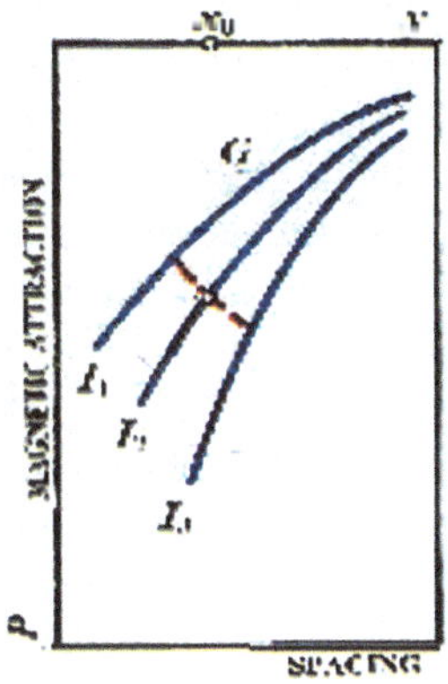

Fig. 15.3 Automatic suspension

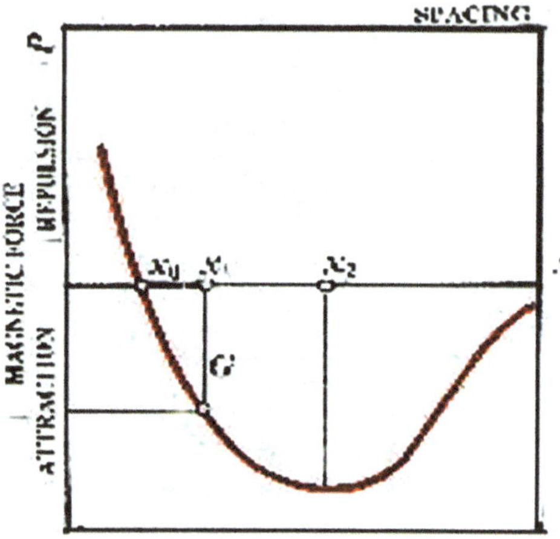

Fig. 15.4 MPW-force

force between two magnets remains repulsive in nature up to the spacing x_0, where it is zero. With a following increase in the spacing comparable with x_0, the magnetic repulsion spontaneously changes into the magnetic attraction increasing in the value up to spacing x_2 and decreasing, as the spacing increase at $x > x_2$.

In the course of researches of the first stage, the necessary prerequisites for the analysis of possibilities to obtain the stable free mass suspension on the basis of the magnetic interaction between two closed loops with zero resistance as a model of superconductive magnets operating in the persistent current mode were obtained. A new statement of the work to derive a mathematical model of levitation was proposed. It was formulated as a problem of necessary and sufficient conditions of the equilibrium stability for a non-linear conservative dynamic system with a free body. The last is under actions of the weight, magnetic forces, and force moments generated by the immobile current carried by zero-resistance loops and ones affiliated to the free body, and other forces that are considered as small with respect to two said types of forces.

The key element of the force interaction is the magnet which is a closed loop without electrical resistance. A short-circuited loop free of electrical resistance possesses the well-known property to preserve the total magnetic flux (magnetic linkage) coupled with this loop. They say also that the full flux is frozen. A high degree of the ideal electric conductivity is achievable in superconductors. If the superconductor resistance were non-zero, the tested difference between resistances in the superconductive and normal states would be greater than that between copper and common insulators.

However, the superconductor is not only the ideal conductor possessing zero electric resistance $\rho = 0$. It displays the ideal diamagnetism $\mu = 0$ as well. The manifestation of these two properties depends on the superconductor body shape and is restricted by the critical temperature and critical magnetic field. If the level of the operating magnetic field is less than critical one and the superconductor sample is a simply connected region (an extended body with non-zero volume and without "holes" for the magnetic field penetration), it exhibits only the ideal diamagnetism. On the other hand, if the superconductor is a closed loop, and the ratio between the thickness and the longitudinal extension is zero, properties of the closed loop are fully determined by the condition $\rho = 0$ [2–5].

A magnet model chosen by us depends only on the ideal electrical conductivity and its exhibition in samples like a thin closed loop. It does not depend on the superconductor diamagnetism in the sense to achieve the stability in the free position. This is because the considered closed loop has the intentionally zero thickness. If such a loop is in the superconductivity state but with a part transferred into the resistive state, it does not manifest diamagnetic properties. Moreover, this loop is non-magnetic generally. If the ideal electric conductivity state of the said part is subsequently restored, the loop continues not to exhibit diamagnetic properties, because these properties depend on the body volume that is zero in our approximation. Hence, we must conclude that there is no possibility to provide the stable free body equilibrium for such loop on the basis of the Earnshaw–Braunbeck's principle that is the single well-grounded scientific principle of magnetic levitation.

But the considered loop exhibits the property to preserve its magnetic linkage (full flux) as any closed loop with zero electrical resistance. The linkage constancy remains valid irrespective of any change in the position of the loop and is a well-known consequence of Faraday's law in the case of closed loops with $\rho = 0$. The last condition can be realized due to superconductivity. Availability of zero resistance in the closed loop is only one bridge connecting our magnet model with superconductivity.

Superconductors made in the form of a thin closed loop with zero electrical resistance are in abundance from Kunzler's discovery of abnormally high critical currents in Nb_3Sn (1961). Critical magnetic fields are for this material 24.5 Tesla at $T = 0$ K and 14 Tesla at $T = 0$ K for NbTi. These levels of the magnetic field were the basis to create a huge amount of superconductive magnets of different shapes and sizes generating and operating in magnetic fields of the interval 10–20 Tesla. Therefore, the closed ideally electrically conducting loop model is applicable to the low-temperature superconductive magnets operating in the persistent current mode. This model can be applicable also to high-temperature superconductors (HTS) operating under the condition of magnetic field penetration and if such material can be represented as a set of zero-resistance closed loops.

Finally, the recent discovery of carbon nanotubes (fullerens) and their properties to conduct persistent currents and to freeze a magnetic flux in high magnetic fields indicate the applicability of our model to such a substance, if it can be manufactured as a closed loop.

It should be noted that our model of the magnet does not account for the quantization of the magnetic flux habitual for electronic superconductive circuits, because we consider high current applications mainly. We simplify also a picture of the magnetic field penetration into samples like a closed loop. In the conventional current carrying model, the current density is usually the same all over its cross section. We consider that the magnetic field and the current are concentrated at a small depth of the loop. The depth of penetration λ is assumed to be corresponding to London's one i.e. it is of the order of $(4\div6) \times 10^{-8}$ m. This circumstance leads to different formulae for the internal inductance responsible for the field inside wires. The magnetic field of ideally electrically conducting loops and that of superconductive loops are distinguished by the field inside the loop wire. This difference vanishes, if the parameter $\tau = da^{-1}$ (d and a are the thickness and longitudinal size of a loop, respectively) tends to zero. We study the case of very small τ. Then the general formula for the vector potential can be expressed as a linear integral, and, for the sake of convenience, the loop self-inductance and the mutual inductance come into play.

15.3 Conception

A conception of the cryogenic-optical sensor based on a competitive adaptive sensitive element and applicable to a gravity meter-sensor is proposed. The sensor element is based on the new magnetic levitation phenomenon, high-precision optical registration of mechanical coordinates of a levitating body, and the robust signal processing tools. The focus of this paper is to equilibrate the gravity forces acting on a free body by magnetic forces so that all six degrees of freedom of a free body are stable. The registration of free body mechanical coordinates is based on an optical system.

15.4 Methods

A variety of research methods and approaches have been used to investigate the main problem (Fig. 15.5):

- Synergetic methods
- Methods of a superconducting probe suspension
- Dynamical-information methodology
- Dynamical stability theory
- Methods of chaotic dynamics
- Global optimization methods [6]
- Risk analysis [7]
- Quantum neural networks

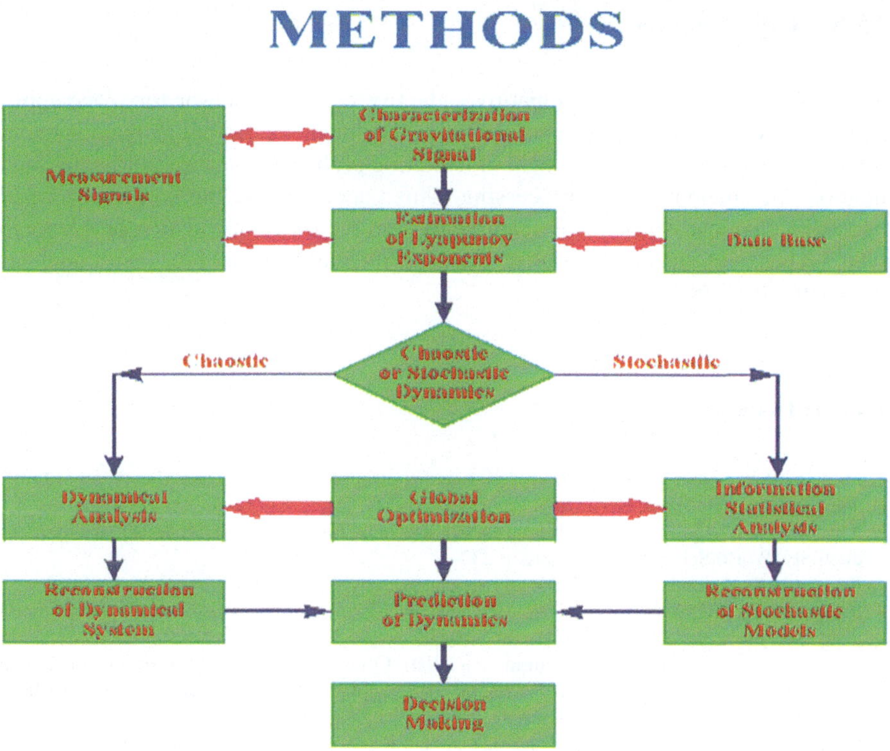

Fig. 15.5 MPW-force

15.5 Estimation of the Gravitation Perturbation by a Biological or Quantum Neural Chip

Over the past decade, deep artificial neural networks (DNNs) have become the state-of-the-art algorithms in Machine Learning (ML), speech recognition, computer vision, natural language processing, and many other tasks. This was made possible by the advancement in Big Data, Deep Learning (DL) and drastically increased chip processing abilities, especially general-purpose graphical processing units (GPGPUs). All this has created a growing interest in making the most of the potential offered by DNNs in almost every field. We propose to use a biological or quantum neural chip for a superconducting gravimeter. The characteristic applications are: signal processing, detection of gravitation perturbations, and other data mining. The Deep Artificial Neuron–Astrocyte Networks (DANAN) could overcome the difficulties in the architecture design, learning process, and scalability of the current ML methods. We also propose to use Hopfield Networks, Probabilistic Neural Network, and Boltzmann Machine.

15.6 Conclusions

This paper proposes a new competitive adaptive cryogenic sensor for space applications based on advanced nanomaterials, the magnetic levitation phenomenon, high-precision optical measurement of levitating probe mechanical coordinates, and methods of quantum signal processing. This paper reviews the construction and operating characteristics of the instruments, and the range of research problems to which it has been and can be applied. An experimental version of the instrument has now being realized.

References

1. Pardalos P, Yatsenko V (2008) Optimization and control of bilinear systems: theory, algorithms, and applicants. Springer, Berlin
2. Kruchinin S, Nagao H, Aono S (2010) Modern aspect of superconductivity: theory of superconductivity. World Scientific, Singapore, p 232
3. Kruchinin S, Nagao H (2012) Nanoscale superconductivity. Int J Mod Phys B 26:1230013
4. Kruchinin S, Klepikov V, Novikov VE (2005) Nonlinear current oscillations in a fractal josephson junction. Mater Sci 23(4):1009–1013
5. Dzhezherya Y, Novak IY, Kruchinin S (2010) Orientational phase transitions of lattice of magnetic dots embedded in a London type superconductors. Supercond Sci Technol 23:1050111–105015
6. Pardalos P, Yatsenko V, Butenko S (2002) Robust recursive estimation and quantum minimax strategies. In: Pardalos P, Murphey R (eds) Cooperative control and optimization. Kluwer Academic Publish, Dordrecht/Boston/London, pp 213–230
7. Yatsenko V et al (2010) Space weather influence on power systems: prediction, risk analysis, and modeling. Energy Syst 1(2):197–207

Chapter 16
Replicated Computer Generated Microstructure onto Piezoelectric Nanocomposite and Nanoporous Aluminum Oxide Membranes Usage in Microfluidics

Y. Patel, C. Justas, V. Naginevičius, and A. Palevicius

Abstract The main aim of the paper is to present replication of computer-generated microstructure on the piezoelectric nano composites with PZT and PMMA (Polymethyl methacrylate) and fabrication of the nanoporous aluminum oxide membrane with nanopores. In this paper computer-generated hologram is produced and replicated on the piezoelectric nano composites by using thermal replication process in the form of micro channels. Nanoporous aluminum oxide membrane fabricated with two step anodization by using custom experimental setup and hexagonal pores with 80 nm diameter and 110 nm interpore distance.

Keywords Microfluidics · Piezoelectric nanocomposite · Aluminum oxide membrane · Computer generated microstructure

16.1 Introduction

Miniaturization of analytical instruments utilizing microfabrication technology has increased the broad usage of analytical chemistry. Focus was an increasing demand for low-cost instruments able to analyze compounds of very small volume with a high level of automation and precision. The terms 'Micro-Total Analysis Systems (μTAS)', and 'Lab-on-a-chip', aims to develop integrated micro-analytical systems [1]. which could complete analysis cycles (e.g. sample pre-treatment,

Y. Patel · C. Justas · A. Palevicius (✉)
Faculty of Mechanical Engineering and Design, Kaunas University of Technology, Kaunas, Lithuania
e-mail: arvydas.palevicius@ktu.lt

V. Naginevičius
Kaunas University of Applied Sciences, Kaunas, Lithuania

J. Bonča, S. Kruchinin (eds.), *Advanced Nanomaterials for Detection of CBRN*, NATO Science for Peace and Security Series A: Chemistry and Biology,
https://doi.org/10.1007/978-94-024-2030-2_16

chemical reactions, analytical separation, detection, and data handling) on the same micro device [2]. At first, the microchannel type of structure was built in solid material substrates such as silicon, glass, etc. [3–7]. but by the evolution of the innovative plastic materials and their development methods poly(methyl methacrylate) (PMMA), polystyrene (PS), polycarbonate (PC), and cyclic olefin copolymer (COC) become more popular in use because the fabrication is progressively effective and compactible with biological substance [8–14]. In this paper, replication of computer generated microstructure has been replicated on the piezoelectric composite material in order to achieve the replica of the master structure on the composite material in the form of the micro channels. The advantage of the piezoelectric nanocomposite provides the better replicated structure because of the polymer composition and the piezoelectric material has a functionality in the channel to travel the liquid by adjusting the shape of the inner geometry of microchannel. Acoustophoresis, a technique dependent on the use of acoustic radiation force for exact control of microscale particles, is known. The employment of surface acoustic waves is particularly assuring. Because of the impact of standing acoustic waves, particles are forming over the microchannel, and because of the surface acoustic waves effect, escalating of the particles can be acquired [15–18]. Previously the research was carried out but in that case separately the microchannel was formed on the PZT [15, 18] and directly on the nonactive glass or silicon [17, 18]. On the other side nanoporous aluminum oxide membrane gained the attraction in the field of the microfluidics for the separation of particles with micro and nano scale as well as acting as a filtration component. There are a lot of investigation going on in the field of the microsystems and their applications in the biomedical field of research [19]. Realtime tracking of optical constants, physical, chemical and organic procedures [20, 21], organically two and three-dimensional miniaturized scale development (microstructures) [22, 23], direct laser recording, in view of multi photon polymerisation [24–26] or 3D printing [27]. They empower shaping of smaller scale structures of the internal geometry yet just in small volume and gradually. Different advancements, for example, particle filtering [28], phase seperation [29] or gas frothing [30] empower forming small scale developments (miniaturized scale structures) very quick and in large volumes yet control of inner geometry of microstructure cast by these methods is extremely poor. therefore, creating a nano pore in the aluminum metal sheet and by integrating the aluminum oxide membrane with the PZT cylinder in hydrodynamic system it is possible to control the fluid flow more efficiently inside the microchannels. Therefore, in this paper nanoporous aluminum oxide membrane with predefined dimensions and geometry crated and could be used as a vibro active nano filter in microfluidics.

16.2 Computer Generated Microstructure and Replication

The replication of the computer-generated microstructure performed after determining the sets of the appropriate process parameters. The Gerchberg-Saxton algorithm with FFT is applied in CGH computation. When discrete data is involved, DFT for digital signal processing is employed. The transformation converts space/time-based data into frequency-based data and the conversion is described with the following formula:

$$F(u,v) = \frac{1}{NM} \sum_{x=0}^{M-1} \sum_{y=0}^{M-1} f(x,y) e^{-i2\pi\left(\frac{xu}{M} + \frac{yn}{N}\right)}. \tag{16.1}$$

Where: u and v – discrete spatial frequencies, M and N – the quantity of sections in x and y directions of space and frequency domains, $F(u,v)$ – 2D discrete $f(x,y)$ spectra.

Initially, an arbitrary number generator creates phase scattering $\varphi[-\pi,\pi]$ or it can be equal to zero. The steps of Gerchberg-Saxton algorithm are the following:

$$\varphi_0^H = 0. \tag{16.2}$$

The following equations describe one iteration of the algorithm, which calculates the improved phase distribution in the hologram plane

$$\begin{aligned} u_n^H &= A(I_H)\exp\left[i\varphi_{n-1}^H\right], \\ \varphi_n^T &= P(FFT(u_n^H)), \\ u_n^T &= A(I_T)\exp\left[i\varphi_n^T\right], \\ \varphi_n^H &= P(FFT^{-1}(u_n^T)), \end{aligned} \tag{16.3}$$

where u_n^H and u_n^T is a light field depicted in a 2D complex plane, I_H – light distribution in the hologram plane, φ_n^T – phase distribution, I_T – object intensity, φ_n^H – hologram phase distribution. The solution of the iteration algorithm is repeated until the difference between the estimated intensity and the predicted intensity in the target plane is lower than the defined limit [31, 32].

Raith e-LiNEplus high resolution electron beam lithography system was used for computer generated hologram. Silicon wafer was exposed to oxygen plasma for 5 min and heated for 30 min at 150 °C on a hotplate. Then PMMA resist of 35 K molecular weight was spin coated at 2000 rpm and dried at 200 °C on a hotplate. E-beam patterning was performed with acceleration voltage of 10 kV, using a 30 μm aperture. PMMA resist was developed in 1:3 methyl isobutyl ketone (MIBK) and isopropyl alcohol (IPA) solvent. Development was finished in IPA and the holograms were washed with water and dried with a compressed air flow. 3D microstructure created in PMMA was metallized with nickel film using electron beam evaporation. profile view of the Computer-generated microstructure and SEM image is presented in Fig. 16.1.

Fabricated computer-generated microstructure was replicated on the piezoelectric nanocomposite material multilayer specimen which was fabricated by using PZT nano powder 80% and binding material PMMA 20%. The synthesis and fabrication of multilayer nanocomposite was done according to technology developed

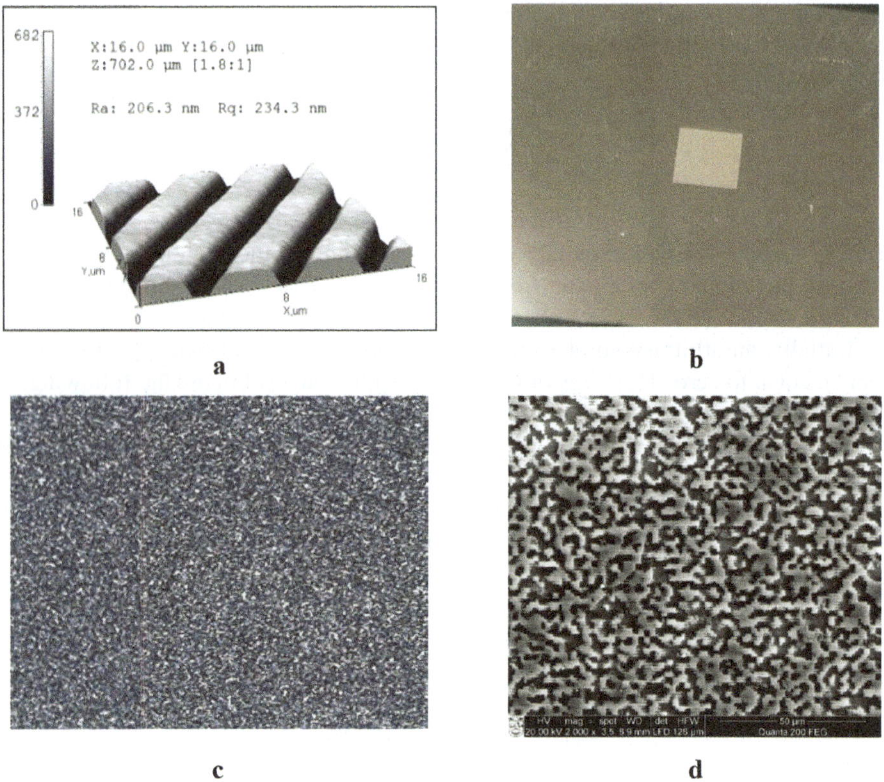

Fig. 16.1 CGH image of periodic microstructure: 3D (**a**) profile view (**b**) photo of produced microstructure (**c**, **d**) SEM image of CGH

and presented in [10]. For the replication of the computer-generated microstructure developed thermal imprinting technology [33] was used which allows to duplicate the CGM in the fabricated piezoelectric nanocomposite. The results obtained from the replication process optical microscope images are presented in the Fig. 16.2.

The developed microchannels on the piezoelectric nanocomposites has a microfluidic structure of period, 4 μm; depth, 0.56 μm. As discussed in the literature it is possible to transport the fluid in microchannel by using the surface acoustic wave phenomenon; standing acoustic wave and travelling acoustic wave. An example of the particle transportation in the microchannel is presented in the Fig. 16.3.

As, shown in the Fig. 16.3 the distilled water drops are passing through the microchannels indicated with arrows. So, by using the piezoelectric nanocomposites it is possible to control the inner geometry of the microchannel.

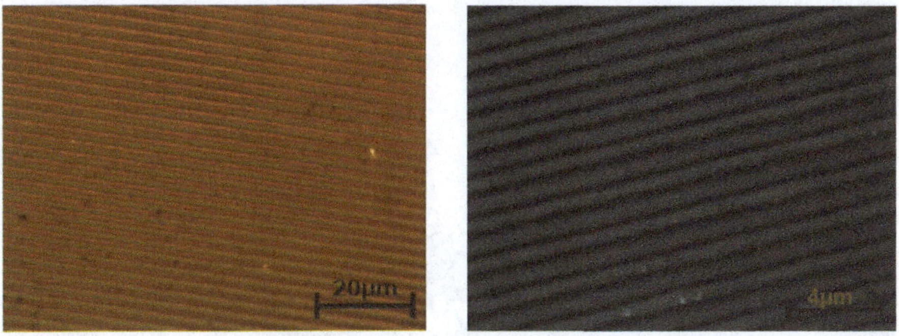

Fig. 16.2 Replica of the microstructure on the piezoelectric nano composite with different magnification

Fig. 16.3 Example of the transportation of fluid through microchannel

16.3 Nanoporous Aluminum Oxide Membrane

16.3.1 Fabrication of Nanoporous Aluminum Oxide Membrane

Anodization process for the nanoporous aluminum oxide membranes were performed in two steps. A natural aluminum smooth sheet used for the anodization process. Before anodization process, the aluminum sheet was ultrasonically cleaned in acetone and phosphoric acid for 5 and 3 min respectively. The first annotation took 1 h in oxalic acid. During the process non ordered holes were observed. This layer was removed by chromium acid at 55 °C temperature for 30 min. Second step anodization takes 24 h in oxalic acid and then unrated aluminum was removed by chloride acid at 55 °C for 10 min. According to literature distance between pores and

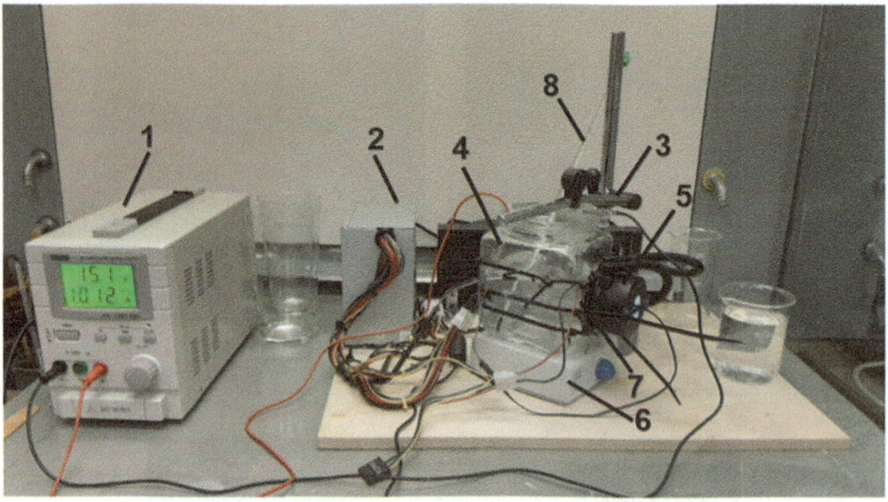

Fig. 16.4 Experimental setup for anodization process

diameter depends on the electrolyte used for anodization; also, 40–100 nm diameter with the inter pore distance of 80–200 nm was achieved by using oxalic acid [34].

Experimental setup of anodization process is presented in Fig. 16.4 consist of DC power unit (1), Power supply unit for peltier element (2), holder (3), glass jar for electrolyte bath (4), cooling device for peltier element (5), electromagnetic stirrer (6), peltier element for cooling inside the jar (7) and thermometer (8) for constant temperature measurement.

During the anodization process constant 60 V DC power was applied. Temperature inside the glass jar during the electrolyte bath was maintained by using peltier element (TEC112715) and to maintain the temperature of the hot side of peltier element master liquid cooling system (MLW-D24M) was attached.

The photo of the fabricated nanoporous aluminum oxide membrane is presented in Fig. 16.5.

Surface morphology of the fabricated nano membrane was investigated using AFM (atomic force microscopy). 2D and 3D view of the selected area $1.4 \times 1.4\,\mu m$ is presented in Fig. 16.6a, b.

Two step anodization using oxalic acid for fabrication of nano porous aluminum oxide membrane pore diameter 70 ± 20 nm, distance between pore 110 ± 10 nm, and hexagonal structure achieved. The obtain results are within the range of the theoretical values as mentioned in the literature [34] that means variation in the result does not exceed 10%.

Fig. 16.5 Photo of nanoporous aluminum oxide membrane formed using two step anodization

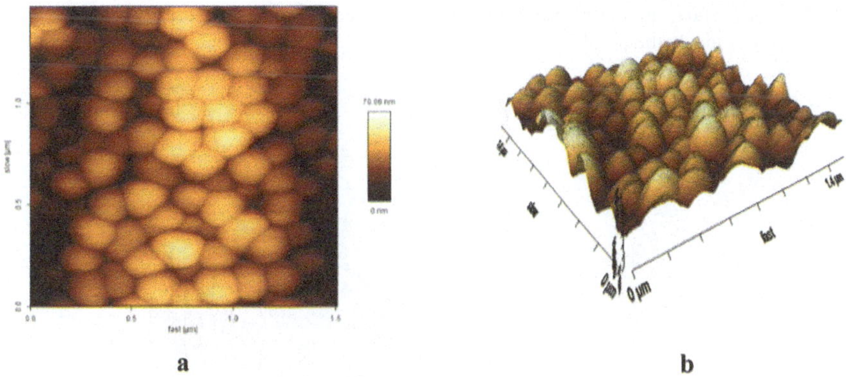

Fig. 16.6 Aluminum oxide nano membrane 2D view (**a**), and 3D view (**b**)

16.3.2 Simulation of Peltier Element and Nanoporous Aluminum Oxide Membrane

To understand the effect of temperature distribution using peltier element inside the glass jar simulation was performed using COMSOL Multiphysics. Model of the glass jar is presented in the Fig. 16.7.

Simulation was performed in isolated environment temperature 20 °C. In Fig. 16.7 peltier element (1) outer surface temperature 70 °C and inner surface 5 °C. Time dependent study was performed, the simulation results and the graph of the temperature vs time is presented in the Figs. 16.8 and 16.9 respectively.

From the simulation observed that temperature inside the jar after 30 s is 11 °C and after 10 min observed decrement in temperature up to 9 °C. Simulation graph is presented in the Fig. 16.9.

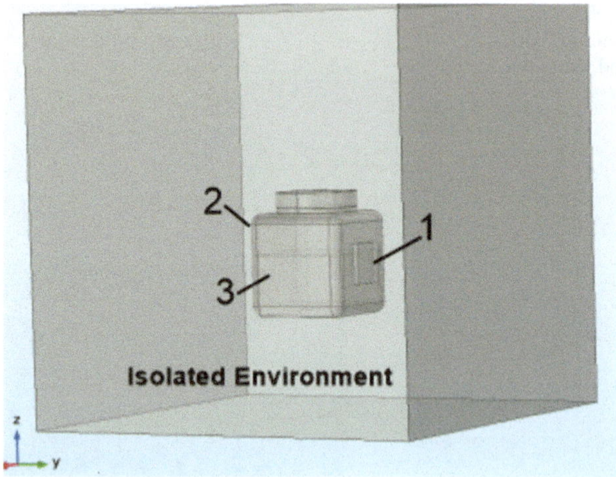

Fig. 16.7 Model of the simulation process for peltier element consists of: peltier element (1), glass jar (2), and water inside the jar (3)

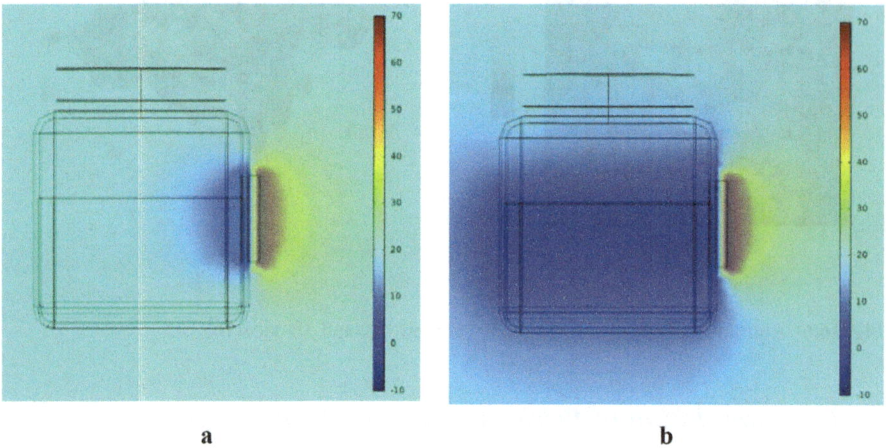

Fig. 16.8 Simulation results for the peltier element inside the jar: after 30 s (**a**) and after 10 min (**b**)

A single tube of these nanoporous aluminum oxide membrane was simulated by applying the acoustic pressure. Dimension of the tube diameter 40 nm and height 200 nm. as a liquid medium water assumed inside the tube and 1 Pa acoustic pressure was applied. Pressure distribution inside the channel presented in the Fig. 16.10.

From the result of the simulation of nanotube clearly shows that frequency 7.62 MHz (Fig. 16.10a) distribution of the pressure at starting of tube is higher than the lower side. on the other side frequency increased to 15.1 GHz (Fig. 16.10b) shows the tube has more deformed shape and equal and more stable distribution of the pressure throughout the nanotube. which may reduce the risk of the damaging

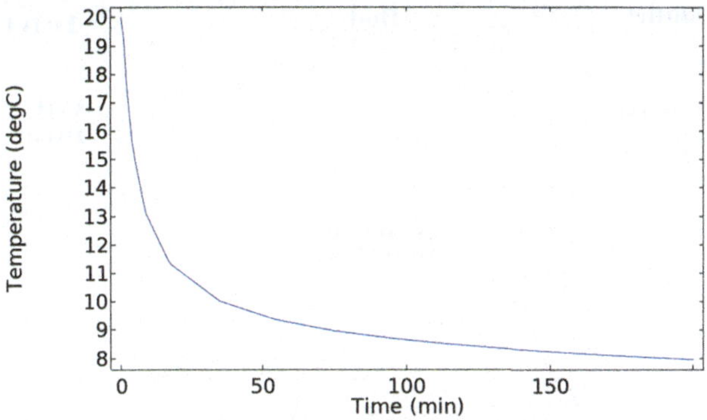

Fig. 16.9 Temperature decrement vs time graph for the peltier element

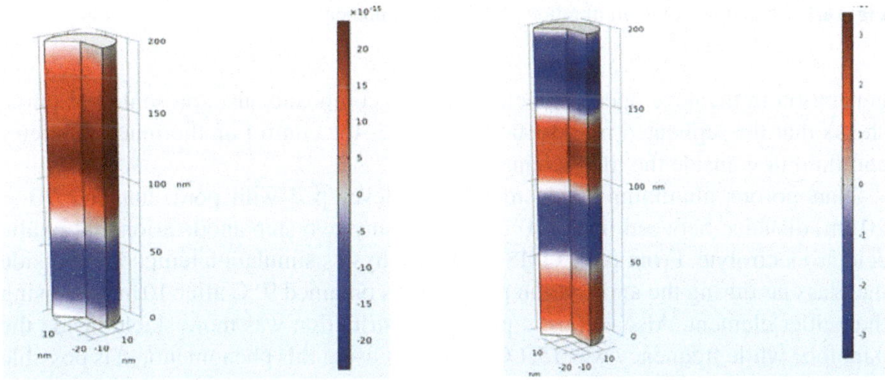

a b

Fig. 16.10 Pressure distribution inside the nanotube: frequency 7.62 MHz (**a**), and frequency 15.1 GHz (**b**)

cells during the separation or filtration process i.e. risk of the mechanical damage of cell during the filtration can be reduced.

Example of the oil filtration using nanoporous aluminum oxide nanomembrane is presented in Fig. 16.11. It consists of rigid body connected with pairs of nano membrane for oil filtration and separation of nano particle.

16.4 Conclusion

Computer generated microstructure was formed on the silicon and by using the thermal replication process microstructure was replicated on the piezoelectric nanocomposites. Used replication technique allow us to duplicate the microstructure (microchannel) formation with necessary and precise geometry for acoustic wave

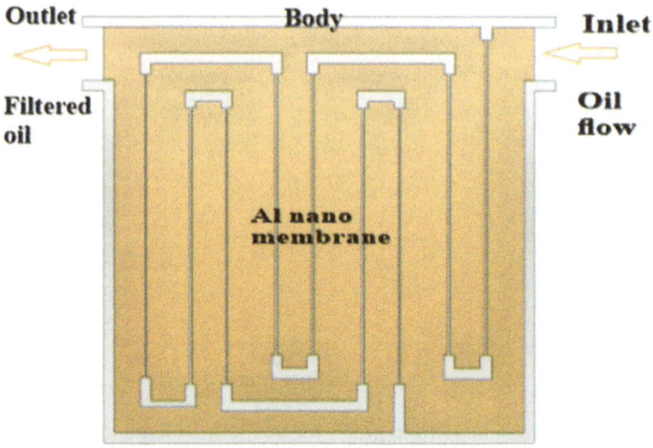

Fig. 16.11 Example of the oil filtration using nanomembrane

generation in the developed piezoelectric nano composite. also, presented example shows that the replicated microstructure enables the control on the inner geometry and fluid flow inside the microchannel.

Nanoporous aluminum oxide membrane developed with pore diameter 70 ± 20 nm, distance between pore 110 ± 10 nm using two step anodization and oxalic acid as electrolyte. From the COMSOL Multiphysics simulation temperature inside the glass jar during the anodization process was obtained 9 °C after 10 min by using the peltier element. Also, acoustic pressure distribution was more stable inside the nanotube while frequency was 15.1 GHz and by using this phenomenon it is possible to reduce the risk of the mechanical damage of the cells during filtration process in microfluidics.

Acknowledgments This research was funded by a Grant S-MIP-17-102 from the Research Council of Lithuania.

References

1. Hardt S, Schönfeld F (2007) Microfluidic technologies for miniaturized analysis systems. https://doi.org/10.1007/978-0-387-68424-6
2. Manz A, Graber N, Widmer HM (1990) Miniaturized total chemical analysis systems: a novel concept for chemical sensing. Sensors Actuators B Chem 1(1–6):244–248. https://doi.org/10.1016/0925-4005(90)80209-i
3. Ding X, Lin SCS, Lapsley MI, Li S, Guo X, Chan CYK, Chiang IK, L. Wang, McCoy JP, Huang TJ (2012) Standing surface acoustic wave (SSAW) based multichannel cell sorting. Lab Chip 12:4228–4231
4. Reichert P, Deshmukh D, Lebovitz L, Dual J (2018) Thin film piezoelectrics for bulk acoustic wave (BAW) acoustophoresis. Lab Chip 18:3655–3667

5. Nava G, Bragheri F, Yang T, Minzioni P, Osellame R, Cristiani I, Berg-Sørensen K (2015) All-silica microfluidic optical stretcher with acoustophoretic prefocusing. Microfluid Nanofluid 19:837–844
6. Ermakov V, Kruchinin S, Fujiwara A (2008) Electronic nanosensors based on nanotransistor with bistability behaviour. In: Bonca J, Kruchinin S (eds) Proceedings NATO ARW "Electron transport in nanosystems". Springer, Berlin, pp 341–349
7. Ermakov V, Kruchinin S, Hori H, Fujiwara A (2007) Phenomena in the resonant tunneling through degenarate energy state with electron correlation. Int J Mod Phys B 11:827–835
8. Nama N, Barnkob R, Mao Z, Kähler CJ, Costanzo F, Huang TJ (2015) Numerical study of acoustophoretic motion of particles in a PDMS microchannel driven by surface acoustic waves. Lab Chip 15:2700–2709
9. Belovickis J, Ivanov M, Samulionis V, Banys J, Solnyshkin A, Gavrilov SA, Nekludov KN, Shvartsman VV, Silibin MV (2018) Dielectric, ferroelectric, and piezoelectric investigation of polymer-based P(VDF-TrFE) composites. Phys Status Solidi (b) 255:1700196
10. Janusas G, Ponelyte S, Brunius A, Guobiene A, Vilkauskas A, Palevicius A (2016) Influence of PZT coating thickness and electrical pole alignment on microresonator properties. Sensors 16:1893
11. Baek C, Yun JH, Wang JE, Jeong CK, Lee KJ, Park KI, Kim DK (2016) A flexible energy harvester based on a lead-free and piezoelectric BCTZ nanoparticle–polymer composite. Nanoscale 8:17632–17638
12. Repetsky P, Vyshyvana IG, Nakazawa Y, Kruchinin SP, Bellucci S (2019) Electron transport in carbon nanotubes with adsorbed chromium impurities. Materials 12:524
13. Ermakov V, Kruchinin S, Pruschke T, Freericks J (2015) Thermoelectricity in tunneling nanostructures. Phys Rev B 92:115531
14. Filikhin I, Peterson T, Vlahovic B, Kruchinin SP, Kuzmichev YB, Mitic V (2019) Electron transfer from the barrier in InAs/GaAs quantum dot-well structure. Phys E Low Dimens Syst Nanostruct 114, Article 113629
15. Darinskii AN, Weihnacht M, Schmidt H (2017) Acoustomicrofluidic application of quasi-shear surface waves. Ultrasonics 78:10–17
16. Guo J, Kang Y, Ai Y (2015) Radiation dominated acoustophoresis driven by surface acoustic waves. J Colloid Interface Sci 455:203–211
17. Sazan H, Piperno S, Layani M, Magdassi S, Shpaisman H (2019) Directed assembly of nanoparticles into continuous microstructures by standing surface acoustic waves. J Colloid Interface Sci 536:701–709
18. Zheng T, Wang C, Xu C, Hu Q, Wei S (2018) Patterning microparticles into a two-dimensional pattern using onecolumn standing surface acoustic waves. Sensors Actuators A 284:168–171
19. Ostasevicius V, Janusas G, Palevicius A, Gaidys R, Jurenas V (2017) Biomechanical microsystems – design, processing and applications. Springer, Switzerland
20. Juknius T, Ružauskas M, Tamulevičius T, Šiugždinieté R, Juknienė I, Vasiliauskas A, Jurkevičiūté A, Tamulevičius S (2016) Antimicrobial properties of diamond-like carbon/silver nanocomposite thin films deposited on textiles: towards smart bandages. Materials 9(5):1–15
21. Bružauskaitė I, Bironaitė D, Bagdonas E, Skeberdis VA, Denkovskij J, Tamulevičius T, Uvarovas V, Bernotienė E (2016) Relevance of HCN2-expressing human mesenchymal stem cells for the generation of biological pacemakers. Stem Cell Res Ther 7:1–15
22. Danilevicius P, Rekstyte S, Balciunas E, Kraniauskas A (2012) Micro-structured polymer scaffolds fabricated by direct laser writing for tissue engineering. J Biomed Opt 17(8):81405
23. Stankevičius E, Gedvilas M, Voisiat B, Malinauskas M, Račiukaitis G (2013) Fabrication of periodic micro-structures by holographic lithography. Lith J Phys 53(4):227–237
24. Sun HB, Kawata S (2004) Two-photon photopolymerization and 3D lithographic microfabrication. Adv Polym Sci 170:169–273
25. Wu S, Serbin J, Gu M (2006) Two-photon polymerisation for three-dimensional microfabrication. J. Photochem Photobiol A 181:1–11
26. Malinauskas M, Gilbergs H, Žukauskas A, Purlys V, Paipulas D, Gadonas R (2010) A femtosecond laser-induced two-photon photopolymerization technique for structuring microlenses. J Opt 12(3):1–8

27. Silva DN, De Oliveira MG, Meurer E, Meurer MI, Da Silva JV, Santa Bárbara A (2008) Dimensional error in selective laser sintering and 3D printing of models for craniomaxillary anatomy reconstruction. J Cranio-Maxillofac Surg 36:443–449
28. Sin D, Miao X, Liu G, Wei F, Chadwick G, Yan C, Friis T (2010) Polyurethane (PU) scaffolds prepared by solvent casting/particulate leaching (SCPL) combined with centrifugation. Mater Sci Eng C 30:78–85
29. Liu X, Ma PX (2009) Phase separation, pore structure, and properties of nanofibrous gelatin scaffolds. Biomaterials 30:4094–4103
30. Salerno A, Oliviero M, Di Maio E, Iannace S, Netti P (2009) Design of porous polymeric scaffolds by gas foaming of heterogeneous blends. J Mater Sci Mater Med 20:2043–2051
31. Palevicius A, Janusas G, Ragulskis M, Palevicius P, Sodah A (2018) Design, analysis and application of dynamic visual cryptography for visual inspection of biomedical systems. In: Nanostructured materials for the detection of CBRN. Springer, Dordrecht, pp 223–232
32. Kruchinin SP, Klepikov VF, Novikov VE, Kruchinin DS (2005) Nonlinear current oscillations in a fractal josephson junction. Mater Sci 23(4):1009–1013
33. Sodah A, Palevicius A, Janusas G, Jurenas V, Patel YR (2018) Sonotrode for formation of piezocomposite functional elements. In: Mechanika 2018: proceedings of the 23rd international scientific conference, pp 150–153
34. Cazorla PH, Fuchs O, Coschet M, Maubert S, Le Rhun G, Foullet Y, Defay E (2014) Integration of PZT thin films on a microfluidic complex system. In: IEEE international ultrasonics symposium proceedings, pp 491–494

Chapter 17
Nanopore-Penetration Sensing Effects for Target DNA Sequencing via Impedance Difference Between Organometallic-Complex-Decorated Carbon Nanotubes with Twisted Single-Stranded or Double-Stranded DNA

A. S. Babenko, H. V. Grushevskaya, N. G. Krylova, I. V. Lipnevich, V. P. Egorova, and R. F. Chakukov

Abstract We offer a highly sensitive and reproducible dielectric-spectroscopy assay of deoxyribonucleic acid (DNA) sequence on a platform of quantum graphene-like structures arranged on nanoporous alumina to correctly identifying an infectious agent in a native double-stranded (ds) DNA. The hybridization of complementary target DNA with probe DNA in the sensor sensitive layer leads to penetration of the formed single-stranded (ss) target DNA into the underlayer nanoporous anodic alumina through the nanocavities of LB-film from organometallic complexes. This results in linking of MWCNT ends, shielding of Helmholtz double layer and following decrease of electrical capacitance of the sensor. The novel electrochemical impedimetric DNA sensor with self-organized multi-walled carbon nanotube (MWCNT) bundles decorated by organometallic complexes as transducer has been utilized to detect the viral DNA in the biological samples of patients with virus infection at DNA concentration as low as 1.0–1.3 ng/μL.

A. S. Babenko
Bioorganics Department, Belarusian State Medical University, Minsk, Belarus

H. V. Grushevskaya (✉) · N. G. Krylova · I. V. Lipnevich · R. F. Chakukov
Physics Department, Belarusian State University, Minsk, Belarus
e-mail: grushevskaja@bsu.by; nina-kr@tut.by; lipnevich@bsu.by

V. P. Egorova
Belarusian State Pedagogical University, Minsk, Belarus

J. Bonča, S. Kruchinin (eds.), *Advanced Nanomaterials for Detection of CBRN*,
NATO Science for Peace and Security Series A: Chemistry and Biology,
https://doi.org/10.1007/978-94-024-2030-2_17

Keywords Multi-walled carbon nanotube (MWCNT) · Nanopore-penetration sensing effect · Double-stranded DNA · Single-stranded DNA

17.1 Introduction

Advances in molecular biology in recent decades are connected with utilizing third-generation DNA-nanosequencing label-free methods [1]. But, the sensitivity of these methods appears to be not enough to recognize a viral infection at the stage of fewness lesions of body cells (so called the window period, or the serologic window). In addition the modern DNA-sequencing methods take quite a lot of time for target viral genome identification. Human parvovirus infection leads to the serious complications, including transient aplastic crisis, chronic anemia, and fetal death. In such cases effectiveness of the infectious diseases treatment often depends on the correct identifying an infectious agent, namely on the performance of medical diagnostic methods. High-sensitive methods to detect viral infection is challenge.

To reveal human parvovirus infection for medical practice we offer a dielectric spectroscopy method based on highly selective hybridization interactions of noncovalent complementary single-stranded ss-DNA molecules. Interacting probe ss-DNA and target genomic linear ss-DNA of samples under investigation form a ds-DNA helix of homoduplex on sensor surface for detection of parvovirus infection. The novel high-sensitive method reliably detects the presence or absence of parvovirus in a sample and does not demand expensive consumable materials.

The goal of the paper is to study effects of penetration of DNA in nanopores at electrochemical DNA sensing on organometallic-complex-decorated MWCNTs deposited on a nanoporous surface. We will utilize the novel label-free electrochemical DNA-nanosensor based on carbon nanotubes (CNTs) to identify viral status of native genomic DNA via impedance difference between the metal-decorated MWCNTs with twisted ss- or ds-DNA.

17.2 Materials and Methods

17.2.1 Reagents

We utilized two types of ss-DNA probes to recognize parvovirus sequences: direct primer sequence to $3' \rightarrow 5'$–ss-DNA-chain and revers one to $5' \rightarrow 3'$–ss-DNA-chain. The direct and revers primers are denoted through $B19VF4$ and $B19VR4$, respectively. The ds-DNA samples have been obtained from blood serum of patients with parvovirus infection (DNA_{pvi}, $i = 1, 2, 3$) and of practically healthy donors (DNA_{hi}, $i = 1, 2, 3$) as negative control. Spectrophotometric data for the infection DNA samples are presented in Table 17.1. The spectrofluorimetric method was used

Table 17.1 Optical densities OD_λ and their differences for different wavelengthes λ and concentration C for parvovirus infection DNA samples

Type of sample	$(OD_{260} - OD_{320})/(OD_{280} - OD_{320})$	OD_{320}	C, μg/mL
DNA_{pv1}	1.8	0.05	4.5
DNA_{pv2}	1.9	0.23	3.8

also to measure the concentration of DNA_{hi}, $i = 1, 2, 3$. The concentration of DNA was estimated about 4.5–5.0 ng/μL.

To calibrate a label-free sequencing an oncogene KRAS, native DNA isolated from colon-cancer tumor, and placental DNA were utilized as a marker gene. The tumor tissue of patients with established diagnosis of colon cancer carrying a mutation single nucleotide polymorphism (SNP) in the second KRAS-exon, codon 12, GGT>GAT were used. Probe DNA $KRAS_m$ is a label-free probe oligonucleotide sequence for the KRAS-gene with SNP. RNA and proteins contents in high-purity ds-DNA (1.03 mg/ml in 10^{-5} M Na_2CO_3 buffer medium) isolated from placenta tissue of healthy donors were less that 0.1% (optical density ratio $D_{260}/D_{230} = 2.378$ and $D_{260}/D_{280} = 1.866$, respectively).

All DNA probes were purchased in "Primetech ALC" (Minsk, Belarus). Length of the oligonucleotides does not exceed 20 nucleotides.

To construct an electrochemical transducer MWCNTs are selected from MWCNTs with diameters ranging from 2.0 to 5 nm and length of ~2.5 μm. The original MWCNTs and single walled CNTs (SWCNTs) were obtained by the method of chemical vapor deposition (CVD-method) [2]. MWCNTs were covalently modified by carboxyl groups and non-covalently functionalized by stearic acid molecules. Salts $Fe(NO_3)_3 \cdot 9H_2O$, $Ce_2(SO_4)_3$ (Sigma, USA), hydrochloric acid, deionized water were used to preparate subphases. Iron-containing films were fabricated from an amphiphilic oligomer of thiophene derivatives with chemically bounded hydrophobic 16-link hydrocarbon chain: 3-hexadecyl-2,5-di(thiophen-2-yl)-1H-pyrrole (H DTP, H-dithionilepyrrole). H-dithionilepyrrole was synthesized by a method proposed in [3]. Working solution of H-dithionilepyrrole, 1.0 mM, was prepared by dissolving precisely weighted substances in hexane. All salt solutions have been prepared with deionized water with resistivity 18.2 MΩ·cm.

All used materials belong to class of analytical pure reagents.

17.2.2 Methods

Impedance Measurements. Electro-physical studies have been performed using planar interdigital electrode structures on pyroceramics support. N pairs, $N = 20$ of aluminum electrodes are arranged in an Archimedes-type spiral configuration. Every such pair is an "open type" capacitor. The dielectric coating of the electrodes represents itself a nanoporous anodic alumina layer (AOA) with a pore diameter of 10 nm. To excite harmonic auto-oscillations of electric current (charging-discharging processes in the capacitors), the sensor was connected as the capacitance C into the relaxation resistance (R) – capacitor (C) oscillator (self-excited RC-oscillator) [1, 4]. Operating of such RC-generator is based on the principle of self-excitation of an amplifier with a positive feedback on the quasi-resonance frequency. The capacitance C of the sensor entered in measuring RC-oscillating circuit has been calculated by the formula $C = 1/(2\pi Rf)$, where R is the measuring resistance, f is the frequency of quasi-resonance.

Biosensitive nanostructured layers which transduce hybridization signals have been fabricated by Langmuir–Blodgett (LB) technique. The biosensitive coating consists of five monomolecular LB-layers (LB-monolayers) fabricated from nanocyclic complexes of high-spin octahedral iron with dithionylpyrrole (DTP) ligands [5]. Complexes of carboxylated hydrophilic MWCNT with different DNA-probes have been deposited on the LB-film of metal-containing conducting dithionylpyrrole polymer [4]. The synthesized LB-nanoheterostructures were suspended on the interdigital electrode system. The fabricated capacitive DNA-nanosensors are sensors of non-Faraday type.

The response to the interactions between the DNA samples and the viral DNA probes has been detected on nanosensors F14, F24, F32 with the direct primers as DNA probes and on the nanosensors R0, R4, R18 with the revers primers as DNA probes. A DNA-nanosensor K11 with the mutant DNA probe $KRAS_m$ recognized SNP in the DNA samples of colon cancer tissue. All electrochemical measurements are performed in deionized water.

All results were confirmed by the method of sequencing by Sanger. We have discriminated mutation and wild type in 100% of 20 samples.

Langmuir–Blodgett technique. A fabrication of the LB-monolayers was carried out on an automated hand-made Langmuir trough with controlled deposition on a substrate, and with computer user interface working under Microsoft Windows operational system. Control of the surface tension has been performed by a highly sensitive resonant inductive sensor. The Y-type transposition of monolayers on supports was performed by their vertical dipping. The complexes $Fe(II)DTP_3$ of high-spin Fe(II) with DTP ligands were synthesized by LB-method at compression of H-dithionilepyrrole molecules on the surface of subphase with salts of three-valence Fe [5]. Horizontally and vertically arranged LB-MWCNT-bundles can be fabricated from the carboxylated multi-walled CNTs [6–11].

We use the LB-technique to deposit two LB-monolayers of stearic acid micelles with DNA/MWCNT complexes inside on five-monolayer LB-film of the organometallic Ce-containing Fe(II)DTP-complexes.

Fabrication of micellar DNA/MWCNT complexes. The micellar complexes ds-DNA/MWCNT and oligonucleotide/MWCNT were obtained by means of ultrasonic treatment of alcoholic solution of ds-DNA or oligonucleotide with MWCNT [6]. Then, the complexes were mixed with a solution of stearic acid in deionized water or in hexane. The resulting mixtures were homogenized by ultrasonic treatment to form hydrophilic or hydrophobic (reverse) micelles of stearic acid with complexes ds-DNA/MWCNT or oligonucleotide/MWCNT inside them [12].

Raman spectroscopy studies. Spectral studies in visible range were carried out using a confocal micro-Raman spectrometer Nanofinder HE ("LOTIS-TII", Tokyo, Japan–Belarus) by laser excitation at wavelengths 355, 473 and 532 nm with power in range from 0.0001 to 20 mW at room (RT) and low temperatures.

17.3 Results

17.3.1 Transducer Characterization

A transducer of electrochemical DNA-sensor is a layered nanoheterostructure consisting of two LB-monolayers of complexes MWCNT/DNA-probe which are deposited on five LB-monolayer of nanocyclic complexes of octahedral high-spin Fe(II) with the dithionilepyrrole ligands. Pyrrole rings of conducting polymer are able to reversible oxidation and reduction [13]. Cation-active (cationic) oxidized pyrrole rings holds dsDNA molecules fixed [14]. This self-redox activity provides DNA fixation during enough time for complementary hybridization, electrostatic repulsion of non-specific-bounded target DNA molecules and leaving from sensor surface in the phase of pyrrole reduction.

Characteristic frequencies of Raman light scattering for the CNTs, DNA, complexes DNA/MWCNT, and MWCNT-bundles arranged in stearic-acid micelles and thin LB-films are presented in Tables 17.2, 17.3, and 17.4. Comparison of Raman spectra of SWCNTs, original MWCNTs, and micellar MWCNTs shown in Table 17.2 demonstrates that the sensitive layer contains CNTs with some number of walls (about two walls). MWCNTs, which has been selected for transducer, practically do not contain impurities, as characteristic vibration mode D are absent in Raman spectrum of micellar MWCNTs. Since a part of MWCNT charge carriers is localized on the support defects the peak D appears in the Raman spectrum of LB-MWCNT-film at laser excitation with wavelength $\lambda = 532$ nm (see Table 17.2). However, decreasing in amplitude the peak D shifts to low frequencies at laser excitation with wavelength $\lambda = 355$ nm (Table 17.2). It testifies that the charge carriers confined on the MWCNT surface can not participate in charge transport onto high-excited impurity levels. Presence of them on the lower excited impurity DNA-levels due $\pi - \pi$-interactions quenches light scattering with wave length $\lambda = 532$ nm in both ds-DNA and oligonucleotides (Tables 17.3 and 17.4).

Quinone can direct interact with DNA by forming covalently bonds with DNA bases as menadione, p-benzoquinone and mitomycin C [23–25] or by intercalating into DNA helix as anthracyclines [26]. Quinones can also interact with CNTs due to $\pi - \pi$ stacking onto the polyaromatic surface of nanotubes [27]. We utilized such a quinone as thymoquinone to prove the presence of recognized complementary target ss-DNA in the senor covering. According data in Fig. 17.1a, thymoquinone interacts with MWCNTs functionalized by DNA because decrease in the sensor capacity occurs at addition of thymoquinone at concentrations 1–50 μmol/L. Since the

Table 17.2 Characteristic molecular vibrations, observed in the Raman spectra of different samples with CNTs at RT, laser excitation with different wavelengths λ

Sample	Bands, cm^{-1}	Assignment or vibration peaks of similar molecular groups, cm^{-1}; λ
Dry SWCNTs*	179(RBM), 1343(D), 1592(G), 2444(D″+D), 2680(2D), 2930(D″D), 3184(2G).	λ = 532 nm
	176.6(RBM), 1352(D), 1567, 1591(G), 2446 (D″+D), 2704(2D), 2930(D″D), 3182(2G)	λ = 473 nm
Dry MWCNTs	1328(D), 1566(G), 2450 (D″+D), 2660(2D), 2890 (D′+D)	[15]; λ = 532 nm
SA micelles	1064.5, 1180, 1298.6, 1440.2, 1459.3, 2847, 2882	Pure SA: 863.2, 885.3, 1065.8, 1079.1, 1121.4, 1180.5, 1299.0, 1422.9, 1440, 1459, 1664.6, 2852.4, 2873.9, 2898.3, 2925.4 [16, 17]; λ = 532 nm
MWCNTs inside SA-micelles	163(RBM), 1575(G), 2430(D″+D), 2648(2D), 3152(2G)	CNT with diameter 2–2.5 nm, pick *D* is practically absent [18]; λ = 532 nm
Thin LB-film from MWCNTs and SA	1350(D), 1573(G), 2406.68(D″+D), 2683(2D), 2923(D′+D).	Two LB-monolayers; λ = 473 nm
	1403(D), 1586(G), 2450 (D″+D), 2810(2D), 2983(D′+D)	Three LB-monolayers; λ = 355 nm

*Vibration modes of CNT are labeled in the following way. RBM is a radial breathing mode; peaks D, 2D, D′+D, D″+D, and D″D are vibrations near $K(K')$-point in a graphene Brillouin zone; peaks G, 2G are vibrations near Γ-point in the zone

MWCNTs are covered by a dense layer of the ss-DNA probe $KRAS_m$ this screening testifies that the quinone penetrates into DNA/MWCNT complexes. Hence, the homoduplexes between $KRAS_m$ and complementary target ss-DNA of colon cancer tumor with SNP are formed on the sensor surface K11.

17.3.2 Electrochemical Impedance Spectroscopy Analysis

The DNA-nanosensors were placed into deionized water. An electrical double (Helmholtz) layer is formed on the interface. Typical frequency dependencies of the sensor capacity are presented in Fig. 17.1b, c. The principe of target DNA sequence detection is based on shielding near-electrode Helmholtz layer that leads to decrease electric capacitance of double layer in a case of complementary target

Table 17.3 Characteristic molecular vibrations, observed in the Raman spectra of native DNA and complexes native DNA/MWCNT at different temperatures *T*, laser excitation with different wavelengths λ

Sample	Bands, cm^{-1}	Assignment or vibration peaks of similar molecular groups, cm^{-1}; λ; *T*
Dry placental DNA*	663.6(T,Gu,A), 728.2(A), 783(C), 802, 881.6, 963.6, 1014(d), 1060.64(d(CO)), 1100(DP), 1141.25, 1181.3(T,C), 1208.9, 1245.67(T), 1303.53(A), 1334(A), 1373.1(T,A,Gu), 1418.2(T,d(CH2)), 1442, 1460(d(CH2)), 1484(A,Gu), 1504.8(A), 1528.6(C), 1573(A,T), 1661(T,Gu(C=O)), 2747(d(CH)), 2894(d(CH)), 2955.36(d(CH))	DNA in A-form DNA; λ = 473 nm, RT
Dry calf DNA	663(T,Gu,A), 682(G), 727(A), 743(T,d), 783(C), 1012(d), 1060(d(CO)), 1100 (DP), 1181(T,C), 1243(T), 1308(A), 1335(A), 1372(T,A,Gu), 1418(T, d(CH2)), 1460(d(CH2)), 1484(A,Gu), 1508(A), 1528(C), 1574(A,T), 1664(T,Gu(C=O)), 2894(d(CH)), 2950(d(CH)), 2957(d(CH))	DNA in A-form [19, 20]
Placental DNA inside SA-micelles	835, 880, 920, 1020, 1050(d(CO)), 1100(DP), 1140, 1185(T,C), 1245(T), 1270(C,A), 1300(A), 1340(A), 1370(Gu,T,A), 1420(A,Gu), 1438(Gu), 1480(A,C), 1490(A,Gu), 1520(A), 1580(Gu,A), 1607(A), 1641(Gu,T), 2721(d(CH)), 2845.6, 2880.7(d(CH)), 2921(d(CH))	Deprotonated DNA at RT; λ = 532 nm. 1605 and 1609 are the characteristic vibrations of the group C=N in adenine of DNA and of the group NH_2 of ds-DNA [12, 18, 21, 22]; 2845.6 is the characteristic frequency of stearic acid
Placental-DNA-encased MWCNTs inside SA-micelles	1343(D), 1573(G), 2460(D″+D), 2686(2D), 2930(D′+D)	λ = 532 nm, RT
	1358(D), 1582(G), 2446(D″+D), 2710(2D), 2927(D′+D)	λ = 473 nm, RT
	1400(D), 1581(G), 2361(D″+D), 2810(2D), 2970(D′+D)	λ = 355 nm, RT
	1350(D), 1587(G), 2467(D″+D), 2698(2D), 1091.1(DP), 1124.5, 1175(T,C), 1621(Gu(C=0),T)	Red shift of frequencies for guanine upon cooling [20] λ = 532 nm, 30 K
	1400(D), 1583(G), 2415(D″+D), 2800(2D), 3190(2G), 1049(d(CO)), 1100(DP), 1294(A), 1327(A), 1747, 2090(d(CH)), 2701(d(CH)), 2949(d(CH))	Red shift of frequencies for Adenine upon cooling [20]; Red shift of band d(CO); λ = 355 nm, 26 K

*Molecular groups of DNA are labeled in the following way. A, Gu, C, and T are the nucleobases adenine, guanine, cytosine, and thymine, respectively; d is deoxyribose; DP is phosphodiester bond

Table 17.4 Characteristic molecular vibrations, observed in the Raman spectra of complexes oligonucleotide/MWCNT at laser excitation with wavelengths $\lambda = 532$ nm and RT

Sample	Bands, cm^{-1}	Assignment
Oligonucleotide-encased MWCNTs inside SA-micelle	1340(D), 1572(G); 1605(A)	MWCNTs adenine
LB-film from oligonucleotide-encased MWCNTs and SA	1340(D), 1571(G); 1605(A)	MWCNTs adenine

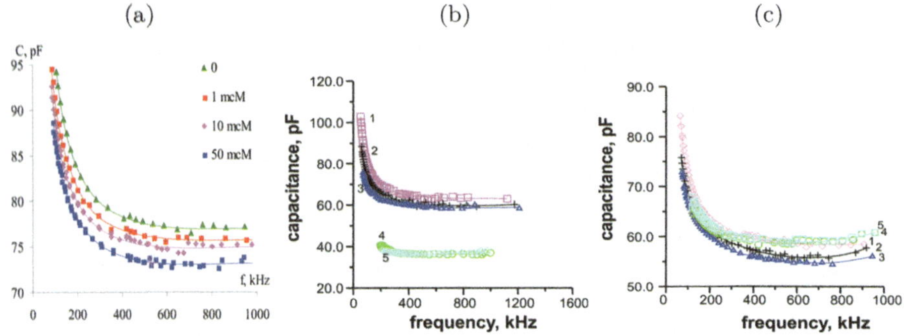

Fig. 17.1 Bode capacity plots for DNA-sensors. (**a**) DNA-sensor K11 with homoduplexes between probe DNA KRAS*m* and recognized colorectal cancer tumor DNA after 30 min exposure of thymoquinone at different concentrations. (**b–c**) DNA-sensor R18 (**b**) and F32 (**c**): 1 – pure sensor, 2 – sensor with electrodes coated by LB-film from nanocyclic organometallic complexes, 3 – sensor with electrodes coated by LB-film from DNA probe allocated on MWCNTs decorated by metal atoms, 4 – deposition of 1.3 ng/μL DNA_{pv1} (**b**) or 0.7 ng/μL DNA_{h1} (**c**), 5 – deposition of 2.6 ng/μL DNA_{pv1} (**b**) or 1.4 ng/μL DNA_{h1} (**c**)

and probe ss-DNAs, as in the case of ds-DNA isolated from blood of patients with parvovirus infection (ds-DNA_{pv1}) (Fig. 17.1b). The hybridization of DNA probe molecule with target DNA molecule that is ss-DNA released at denaturation of the original ds-DNA proceeds on the sensor surface. The complementary target ss-DNA bound with the ss-DNA probe $B19VR4$ of sensor penetrates in the sensitive layer of transducer through the nanocavities of LB-DTP-film into nanopores of AOA with subsequent binding of MWCNTs ends and a shielding effect emerges. In the individual case the capacitance decrease has been registered for DNA sample with parvovirus only at DNA concentration 3 ng/μL. The decrease of sensor capacitance indicates the complementary interaction of revers-primer probe DNA with viral DNA from patient's DNA samples.

A target ss-DNA which is noncomplementary to the ss-DNA probe reacts very quickly with another complementary ss-DNA forming the original target ds-DNA. Since a diameter of this ds-DNA is larger than the nanocavities, the shielding effect is absent and a capacitance of sensor increases (Fig. 17.1c).

The dielectric loss is measured as the inverse capacity C^{-1} of sensor. Spectra of Cole-Cole plots are dependencies of the dielectric losses on signal power W. These dependencies correspond to the dependencies of dielectric loss constant on the real part of complex dielectric permeability (dielectric dispersion). As one can see in Fig. 17.2, the Cole-Cole plots of pure sensors without sensitive coating are characterized by the presence of three Cole-Cole plots with characteristic frequencies $\lambda_0, \lambda_1, \lambda_2$ of dipole relaxation in the range in signal power value W from 2 to 25 V*V. In addition to a capacitance of electrically charged Helmholtz double layer, there is a Warburg impedance element of diffusion layer at signal power more than 40 V*V.

Deposition of sensor coating leads to appearing of additional frequencies λ_p and λ_{ON} of relaxation oscillations of dipoles in the LB-DTP-film and oligonucleotide-MWCNT-LB-film, as Fig. 17.2 demonstrates. In Fig. 17.2c, d the data of electrochemical response of the sensors on hybridization of probe oligonucleotides with DNA isolated from the blood of patient with parvovirus infection are presented. As one can see, hybridization of oligonucleotide with complementary viral DNA results in appearing additional frequency λ_{DNA} of dipole relaxation at signal power 17–19 V*V. For revers primer $B19VR4$ along with the appearing the Cole–Cole plots λ_{DNA} dielectric losses increase (screening effect).

The characteristic Cole–Cole plots λ_{DNA} are absent at non-complementary hybridization between target DNA of all samples obtained from healthy donors and both the DNA probes $B19VR4$ and $B19VF4$ (see Fig. 17.2a, b). Since dielectric losses decrease and accordingly the screening effect is absent, an increase of capacity emerges after the noncomplementary hybridization.

17.4 Discussion and Conclusion

The Raman and impedance spectroscopic assays demonstrate that charge CNT-carriers are confined on CNT-surface. Quinone, intercalating into ds-DNA helix, keeps sterically its nucleosides away from CNT-surface. Since the CNT charge carriers can not be transported (localized) on the remote impurity defects, a number of free CNT charge carriers increases by the number of charge carriers localized before. Meanwhile $\pi - \pi$-bonds between nucleosides and CNT-surface break. Due to attenuation of $\pi - \pi$-interactions conformational mobility of DNA increases and a conformation of DNA molecules attaching to CNT-ends that MWCNTs linked with DNA form a network is an energy-efficient DNA configuration. Hopping conduction of DNA appears after doping in the sites of contact between the DNA molecules and the end groups. The number of contacts increases with thymoquinone concentration (see Fig. 17.1a). Now, a transport of electrical charge occurs along two systems. Since both of systems are high-conductivity ones, screening of near-electrode Helmholtz layer by the DNA–CNT network is more effective than by the MWCNT LB-bundles. The native-DNA sequencing performed and presented in the paper is based on this double shielding of external electrical fields. The DNA doped by CNT-ends on sensor surface results from conformal wandering of complementary ss-DNA which penetrates through nanocavities in the LB-film of nanocyclic compound Fe(II)DTP in the nanoporous AOA.

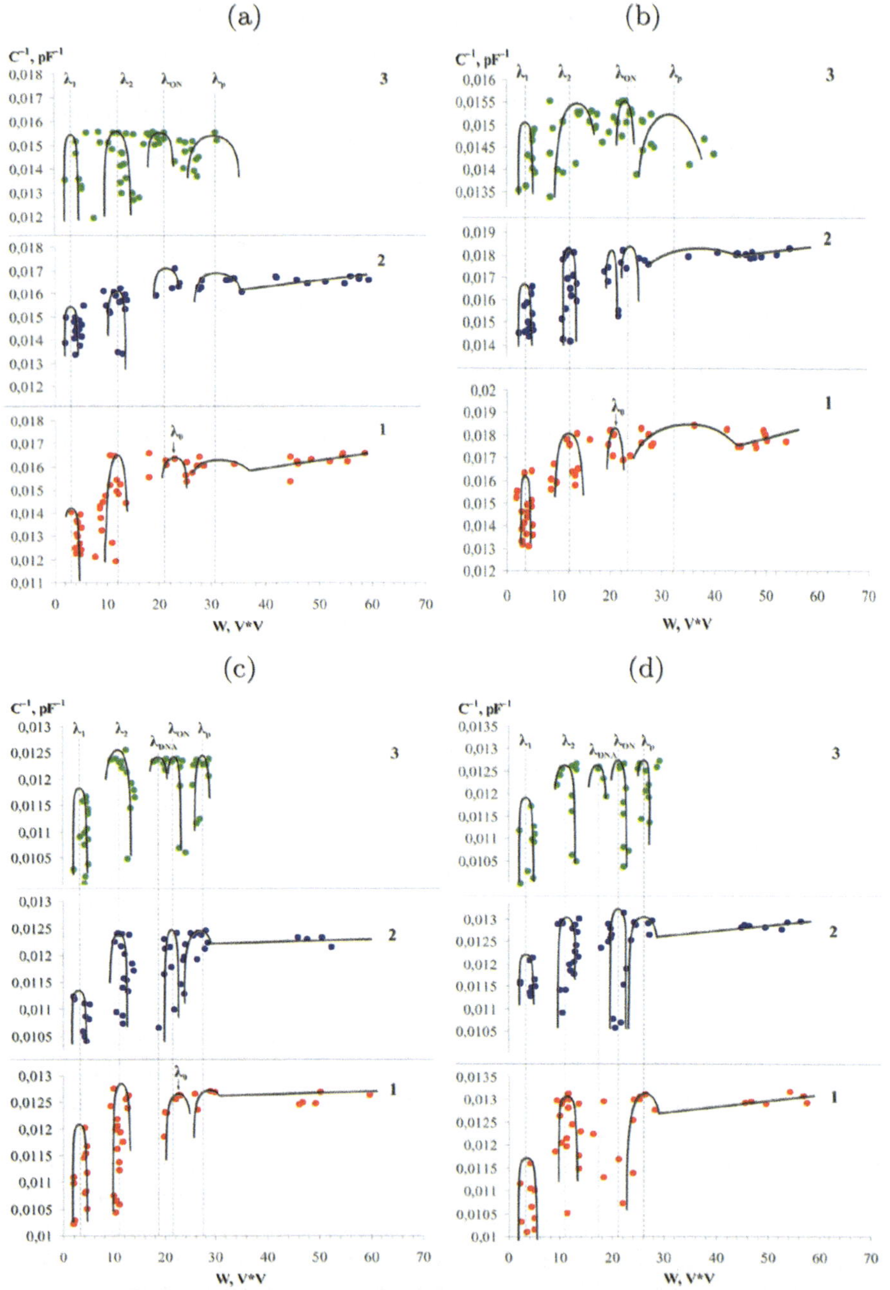

Fig. 17.2 Dielectric spectra with Cole–Cole plots for sensors with LB-DTP-film (curve 1), coated with LB-CNT-film with oligonucleotide before (curve 2) and after hybridization with DNA (curve 3). (**a**, **b**) DNA isolated from healthy donors: control DNA_{h2} (**a**) on sensor R0 and DNA_{h3} (**b**) on sensor F24; deposition of 0.7 ng/μL. (**c**, **d**) DNA isolated from the blood of patients with parvovirus infection, deposition of 1.0 ng/μL DNA_{pv2} on sensor R4 (**c**) and sensor F14. Characteristic maxima of the plots are λ_0, λ_1, λ_2 for AOA; λ_p for LB-DTP-film, λ_{ON} for DNA probe, λ_{DNA} for target DNA

So, the proposed dielectric-spectroscopy method allows to detect the presence or absence of target viral infection in DNA samples and can be used as an alternative laboratory diagnostic method.

References

1. Grushevskaya HV et al (2019) Single-molecule EIS-sequencing of DNA on composite nanoporous structures: advances and perspectives. Sci Innov 4(194):23 (in Russian)
2. Labunov V, Shulitski B, Prudnikava A, Shaman YP, Basaev AS (2010) Composite nanostructure of vertically aligned carbon nanotube array and planar graphite layer obtained by the injection CVD method. Quantum Electron Optoelectron 13:137
3. Kel'in A, Kulinkovich O (1995) A new synthetic approach to the conjugated five numbered heterocycles with long-chain substituents. Folia Pharm Univ Carol (supplementum) 18:96
4. Grushevskaya HV, Krylova NG, Lipnevich IV, Babenka AS, Egorova VP, Chakukov RF (2018) CNT-based label-free electrochemical sensing of native DNA with allele single nucleotide polymorphism. Semiconductors 52(14):1836
5. Grushevskaya HV, Lipnevich IV, Orekhovskaya TI (2013) Coordination interaction between rare earth and/or transition metal centers and thiophene series oligomer derivatives in ultrathin Langmuir–Blodgett films. J Mod Phys 4:7
6. Egorov AS, Krylova HV, Lipnevich IV, Shulitsky BG, Baran LV, Gusakova SV, Govorov MI (2012) Structure of modified multi-walled carbon nanotube clusters on conducting organometallic Langmuir–Blodgett films. J Nonlin Phenom Complex Syst 15:121
7. Repetsky SP, Vyshyvana IG, Nakazawa Y, Kruchinin SP, Bellucci S (2019) Electron transport in carbon nanotubes with adsorbed chromium impurities. Materials 12:524
8. Repetsky SP, Vyshyvana IG, Kruchinin SP, Bellucci S (2018) Influence of the ordering of impurities on the appearance of an energy gap and on the electrical conductance of graphene. Sci Rep 8:9123
9. Filikhin I, Peterson TH, Vlahovic B, Kruchinin SP, Kuzmichev YuB, Mitic V (2019) Electron transfer from the barrier in InAs/GaAs quantum dot-well structure. Phys E Low-dimensional Syst Nanostruct 114:113629
10. Ermakov V, Kruchinin S, Pruschke T, Freericks J (2015) Thermoelectricity in tunneling nanostructures. Phys Rev B 92:115531
11. Kruchinin S, Pruschke T (2014) Thermopower for a molecule with vibrational degrees of freedom. Phys Lett A 378:157–161
12. Egorov AS, Egorova VP, Grushevskaya GV, Krylova NG, Lipnevich IV, Orekhovskaya TI, Shulitsky BG (2016) CNTenhanced Raman spectroscopy and its application: DNA detection and cell visualization. Lett Appl NanoBioSci 5:343
13. Tsai YT, Choi CH, Gao N, Yang EH (2011) Tunable wetting mechanism of polypyrrole surfaces and low-voltage droplet manipulation via redox. Langmuir 27:4249
14. Jeon SH, Lee HJ, Bae K, Yoon K-A, Lee ES, Cho Y (2016) Efficient capture and isolation of tumor-related circulating cell-free DNA from cancer patients using electroactive conducting polymer nanowire platforms. Theranostics 6:828
15. Cooper DR et al (2012) Experimental review of graphene. ISRN Condens Matter Phys 2012:501686
16. Potcoava MC, Futia GL, Aughenbaugh J, Schlaepfer IR, Gibson EA (2014) Raman and coherent anti-Stokes Raman scattering microscopy studies of changes in lipid content and composition in hormone-treated breast and prostate cancer cells. J Biomed Opt 19(11):111605
17. Deepika, Hait SK, Chen Y (2014) Optimization of milling parameters on the synthesis of stearic acid coated $CaCO_3$ nanoparticles. J Coat Technol Res 11(2):273

18. Grushevskaya HV, Krylova NG, Lipnevich IV, Egorova VP, Babenka AS (2018) Single nucleotide polymorphism genotyping using DNA sequencing on multiwalled carbon nanotubes monolayer by CNT-plasmon resonance. Int J Mod Phys B 32(17):1840033
19. Prescott B, Steinmetz W, Thomas GJ Jr (1984) Characterization of DNA structures by laser Raman spectroscopy. Biopolymers 23(2):235
20. Anokhin AS, Gorelik VS, Dovbeshko GI, Pyatyshev AYu, Yuzyuk YuI (2015) Difference Raman spectroscopy of DNA molecules. J Phys Conf Ser 584:012022
21. Zhizina GP, Oleinik EF (1972) Infrared spectroscopy of nucleic acids. Russ Chem Rev 41(3):474
22. Grushevskaya HV, Krylova NG, Lipnevich IV, Orekhovskaja TI, Egorova VP, Shulitski BG (2016) Enhancement of Raman light scattering in dye-labeled cell membrane on metal-containing conducting polymer film. Int J Mod Phys B 30:1642018
23. Gutierrez PL (2000) The metabolism of quinone-containing alkylating agents: free radical production and measurement. Front Biosci 5:d629
24. Esteves-Souza A et al (2007) Cytotoxic and DNA-topoisomerase effects of lapachol amine derivatives and interactions with DNA. Braz J Med Biol Res 40:1399
25. Hasinoff BB et al (2006) Structure-activity study of the interaction of bioreductive benzoquinone alkylating agents with DNA topoisomerase II. Cancer Chemother Pharmacol 57:221
26. Martinez R, Chacon-Garcia L (2005) The search of DNA-intercalators as antitumoral drugs: what it worked and what did not work. Curr Med Chem 12:127
27. Grushevskaya HV, Krylova NG (2018) Carbon nanotubes as a high-performance platform for target delivery of anticancer quinones. Curr Pharm Des 24(43):5207

Chapter 18
Functional Nanocomposites Based on Quantum Dots

V. Smyntyna, V. Skobeeva, and G. Skobeeva

Abstract The optical and luminescent properties of composites based on cadmium sulfide quantum dots (CdS QDs) with dyes methylene blue (MB) and acridine yellow (AY) were studied. An increase in the luminescence intensity of the MB dye in the composite with QDs was found at low dye concentrations, which is explained by the mechanism of energy transfer from the QD (donor) to the dye (acceptor). The quenching of the luminescence of MB and AY dyes at high concentrations is explained by the transition of some of the dye molecules to aggregates that do not luminesce.

Keywords Nanocomposites · Energy transfer · Quantum dots of cadmium sulfide · Methylene blue dye · Acridine yellow dye · Luminescence · Aggregates of dyes

18.1 Introduction

One of the relevant areas of nanotechnology the last decade is the creation and study of functional nanomaterials, which opens up new opportunities for the development of information technology, energy, electronics, sensors, nanomedicine [1–9]. A key element of the new functional materials are nanomaterials whose functional properties are determined by the nanostructure consisting of nanoparticles ranging in size from 1 to 100 nm. The purpose of creating functional materials is to obtain and

V. Smyntyna
Odesa I.I. Mechnikov National University, Odessa, Ukraine
e-mail: Smyntyna@onu.edu.ua

V. Skobeeva (✉)
Research Institute of Physics, I.I. Mechnikov National University, Odessa, Ukraine
e-mail: v_skobeeva@ukr.net

G. Skobeeva
I.I. Mechnikov National University, Odessa, Ukraine

J. Bonča, S. Kruchinin (eds.), *Advanced Nanomaterials for Detection of CBRN*,
NATO Science for Peace and Security Series A: Chemistry and Biology,
https://doi.org/10.1007/978-94-024-2030-2_18

study new nanomaterials with properties that will correspond to a specific practical application. Thus, the relevance of research into hybrid nanosystems is due to both the interest in the issue of the interaction mechanism between components and caused by the prospect of practical application.

Composites based on colloidal semiconductor QDs and organic molecules are the most common hybrid structures in which QDs act as a photoexcitation energy donor for the second component of the structure. The optimal conditions under which efficient energy conversion is achieved depends on many factors. These include the degree of overlap of the luminescence spectra of the donor and the absorption of the acceptor, the distances between them, the brightness and stability of the luminescence of the QD, the shape and size of the QD. For the functioning of nanostructures in biological media, it is desirable to synthesize complexes in aqueous solutions to obtain hydrophilic QDs [10–13].

In this work, it are study composites based on semiconductor quantum dots of cadmium sulfide (CdS QDs) and dyes methylene blue (MB) and acridine yellow (AY). Each of the components, individually, has unique properties. CdS QDs have generated considerable interest due to their unique properties associated with dimensional effects. CdS luminescent QDs are promising materials for creating biomedical markers and optical sensors [14].

One of the most promising materials for creating hybrid structures is MB dye. The properties of this dye allow it to be used for various applications, including chemical indicators, markers, and spectral-sensitizing systems for photoelectric and photocatalysis of elements [15–19].

AY dye is used as indicators in medicine, biology, microbiology, and pharmaceuticals. In microbiology, it is used as a mutagen when working with chromosomes. AY dye is also used as a fluorescent dye and anti-inflammatory agent [20].

18.2 Experimental

Cadmium sulfide nanocrystals were grown by the colloid-chemical method from solutions of cadmium and sulfur salts in an aqueous solution (5%) of gelatin. The concentration of cadmium salt $(CdNO3)_2$ and Na_2S was 0.5 M and 0.25 M, respectively. QDs of cadmium sulfide occurred as a result of an exchange-chemical reaction at a temperature of 40 °C. The average radius of cadmium sulfide nanoparticles was calculated from the optical absorption spectra according to the theory of interband absorption [21] and was 3.1 ± 0.3 nm. Dyes of different concentrations were added to the obtained colloidal solution with CdS QDs.

Optical absorption spectra were measured on a spectrophotometer SF-26 in the wave length region from 320 to 600 nm. To reduce the error associated with the effect of light scattering in the short-waveregion (320–360 nm) we used the UVLF-2 light filter. The measurement error did not exceed ±1%. The luminescence was excited by a pulsed LCS-DTL-374QT laser with a wave length of light 355 nm (power is 35 mW).

18.3 Results and Discussion

Taking into account the results of our studies of the thermodynamics and kinetics of the chemical synthesis reaction, CdS QDs with a controlled emission spectrum were synthesized [21, 22]. For this study, CdS QDs were synthesized, which showed a broad luminescence band in the region of 400–800 nm with a maximum of $\lambda_{max} = 600$ nm (curve 1, Fig. 18.1). The same figure shows the luminescence spectra of CdS QDs with MB dye of different concentrations. The luminescence spectrum of a composite differs significantly from the luminescence of its components. Curve 1 in the figure corresponds to the spectrum of CdS nanocrystals. It was found that upon the introduction of the dye, the luminescence of CdS nanocrystals in the wavelength range of 600–650 nm is quenched and a narrower dye band appears with a maximum in the region of 700 nm (curves 3–7).

The observed dependency can be explained by the luminescence energy transfer of CdS quantum dots to the dye in the spectral region of dye absorption. The absorption spectra of the composite are shown in Fig. 18.2. Here, curve 1 corresponds to the absorption of CdS QDs, and curves 2–5 relate to the absorption of the CdS QDs

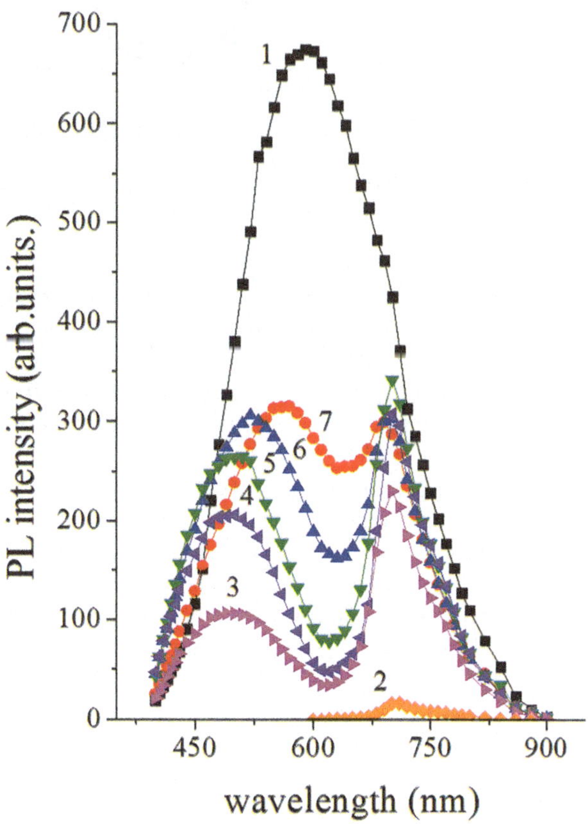

Fig. 18.1 Luminescence spectra of QDs CdS (1), dye MB concentration 5×10^{-6} M (2) and QD CdS with dye MB, concentration, M: 27×10^{-6} (3), 21×10^{-6} (4), 16×10^{-6} (5), 11×10^{-6} (6), 5×10^{-6} (7)

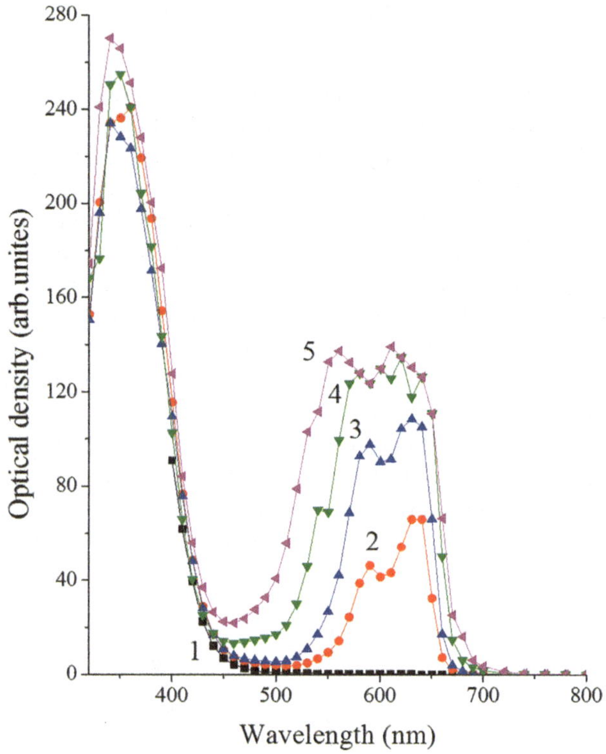

Fig. 18.2 Absorption spectra of QDs CdS (1) and QDs CdS with MB dye of various concentrations, M: 5×10^{-6} (2), 11×10^{-6} (3), 16×10^{-6} (4), 27×10^{-6} (5)

composite with MB dye. The dye itself at these concentrations absorbs in the range 450–650 nm, and luminesces in the longer wavelength region ($\lambda_{max} = 700$ nm). It is characteristic that the absorption spectra of CdS QDs and the MB dye do not overlap, and therefore light with a wavelength from the CdS absorption region (in our case, the laser wavelength $\lambda = 355$ nm) will not excite the luminescence of the dye.

The exciting light for the dye is the radiation energy of the quantum dots, whose spectrum coincides with the absorption spectrum of the dye. Indeed, the luminescence intensity of QDs decreases (curves 1 and 7), and the dye increases by almost 2 orders of magnitude (curves 2 and 7). However, it is noticeable that a slight increase in the luminescence of the dye with an increase in its concentration is replaced by quenching (curves 3 and 4). This fact is explained in the structuring of MS dye molecules, which is well confirmed by the above absorption spectra (Fig. 18.2), which contain both the molecular (630 nm) band and the aggregated (580 nm) dye band. With increasing dye concentration, an increasing number of molecules aggregate (curves 3–5) and in this state they do not luminesce. This leads to quenching of the dye. As can be seen from Fig. 18.2, the dye absorption spectra

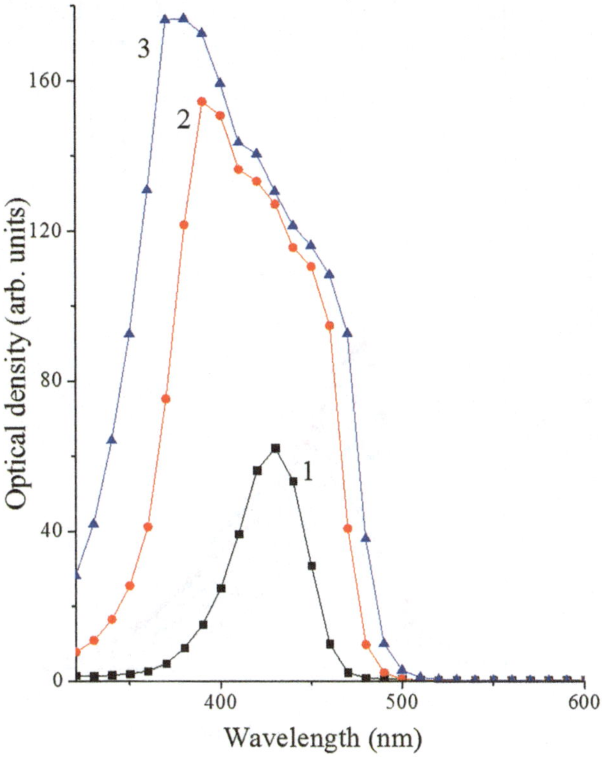

Fig. 18.3 Absorption spectra of dye AY, concentration, M: 7×10^{-6} (1), 37×10^{-6} (2), 183×10^{-6} (3)

broaden due to the absorption of molecular aggregates. This increases the overlap of the luminescence bands of CdS QDs as a donor and dye absorption as an acceptor. The consequence of this is a decrease in the luminescence intensity of with a shift of the band maximum to the short-wavelength region.

AY dye has a wide absorption band in the region from 300 to 500 nm. The absorption bands of QDs and AY dye overlap (Fig. 18.3). The position of the maximum of the molecular form of the dye is localized at $\lambda = 430$ nm (curve 1, Fig. 18.3). With an increase in the dye concentration, bands characteristic of the aggregated form of molecules appear in the spectrum, the spectrum shifts to a shorter wavelength region, and the overlap area with the absorption spectrum of quantum dots increases.

In contrast to composites with MB dye, luminescence of a composite with AY dye should appear in the shorter wavelength region of the spectrum. This is illustrated in Fig. 18.4. Curves 2 and 3 in this figure refer to the luminescence spectra of a dye without quantum dots. It is noticeable that, at high dye concentrations, no enhancement of its luminescence is observed, which is explained by the transition of some of the molecules to a non-luminescent aggregated form.

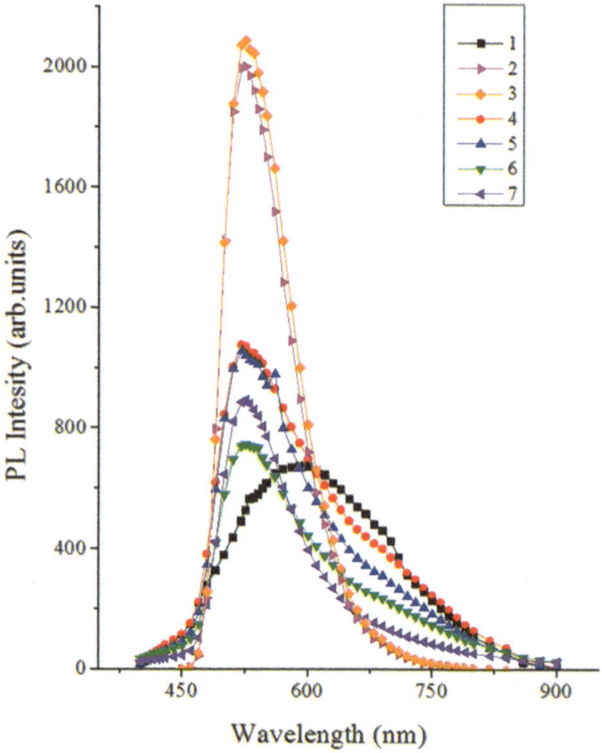

Fig. 18.4 Luminescence spectra of QD CdS (1), dye of AY, concentration, M: 37 × 10^{-6} (2), 183 × 10^{-6} (3) and QD of CdS with dye of AY, concentration, M: 7 × 10^{-6} (4), 22 × 10^{-6} (5), 37 × 10^{-6} (6), 183 × 10^{-6} (7)

A comparison of spectra 2 and 3 with spectra 6 and 7 shows that luminescence quenching occurs in a composite with CdS quantum dots at the same concentrations. The maximum of the luminescence band has shifted from red to green.

18.4 Conclusion

Hydrophilic CdS QDs were synthesized in a gelatinous colloid with an intense wide luminescence band, which allows the formation of CdS QDs composites – dyes with different absorption spectra.

An increase in the luminescence intensity of the MB dye in the composite with QDs was found at low dye concentrations, which is explained by the mechanism of energy transfer from the QDs (donor) to the dye (acceptor). The quenching of the luminescence of MB and AY dyes at high concentrations is explained by the transition of some of the dye molecules to aggregates that do not luminesce.

The resulting composites are stable and can be promising as luminescent biological labels.

References

1. Mansur H, Ancelmo A, Mansur P (2012) Fluorescent nanohybrids: quantum dots coupled to polymer recombinant protein conjugates for the recognition of biological hazards. J Mater Chem 22:9006–9018. https://doi.org/10.1039/C2JM31168B
2. Ramanery F, Mansur A, Mansur H (2013) One-step colloidal synthesis of biocompatible water-soluble ZnS quantum dot/chitosan nanoconjugates. Nanoscale Res Lett 8(1):512. https://doi.org/10.1186/1556-276X-8-512
3. Moulick A, Milosavljevic V, Vlachova J, Podgajny R, Hynek D, Kopel P, Adam V (2017) Using CdTe/ZnSe core/shell quantum dots to detect DNA and damage to DNA. Int J Nanomed 12:1277–1291. https://doi.org/10.2147/IJN.S121840
4. Reyes-Esparza J, Martínez-Mena A, Gutiérrez-Sancha I, Rodríguez-Fragoso P, Gonzalez de la Cruz G, Mondragón R, Rodríguez-Fragoso L (2015) Synthesis, characterization and biocompatibility of cadmium sulfide nanoparticles capped with dextrin for in vivo and in vitro imaging application. J Nanobiotechnol 13:83. https://doi.org/10.1186/s12951-015-0145-x
5. Wang N, Zhang H, Yang X (2016) A facile hydrothermal route for synthesis of ZnS hollow spheres with photocatalytic degradation of dyes under visible light. J Appl Spectrosc 83(6):1007–1011. https://doi.org/10.1007/s10812-017-0398-2
6. Ermakov V, Kruchinin S, Fujiwara A (2008) Electronic nanosensors based on nanotransistor with bistability behaviour. In: Bonca J, Kruchinin S (eds) Proceedings NATO ARW "Electron transport in nanosystems". Springer, pp 341–349, Dordrecht, Netherland
7. Ermakov V, Kruchinin S, Hori H, Fujiwara A (2007) Phenomena in the resonant tunneling through degenarate energy state with electron correlation. Int J Mod Phys B 11:827–835
8. Kruchinin S, Pruschke T (2014) Thermopower for a molecule with vibrational degrees of freedom. Phys Lett A 378:157–161
9. Ermakov V, Kruchinin S, Pruschke T, Freericks J (2015) Thermoelectricity in tunneling nanostructures. Phys Rev B 92:115531
10. Musikhin S, Alexandrova O, Luchinin V, Maksimov A, Moshnikov V (2012) Semiconductor colloidal nanoparticles in biology and medicine. Biotechnosphere 5–6(23–24):40–48. https://www.researchgate.net/publication/280247410
11. Alison, Funston M, Jasieniak J, Mulvaney P (2008) Complete quenching of CdSe nanocrystal photoluminescence by single dye molecules. Adv Mater 20:4274–4280. https://doi.org/10.1002/adma.200703186
12. Sekhar M, Santhosh K, Kumar J, Mondal N, Soumya S, Samanta A (2014) CdTe quantum dots in ionic liquid: stability and hole scavenging in the presence of a sulfide salt. J Phys Chem C 118(32):18481–18487. https://doi.org/10.1021/jp507271t
13. Smirnov M, Ovchinnikov O, Dedikova A, Shapiro B, Vitukhnovsky A (2016) Luminescence properties of hybrid associates of colloidal CdS quantum dots with J-aggregates of thiatrimethine cyanine dye. J Lumin 176:77–85. https://doi.org/10.1016/j.jlumin. 2016.03.015
14. Michalet X, Pinaud F, Bentolila L, Tsay J, Doose S, Li J, Sundaresan G, Wu A, Gambhir S, Weiss S (2005) Quantum dots for live cells, in vivo imaging, and diagnostics. Scince 307(5709):538–544. https://doi.org/10.1126/science.1104274
15. Jockusch S, Lee D, Turro NJ, Leonard EF (1996) Photo-induced inactivation of viruses: adsorption of methyleneblue, thionine, and thiopyronine on Qf3 bacteriophage. Proc Natl Acad Sci U S A 93:7446–7451

16. Gabrielli D, Belisle E, Severino D, Kowaltowski AJ, Baptista MS (2004) Binding, aggregation and photochemical properties of methylene blue in mitochondrial suspensions. Photochem Photobiol 79(3):227–232. https://doi.org/10.1111/j.1751-1097.2004.tb00389.x
17. Tardivo J, Giglio A, Oliveira C, Gabrielli D, Junqueira H, Tada D, Severino D, Turchiello R, Baptista M (2005) Methylene blue in photodynamic therapy: from basic mechanisms to clinical applications. Photodiagn Photodyn Ther 2(3):175–191. https://doi.org/10.1016/S1572-1000(05)00097-9
18. Tuite EM, Kelly JM (1993) Photochemical interactions of methylene blue and analogues with DNA and other biological substrates. J Photochem Photobiol B 21(2–3):103–124, PMID: 8301408
19. Filikhin I, Peterson TH, Vlahovic B, Kruchinin SP, Kuzmichev YuB, Mitic V (2019) Electron transfer from the barrier in InAs/GaAs quantum dot-well structure. Phys E Low-dimensional Syst Nanostruct 114:113629
20. Yurre T, Rudaya L, Klimova I, Shamanin V (2003) Organic materials for photovoltaic and light-emitting devices. Phys Technol Semicond 37:835–842
21. Smyntyna V, Skobeeva V, Verheles K, Malushin N (2019) Influence of technology on the formation of luminescence centers in QDs CdS. J Nano Electron Phys 11(5):05031–05034. https://doi.org/10.21272/jnep.11(5).05031
22. Verheles K, Smyntyna V, Skobeeva V, Malushin N (2019) The dependence of photoluminescence spectra of CdS QDs on stoichiometry. Visnyk Lviv Univ Ser Phys 56:3–10. https://doi.org/10.30970/vph.56.2019.3

Chapter 19
Polymer Lattice and Track Nanostructures to Create Novel Biosensors

T. Kavetskyy, D. Fink, A. Kiv, Yu. Bondaruk, O. Šauša, Y. Kukhazh, K. Zubrytska, O. Smutok, and M. Gonchar

Abstract Continuous environmental monitoring and human health diagnostics are important tasks of modern science. We are developing multifunctional biosensor equipment for solving these problems. It is essential that the appropriate devices are multifunctional, highly sensitive and convenient for practical use in everyday life and industrial conditions. As an example, here we show the results of the development of highly sensitive new generation biosensors for the analysis of some xenobiotics in wastewater and functional health markers in human liquids. The recent results on the construction of laboratory prototypes of amperometric

---

T. Kavetskyy (✉)
Drohobych Ivan Franko State Pedagogical University, Drohobych, Ukraine

The John Paul II Catholic University of Lublin, Lublin, Poland

Y. Kukhazh · K. Zubrytska
Drohobych Ivan Franko State Pedagogical University, Drohobych, Ukraine

D. Fink
Universidad Autónoma Metropolitana-Iztapalapa, México, DF, Mexico

Nuclear Physics Institute, ASCR, Řež, Czech Republic
e-mail: fink@xanum.uam.mx

A. Kiv
Department of Materials Engineering, Ben-Gurion University of the Negev, Beer-Sheva, Israel

South-Ukrainian K.D. Ushynsky National Pedagogical University, Odessa, Ukraine
e-mail: kiv@bgu.ac.il

Yu. Bondaruk
South-Ukrainian K.D. Ushynsky National Pedagogical University, Odessa, Ukraine

O. Šauša
Institute of Physics, Slovak Academy of Sciences, Bratislava, Slovak Republic

O. Smutok · M. Gonchar
Drohobych Ivan Franko State Pedagogical University, Drohobych, Ukraine

Institute of Cell Biology, National Academy of Sciences of Ukraine, Lviv, Ukraine

J. Bonča, S. Kruchinin (eds.), *Advanced Nanomaterials for Detection of CBRN*, NATO Science for Peace and Security Series A: Chemistry and Biology,
https://doi.org/10.1007/978-94-024-2030-2_19

enzymatic biosensors in which we use the gold nanoparticles in combination with different polymers as a host matrix will be discussed. The results of using track polymer structures to create novel biosensors will be also presented. The developed ion track-based nanosensors provide high sensitivity, low power and low cost. They shall be modified further towards portable tools for in-situ chemical analysis in space and terrestrial applications. The generic track-based nanosensors platform is suitable for detection and analysis of gases and volatile organic compounds, specifically of biomolecules, in both gaseous and liquid phases.

Keywords Enzymatic biosensors · Gold nanoparticles · Polymers · Track nanostructures · Human health · Environmental monitoring

19.1 Introduction

Technogenic pressure on the environment, as well as irrational unhealthy eating, has significantly affected human health. To eliminate these negative impacts it is greatly important creation of new highly effective approaches for analysis of the levels of water pollution, food quality and physiological markers in human blood. The most dangerous pollutants of water resources are xenobiotic – products of the chemical and pharmaceutical industry, which negatively impact on the physiological state of living organisms and have carcinogenic properties even at very low concentrations. On the other hand, *L*-lactate is served as an important marker in food technology, clinical diagnostics, and sports medicine. In the present work, we reported the recent results obtained on construction of the new laboratory prototypes of two types of amperometric enzymatic biosensors based on novel polymer matrixes and nanocomposites for xenobiotic analysis in wastewater and *L*-lactate assay in biological fluids. The results of using track polymer structures to create novel biosensors are also reported. The generic track-based nanosensors platform is suitable for detection and analysis of gases and volatile organic compounds, specifically of biomolecules, in both gaseous and liquid phases.

19.2 Materials and Methods

The organic (photopolymers and others) and organic-inorganic hybrid (ureasil-composite and others) polymer matrixes with micro/nanoparticles (microspheres of chalcogenide clusters and gold nanoparticles (AuNPs) were synthesized and used for immobilization of commercial enzymes (laccase from *Trametes versicolor*), as well as recombinant *L*-lactate: cytochrome *c* oxidoreductase (flavocytochrome b_2) from a gene-engineered strain of methylotrophic yeast *Ogataea polymorpha* "tr 1" (*gcr1 catX/prAOX_CYB2*) [1].

In order to estimate the nature of the biosensor's characteristics obtained the structural and physico-chemical properties of the nanocarriers, as well as polymers used for immobilization of the enzymes, were investigated with the most proper and informative experimental techniques such as positron annihilation spectroscopy, swelling, UV-visible spectroscopy, scanning electron microscopy, atomic force microscopy and others.

In the case of track-based biosensors it is important in particular to know the peculiarities of the passage of ion flows through the nanopores. To reveal the mechanisms of the passage of the matter through nanotracks in polymers, we apply a computer simulation of these processes [2].

Using the classical method of molecular dynamics, a cylindrical nanopore model was created and the passage of ionic fluxes through such a pore was simulated. It was found that the important factors determining the magnitude of the ion flux are the density of the adsorption centers (AC) on the inner surface of the nanocylinder and its diameter.

19.3 Results and Discussion

19.3.1 Amperometric Enzyme Biosensors

For the first time, we have reported construction of the innovative amperometric biosensors of the "third generation" for L-lactate analysis in human liquids based on flavocytochrome b_2 (FC b_2) immobilized onto gold nanoparticles (AuNPs) [3] and for xenobiotic determination in water samples based on the organic-inorganic ureasilicate (shorten name "ureasil") composites and commercial laccase [4]. The principal schemes of direct electron transfer in the biorecognition layer of the constructed biosensors are shown in Fig. 19.1.

The developed highly sensitive and selective FC b_2-AuNPs-based bioelectrodes were integrated into an automatic amperometric microcomputer-based analyzer system and non-invasive analysis of the L-lactate in human sweat and saliva samples was done. The results for the analyzed real human liquids obtained by direct non-invasive and reference analysis have demonstrated a high correlation. The new prototype of non-invasive electrochemical biosensor of third generation based on FC b_2 immobilized onto AuNPs is supposed to be applied for the determination of L-lactate in clinical diagnostics and sport medicine [3].

The constructed laccase-based biosensor using ureasil/As_2S_3 composite was characterized by very high sensitivity, but a weak point of the biosensor was a very strong unexpected electrochemical noise at chronoamperometric measurement. At the same time, new perspectives of the ureasil-based polymers for construction of amperometric enzyme biosensors were further found [5–10]. In particular, a correlation between the network properties of the biosensor sensing layers (e.g., free-volume V_h at glass transition temperature T_g and coefficients for the thermal

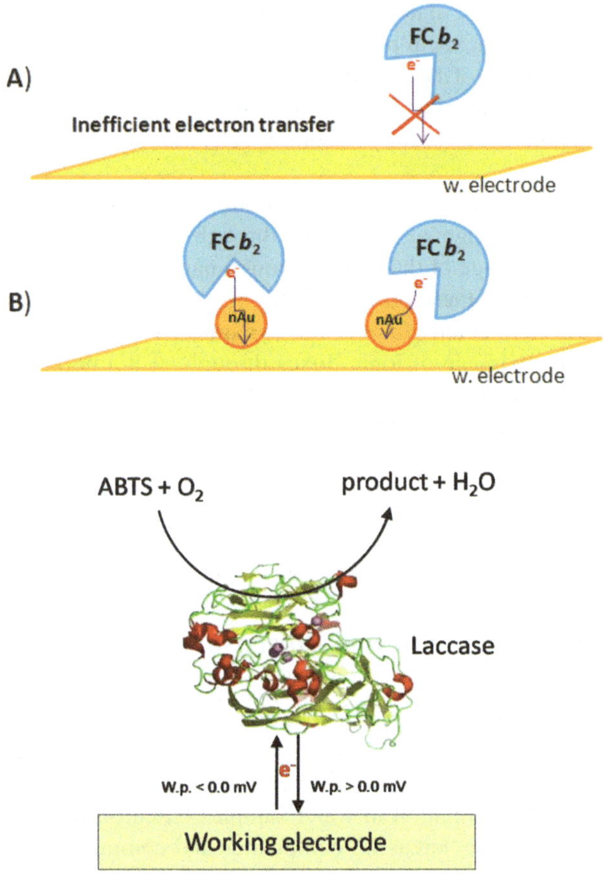

Fig. 19.1 The principal schemes of direct electron transfer in the biorecognition layer of the constructed biosensors: from reduced flavocytochrome b_2 (FC b_2) through AuNPs to the working electrode (top); electrochemical processes on the working electrode's surface of the laccase-based biosensor (bottom). (Adapted from [3, 4])

expansion of free-volume voids α_{F1}, α_{F2} as well as their difference ($\alpha_{F2} - \alpha_{F1}$), and swellability or crosslinked density) based on the pure ureasil and ureasil/As_2S_3 composites of different history (fresh and aged samples) and biosensor characteristics (e.g., a maximal current at substrate saturation I_{max}, apparent Michaelis-Menten constant K_M^{app} to ABTS chosen as a substrate, the slope of the calibration curve, and the sensitivity of bioelectrodes obtained by means of chronoamperometric analysis) was established. Further research showed an enhancement of the sensitivity of the laccase-based bioelectrodes using enzyme-immobilized AuNPs coupling with the ureasil/As_2S_3 polymer as a host matrix [11].

The above mentioned correlation, which was firstly observed in the case of ureasil matrixes, is also proved in the case of photocross-linked polymers [12]. It was suggested that the correlation found may be more general. Summarizing, it was concluded that controlling the network properties of polymer matrixes suitable for construction of amperometric enzyme biosensors by examining ($\alpha_{F2} - \alpha_{F1}$) magnitude in temperature dependent positron annihilation lifetime spectroscopy experiment may be used for improvement of operational parameters of such sensors [13].

19.3.2 Track-Based Biosensors

Two track-based biosensor concepts are presented in [14] (Fig. 19.2). In this Figure at the top is the pore-blocking concept of Siwy [15], illustrated for the system biotin-streptavidin. At the bottom is the product enrichment concept according to the work [16], illustrated for the system glucose oxidase – (GOx)-glucose. Whereas in the left case, the measuring signal is the interruption of a test current through the nanopore,

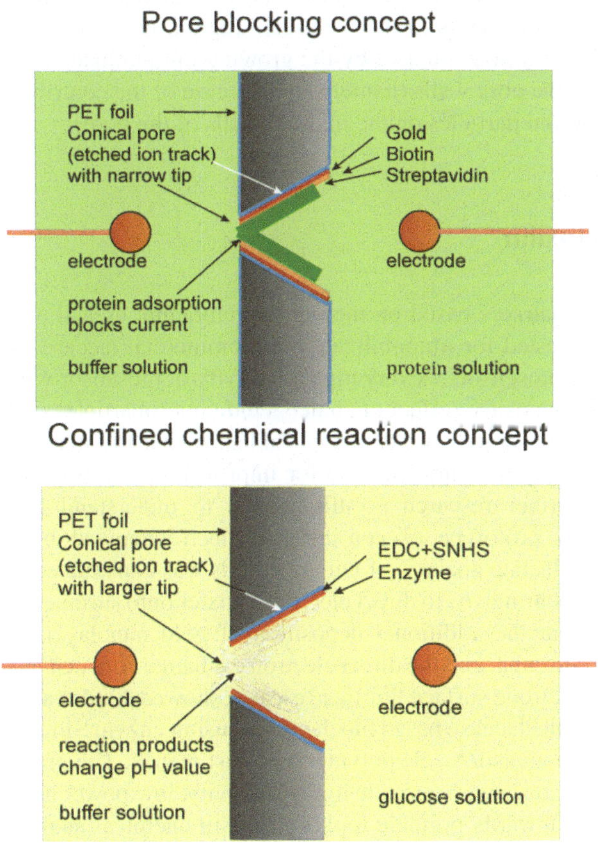

Fig. 19.2 Two concepts of nuclear track-based biosensors. (Adapted from [14])

in the right case it is the increase of the test current. In the latter case, the layer indicated by EDC+SNHS only serves for bonding the enzyme to the polymer wall. Once this is accomplished it does not exist any longer.

Computer simulation showed that the magnitude of the ion flux is significantly affected by the density of AC and the binding energy of ions in AC. As a result, the dependences of the ion current on the diameter of the cylindrical pore were obtained, which are consistent with experimental data [14]. The simulation result obtained indicates that the flux (and hence also the diffusion coefficient) decreases with decreasing pore radius (from 50 to 25 nm). This is due to a sharp increase in the contribution of multiple scattering of particles on the internal pore walls.

Another result was obtained by including a number of "foreign" particles to the composition of the gas passing through the pore. These particles differ in charge from the particles of the main stream. The computer program yields a change of the flux density of particles through the pore depending on the percentage of "foreign" particles.

In addition, it turns out how this dependence is affected by the magnitude of the tube radius. The simulation shows that, despite the decrease in the particle flux with decreasing pore radius in the nanoscale region, the sensitivity of the sensor increases, which can be explained by the growing predominance of the interaction of particles with the pore walls (namely, the increase of the contribution of multiple scattering of foreign particles on the internal walls of the pore).

19.4 Conclusion

Novel polymer matrixes based on the organic-inorganic ureasil and ureasil/As_2S_3 composite were tested for immobilization of commercial laccase and construction of amperometric biosensors. A very high sensitivity of biosensor with ureasil/As_2S_3 composite and the positive effect of chalcogenide microparticles on the biosensor's parameters were observed. It was also established that the network properties of polymer matrix play a significant role for improvement of amperometric enzyme biosensor but further research is still needed to understand better the results obtained. A new prototype of non-invasive electrochemical biosensor of third generation for L-lactate analysis of human liquids was constructed using the ability of yeast flavocytochrome b_2 to direct electron transfer onto surface of gold electrode. It was proven that the additional deposition of gold nanolayer (AuNPs) onto a working electrode enhanced the direct electron exchange between the enzyme active center and the electrode surface. Further research showed similar an enhancement of the sensitivity of the laccase based bioelectrodes using enzyme-immobilized AuNPs coupling with the ureasil/As_2S_3 polymer as a host matrix. The presently developed ion track-based nanosensors provide high sensitivity, low power and low cost. They will be modified towards portable tools for in-situ chemical analysis in space and terrestrial applications.

Acknowledgments This work was financially supported in part by the Ministry of Education and Science of Ukraine (projects #0117U007143; to YK and KZ, #0118U000297 and #0119U100671; to TK, YK, OS, and MG), National Academy of Sciences of Ukraine in the frame of the Scientific-Technical Program "Smart sensor devices of a new generation based on modern materials and technologies" (project #13; to OS and MG), Slovak Grant Agency VEGA (project #2/0157/17; to OŠ), and Slovak Research and Development Agency (project #APVV-16-0369; to OŠ). TK, YK, and KZ also acknowledge the SAIA for scholarship in the IPSAS within the National Scholarship Programme of the Slovak Republic.

References

1. Dmitruk K et al (2008) Construction of flavocytochrome b_2-overproducing strains of the thermotolerant methylotrophic yeast. Hansenula polymorpha (Pichia angusta). Microbiology 77:213
2. Bondaruk YU et al (2020) Simulation of the passage of ion flows through nanotracks. Intern J Adv Comp Techn 9:1
3. Smutok O et al (2017) Development of a new mediatorless biosensor based on flavocytochrome $b_2$immobilized onto gold nanolayer for non-invasive L-lactate analysis of human liquids. Sens Actuators B 250:469
4. Kavetskyy T et al (2017) Laccase-containing ureasil-polymer composite as the sensing layer of an amperometric biosensor. J Appl Polym Sci 134:45278
5. Kavetskyy T et al (2017) Network properties of ureasil-based polymer matrixes for construction of amperometric biosensors as probed by PALS and swelling experiments. Acta Phys Pol A 132:1515
6. Kavetskyy TS et al (2018) Ureasil-based polymer matrices as sensitive layers for the construction of amperometric biosensors. NATO SPS B Phys Biophys 30:309
7. Kavetskyy TS et al (2018) Swelling behavior of organic-inorganic ureasil-based polymers. NATO SPS B Phys Biophys 32:333
8. Ermakov V, Kruchinin S, Fujiwara A (2008) Electronic nanosensors based on nanotransistor with bistability behaviour. In: Bonca J, Kruchinin S (eds) Proceedings NATO ARW "Electron transport in nanosystems". Springer, Dordrecht, pp 341–349
9. Ermakov V, Kruchinin S, Hori H, Fujiwara A (2007) Phenomena in the resonant tunneling through degenarate energy state with electron correlation. Int J Mod Phys B 11:827–835
10. Ermakov V, Kruchinin S, Pruschke T, Freericks J (2015) Thermoelectricity in tunneling nanostructures. Phys Rev B 92:115531
11. Kavetskyy T et al (2019) Improvement of amperometric laccase biosensor using enzyme-immobilized gold nanoparticles coupling with ureasil polymer as a host matrix. Gold Bull 52:79
12. Kavetskyy T et al (2019) Dependence of operational parameters of laccase-based biosensors on structure of photocross-linked polymers as holding matrixes. Eur Polym J 115:391
13. Kavetskyy T et al (2020) Controlling the network properties of polymer matrixes for improvement of amperometric enzyme biosensors: contribution of positron annihilation. Acta Phys Pol A 137:246
14. Fink D et al (2016) Nuclear track-based biosensing: an overview. Radiat Eff Defects Solids 171:173
15. Acar ET et al (2018) Concentration-polarization-induced precipitation and ionic current oscillations with tunable frequency. J Phys Chem C 122:3648
16. Fink D et al (2009) Glucose determination using a re-usable enzyme-modified ion track membrane. Biosens Bioelectron 24:2702

Chapter 20
The Sensitivity Peculiarities of Nanosized Tin Dioxide Films to Certain Alcohols

A. P. Chebanenko, V. S. Grinevych, L. M. Filevska, and V. Smyntyna

Abstract The electrophysical and kinetic characteristics of thin films of nanosized tin dioxide obtained using polymers were investigated in their contact with an alcohol-containing atmosphere at room temperature. A reversible change in the conductivity of SnO_2 films in atmospheres with ethanol and isopropanol vapors at room temperature was established. Differences in the values of sensitivity and adsorption/desorption constants were also established, which ensure the selectivity of thin SnO_2 films to alcohols (ethanol and isopropanol). All this allows using them as basic sensitive elements of gas sensors for ethyl and isopropyl alcohols detections in ambient air, operating at room temperature.

Keywords Nanosized tin dioxide · Thin films · Sensitivity to ethanol · Sensitivity to isopropanol

20.1 Introduction

Tin dioxide in nanoscale form is widely used as a sensitive element for monitoring the air impurity in industrial zones, at enterprises, in the food industry and trading enterprises while monitoring the freshness of products, etc. [1]. Control of alcohol content (in particular) is important in the atmosphere of working areas of chemical and food industry enterprises, in exhaled gases to determine the degree of alcohol poisoning or control the alcohol intoxication of vehicle drivers [2–4]. The effect of various alcohols vapors on the physical characteristics of tin dioxide and the possibility of creating alcohol sensors are widely studied [5–8, 13]. For this case, the main parameter which is sensitive to external atmosphere changes is, most often, the conductivity (resistivity) of a sensitive material [9]. As a rule, sensitive elements operate with a thermoheater which provides the temperature for surface adsorption

A. P. Chebanenko · V. S. Grinevych (✉) · L. M. Filevska · V. A. Smyntyna
Odessa I.I. Mechnikov National University, Odessa, Ukraine
e-mail: grinevich@onu.edu.ua

J. Bonča, S. Kruchinin (eds.), *Advanced Nanomaterials for Detection of CBRN*,
NATO Science for Peace and Security Series A: Chemistry and Biology,
https://doi.org/10.1007/978-94-024-2030-2_20

reactions close to or more than 200 °C and are used to control the composition of the atmosphere [7, 8, 10, 11]. Sensors operating at a temperature of 100–110 °C are considered to be low-temperature [6, 12, 13] ones.

The electrophysical and kinetic characteristics of thin films of nanosized tin dioxide obtained using polymers were investigated in contact with an alcohol-containing atmosphere at room temperature. The results of these studies are presented in this paper.

20.2 Samples and Research Methods

The studied tin dioxide films were obtained by the sol–gel method using polymers (polyvinyl acetate) for structuring [14] and bis(acetylacetonato)-dichlorotin(IV) as a precursor [15, 16]. The presence of a nanoscale structure is confirmed by AFM studies.

Ethanol and isopropanol (propanol-2) were used as detectable alcohols. The studies were carried out at room temperature, since the use of sensitive elements without additional heating simplifies their production and saves expensive structural materials (platinum, iridium, etc.) for the production of heaters.

To measure the electrical characteristics, the studied SnO_2 films were supplied with indium contacts, thermally sprayed in high vacuum on the surface of the films in the form of two parallel strips. The inter electrode distance was 2 mm. The measurement technique for the electrophysical characteristics of SnO_2 nanofilms is a standard I–V method, i.e. recording of the current kinetics with the composition change of the surrounding atmosphere [17].

The sensitivity studies were the comparison of the electrophysical parameters changes of thin SnO_2 films with the atmosphere composition change in which the sample is located (dry air, air with ethanol or propanol-2 vapors).

20.3 Results and Discussion

Prior to measuring the sensitivity of the films, the nature of the current flow in them was studied in different atmospheres. Figure 20.1 shows the current-voltage characteristics measured in vapors of ethyl and isopropyl alcohols. For comparison, the I–V characteristics measured in dry air are also presented there.

As it is seen, all the current-voltage characteristics are superlinear. However, the curvature of the characteristics measured in the atmosphere of ethyl and isopropyl alcohol vapors (curves 2 and 3) is greater than of the I–V characteristics measured in air (curve 1). This behavior can be explained by the increase in intercrystalline barriers due to their decoration by ethanol or isopropanol molecules upon their adsorption on the film surface. As a result, the current flow mechanism to a greater extent becomes a barrier type. This assumption is supported by the fact that the

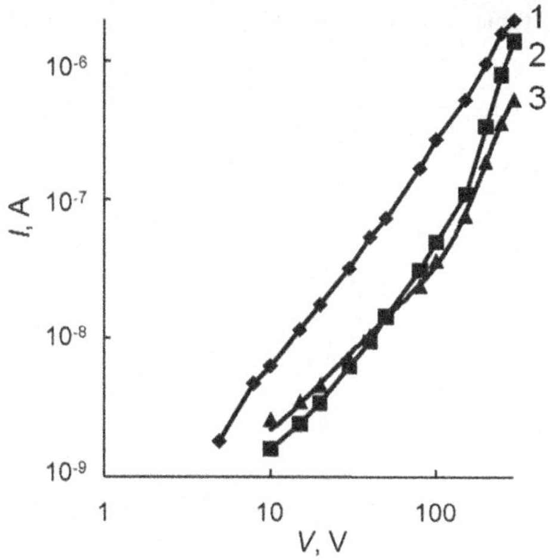

Fig. 20.1 The current–voltage characteristics of the SnO_2 film, measured in an atmosphere of dry air (1), ethyl alcohol vapor (2), isopropyl alcohol vapor (3) ($T = 290$ K)

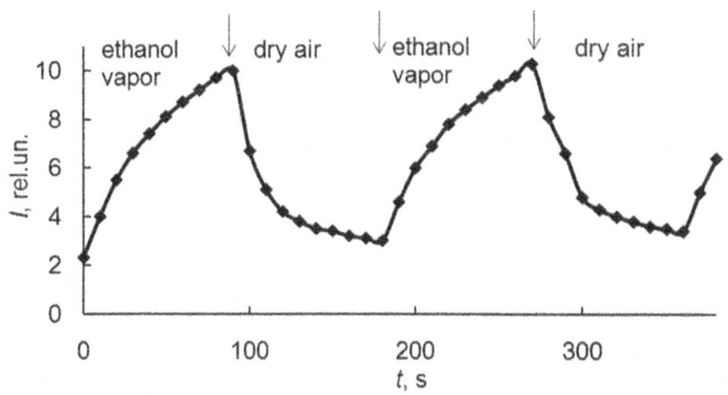

Fig. 20.2 Current relaxation in the SnO_2 film during periodic injection of ethanol vapor or dry air ($V = 250$ V) into the chamber

electrical resistance of SnO_2 films in the atmosphere of ethanol or isopropanol vapors becomes greater than in the atmosphere of air. We also note that in the ethanol atmosphere there is a slightly greater superlinearity of the I–V characteristics compared to the I–V characteristics in the propanol-2-containing atmosphere.

The specificity of the sensitivity of tin dioxide films to alcohols was studied as a kinetic change of the films' conductivity upon periodic admission of alcohol and dry air into the measuring chamber. Figure 20.2 shows the relaxation of current in the studied film at the periodically letting ethanol and dry air vapor into the chamber.

Figure 20.2 shows that at the presence ethanol vapor in the chamber, the current strength decreases during 60 s by an order of magnitude, i.e., the SnO_2 film sample becomes highly resistive. One minute after the ethanol vapor inlet, the current through the SnO_2 sample reaches a minimum, which corresponds to the sensitivity maximum. After having blown a dry air through the chamber, the conductivity of the SnO_2 sample restores to its original values. Reversible changes in conductivity in the presence of ethanol vapors are repeated at the repeating of ethanol/dry air vapor cycles.

Table 20.1 The SnO_2 film sensitivity to ethanol vapor in the dependence on the applied voltage

V, V	80	120	150	200	250	300	320
S, rel. un	0.51	0.54	0.56	0.61	0.68	0.73	0.75

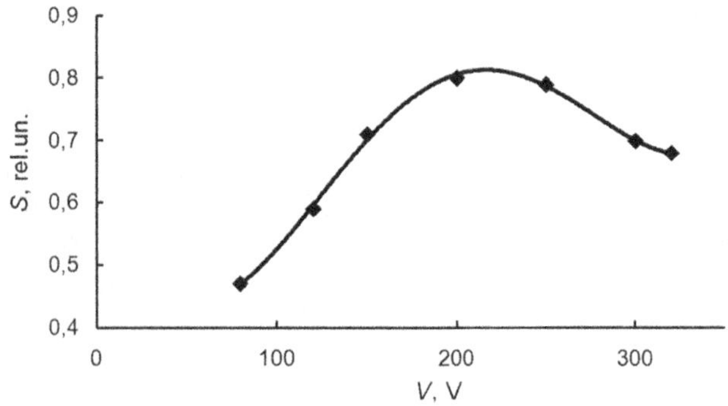

Fig. 20.3 The dependence of the sensitivity of the SnO_2 film to ethanol vapor on the applied voltage

The current relaxation in the SnO_2 film was studied under periodic inflow of ethanol or dry air vapor into the chamber at various voltages applied to the sample. The sensitivity value S at various voltage values was calculated by the formula:

$$S = \frac{I_1 - I_0}{I_0}, \tag{20.1}$$

where I_1 is the current in the presence of ethanol vapor, I_0 is the current in the atmosphere of dry air.

For each voltage, the sensitivity values of the SnO_2 film were calculated, presented in Table 20.1 and for visibility, in Fig. 20.3.

It is seen, that the maximum sensitivity of the SnO_2 film is reached at a voltage of 200 V.

The kinetic dependences of the current strength (Fig. 20.2) upon interaction with ethanol vapors and dry air are close to exponential. Namely, the decrease in current with time during the inlet of ethanol occurs according to the law:

$$I = I_0 e^{-\frac{t}{\tau_d}}. \tag{20.2}$$

The increase in current with time during the inlet of dry air is according to the law:

$$I = I_0 \left(1 - e^{-\frac{t}{\tau_a}}\right). \tag{20.3}$$

Here, τ_a and τ_d are some time constants that characterize the rate of adsorption (τ_a) and desorption (τ_d) processes. In this case, the time dependences of the current decay, rearranged in coordinates $\ln(I_0/I) = f(t)$ and current growth, rearranged in coordinates $\ln(I_0/(I_0 - I)) = f(t)$, should be straight lines.

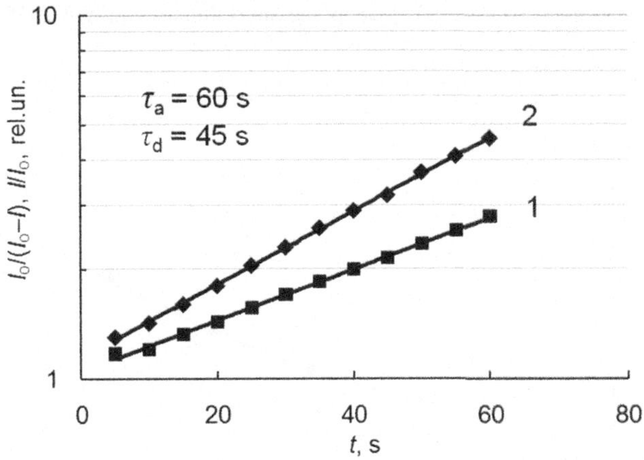

Fig. 20.4 Current relaxation in the SnO_2 film when ethanol vapor (1) and dry air (2) ($V = 200$ V) are introduced into the chamber

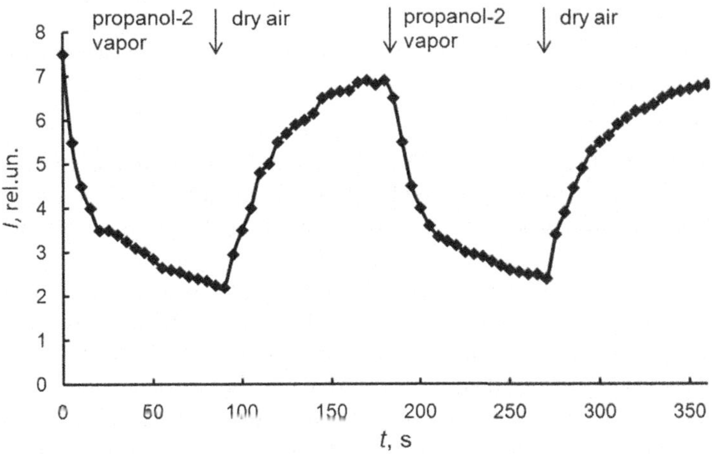

Fig. 20.5 Current relaxation in the SnO_2 film upon periodic injection of vapors of isopropyl alcohol or dry air ($V = 200$ V) into the chamber

The dependences rearranged in a logarithmic scale are shown in Fig. 20.4, of which the adsorption and desorption time constants were calculated. The desorption constant in this case is less than the constant adsorption, which indicates a faster process of desorption of ethanol molecules from the surface of the studied films in comparison with adsorption. This adsorption/desorption ratio is useful for using tin dioxide layers as ethanol sensors at room temperature, since it indicates that adsorption centers can be released from adsorbed molecules without additional heating.

The sensitivity changes of the SnO_2 film during alternate inlet of vapors of isopropyl alcohol and dry air into the measuring chamber is shown at Fig. 20.5.

Table 20.2 The dependence of the sensitivity of the SnO_2 film to isopropyl alcohol vapors on the applied voltage

V, V	80	120	150	200	250	300	320
S, rel. un	0.47	0.59	0.71	0.8	0.79	0.7	0.6

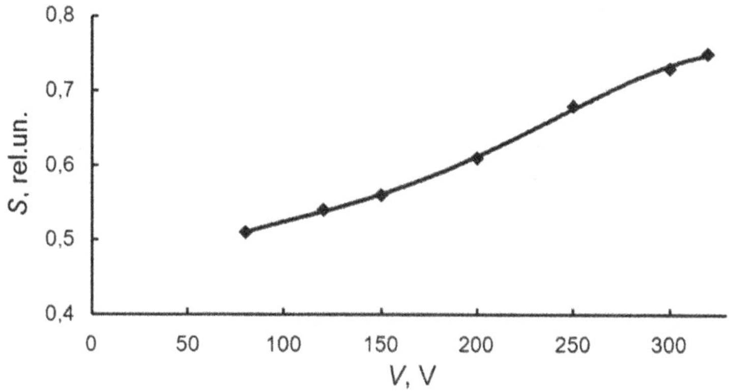

Fig. 20.6 The dependence of the sensitivity of the SnO_2 film to isopropyl alcohol vapors on the applied voltage

Isopropyl alcohol, like ethyl alcohol, makes the SnO_2 film sample highly resistant, and a sensitivity maximum is reached over approximately the same time intervals. When dry air is introduced, the conductivity of the SnO_2 film returns to its original values.

Same as for ethyl alcohol, series of experiments were conducted at various voltages applied to the sample. For each voltage value, the sensitivity of the tin dioxide film was calculated. The results are presented in Table 20.2.

It can be seen (Fig. 20.6) that the SnO_2 film' sensititvity dependence to isopropyl alcohol vapor on the applied voltage is almost linear, and the sensitivity increases in the entire range of given voltages, in contrast to similar results for ethyl alcohol. The dependences (20.2) and (20.3), were plotted in a logarithmic scale (Fig. 20.7), from which the time constants for adsorption and for desorption were calculated, which gave $\tau_a = 27$ s, $\tau_d = 20$ s.

The comparison of these results with the results of Table 20.2 and Fig. 20.2, shows that at the same voltages applied to the SnO_2 film, its sensitivity to ethanol vapor is higher than to isopropyl alcohol one. Besides that, the time constants of adsorption and desorption in vapors of ethyl alcohol are twice as much as in vapors of isopropyl alcohol. This is evidently due to the more complex structure of the isopropyl alcohol molecule compared to ethanol (Fig. 20.8).

The adsorption of isopropyl alcohol molecules is more complicated on the surface of the SnO_2 film in comparison with ethyl alcohol due to the additional CH_3 radical. This structure of the isopropanol molecule provides lower binding energy to the surface of the adsorbent. A consequence of this is the lower values

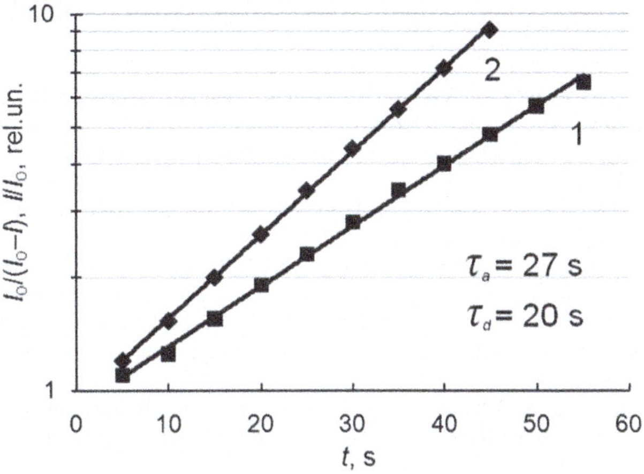

Fig. 20.7 Current relaxation in the SnO_2 film during the inlet of vapors of isopropyl alcohol (1) and dry air (2) ($V = 200$ V) into the chamber

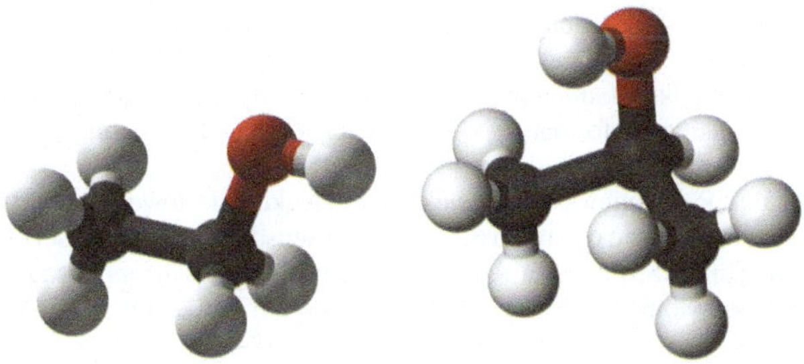

Fig. 20.8 The structure of molecule of ethyl alcohol (ethanol) – left, and isopropyl alcohol (propanol-2) – right

of the adsorption and desorption time constants for isopropanol compared to that of ethanol. Differences in the values of sensitivity and adsorption/desorption constants provide the selectivity of the films to different alcohols.

20.4 Conclusion

Thus, as a result of the studies, a reversible change in the conductivity of tin dioxide films in atmospheres with ethanol and isopropanol vapors at room temperature was established. Differences in the values of sensitivity and adsorption/desorption constants were also established, which ensure the selectivity of thin SnO_2 films to alcohols (ethanol and isopropanol). All this allows using them as basic sensitive elements of gas sensors for ethyl and isopropyl alcohols detections in ambient air, operating at room temperature.

References

1. Grinevych V, Smyntyna V, Filevska L (2018) Nanostructured SnO_2 as CBRN safety material. In: Bonca J, Kruchinin S (eds) Nanostructures materials for the detection of CBRN, NATO science for pease and security series A: chemistry and biology. Springer Science+Business Media, pp 107–127
2. Fell JC, Compton C, Voas RB (2008) A note on the use of passive alcohol sensors during routine traffic stops. Traffic Inj Prev 9(6):534–538
3. Ljungblad J, Hök B, Allalou A, Pettersson H (2017) Passive in-vehicle driver breath alcohol detection using advanced sensor signal acquisition and fusion. Traffic Inj Prev 18(1):S31–S36
4. Ozoemena KI, Musa S, Modise R, Ipadeola AK, Gaolatlhe L, Peteni S, Kabongo G (2018) Fuel cell-based breath-alcohol sensors: innovation-hungry old electrochemistry. Curr Opin Electrochem 10:82–87
5. Tan W, Yu Q, Ruan X, Huang X (2015) Design of SnO2-based highly sensitive ethanol gas sensor based on quasi molecular-cluster imprinting mechanism. Sens Actuators B Chem 212:47–54
6. Zhang W, Yang B, Liu J, Chen X, Wang X, Yang C (2017) Highly sensitive and low operating temperature SnO_2 gas sensor doped by Cu and Zn two elements. Sens Actuators B Chem 243:982–989
7. Guan Y, Wang D, Zhou X, Sun P, Wang H, Ma J, Lu G (2014) Hydrothermal preparation and gas sensing properties of Zn-doped SnO_2 hierarchical architectures. Sens Actuators B Chem 191:45–52
8. Poloju M, Jayababu N, Manikandan E, Ramana Reddy MV (2017) Enhancement of the isopropanol gas sensing performance of SnO_2/ZnO core/shell nanocomposites. J Mater Chem C 5:2662–2668
9. Teterycz H, Halek P, Wiśniewski K, Halek G, Koźlecki T, Polowczyk I (2011) Oxidation of hydrocarbons on the surface of tin dioxide chemical sensors. Sensors (Basel) 11(4):4425–4437
10. Ermakov V, Kruchinin S, Fujiwara A (2008) Electronic nanosensors based on nanotransistor with bistability behaviour. In: Bonca J, Kruchinin S (eds) Proceedings NATO ARW "Electron transport in nanosystems". Springer, Dordrecht, pp 341–349
11. Ermakov V, Kruchinin S, Hori H, Fujiwara A (2007) Phenomena in the resonant tunneling through degenarate energy state with electron correlation. Int J Mod Phys B 11:827–835
12. Ermakov V, Kruchinin S, Pruschke T, Freericks J (2015) Thermoelectricity in tunneling nanostructures. Phys Rev B 92:115531
13. Jawaher KR, Indirajith R, Krishnan S, Robert R, Pasha SKK, Deshmukh K, Sastikumar D, Das SJ (2018) A high sensitivity isopropanol vapor sensor based on Cr_2O_3-SnO_2 heterojunction nanocomposites via chemical precipitation route. J Nanosci Nanotechnol 18(8):5454–5460
14. Filevskaya LN, Smyntyna VA, Grinevich VS (2006) Morphology of nanostructured SnO_2 films prepared with polymers employment. Photoelectronics 15:11–14
15. Ulug B, Türkdemir HM, Ulug A, Büyükgüngör O, Yücel MB, Smyntyna VA, Grinevich VS, Filevskaya LN (2010) Structure, spectroscopic and thermal characterization of bis(acetylacetonato)dichlorotin(IV) synthesized in aqueous solution. Ukr Chem J 76(7):12–17
16. Grinevych V, Smyntyna V, Filevska L, Savin S, Ulug B (2017) Thermogravimetric study of nano SnO_2 precursors. In: Fesenko O, Yatsenko L (eds) Nanophysics, nanomaterials, interface studies, and applications. Springer proceedings in physics, vol 195. Springer International Publishing AG, Springer, Cham, pp 53–61
17. Chebanenko AP, Filevska LM, Grinevych VS, Simanovich NS, Smyntyna VA (2017) The humidity and structuring additives influence on electrophysical characteristics of tin dioxide films. Photoelectronics 26:5–10

Chapter 21
Porous Silica Glasses as a Model Medium for the Formation of Nanoparticles Ensembles: Review

I. K. Doycho, V. S. Grinevych, and L. M. Filevska

Abstract The types of porous silicate glasses, their preparation methods, and structural properties for each type of glass are described. Ways to use glasses as a matrix for obtaining the ensembles of silicon nanoparticles, dyes, CdS, AgBr, and SnO_2 are briefly described. The work may be useful as a reference material for researchers working in the fields of nanoelectronics and sensorics.

Keywords Porous silicate glasses · Ensemble of nanoparticles · Model medium

21.1 Introduction

The quantum confinement effects can significantly affect the optical properties of semiconductor materials. The manifestation of these effects needs the particle sizes of the material to be in the nanometer range. Moreover, the properties of such a system will be stable, if it is a large canonical ensemble of nanoparticles placed in a certain model medium. Some authors use polymers [1, 2] or gelatin [1, 3] as such a medium, which, being enough chemically inert, are able due to their structure to retain the clusters of a necessary substance which have grown inside them. However, from our point of view, the most suitable material as a model medium is porous silicate glass with a columnar structure, i.e., a medium containing through voids with nanometer size. The advantage of this substance as a matrix for the semiconductor clusters embedding is due to the fact that, unlike polymers and gelatin, it is characterized by a highly dispersed distribution of interpenetrating pores, the sizes of which can vary from several to hundreds of nanometers. Since a cluster cannot exceed its size inside the pore, the sizes of the formed semiconductor particles are in the same range. The columnar structure allows one to affect the internal surfaces of pores, saturating the glass with solutions of suitable substances.

I. K. Doycho · V. S. Grinevych (✉) · L. M. Filevska
Odessa I.I. Mechnikov National University, Odessa, Ukraine
e-mail: grinevich@onu.edu.ua

J. Bonča, S. Kruchinin (eds.), *Advanced Nanomaterials for Detection of CBRN*, NATO Science for Peace and Security Series A: Chemistry and Biology,
https://doi.org/10.1007/978-94-024-2030-2_21

21.2 Fabrication of Various Types of Porous Glasses

The model medium used to create semiconductor nanoclusters with given dimensions must satisfy the following requirements:

(1) to be enough rigid and mechanically strong to form particles, so that the particles, arbitrarily growing, cannot form its frame;
(2) to be sufficiently chemically inert not reacting with the working substance;
(3) to have no luminescence in the spectral region of the substance whose particles are formed in this medium.

As can be concluded from a number of works [1, 4–11], silicate porous glass satisfies these requirements in the best way.

The essence of a production technology for silicate porous glass is as follows [12]. The charge for melting sodium-borosilicate two-phase glass, initially brought to the mixture melting point temperature (about 750 °C), was slowly adiabatically cooled. Since the melting temperature of quartz glass is much higher than that of sodium borate, the temperature at which phase separation occurs is chosen such that the silicate phase is simply very heated, whereas the sodium borate phase still melts, forming large bubbles for the silicate framework. At this temperature, 650 °C, the melt, due to its viscosity, is held for several hundred hours until both phases mutually dissolve in each other, then it is slowly cooled to room temperature. The obtained two-phase glass is a fairly large (up to hundreds of nanometers) section of mutually interwoven silicate and sodium borate phases. In addition to differences in the melting points, these phases are also not chemically stable. Using a specially prepared etchant based on hydrofluoric, nitric, and glacial acids, it is managed to etch away the sodium borate phase completely, practically not affecting the silicate phase. In this case, a silicate framework with through voids (or a columnar structure) remains in place of the etched away sodium borate phase. Due to the mutual dissolution, very small sandy-type particles of quartz were also found in the sodium borate phase, which, after etching, precipitated into the formed voids (pores). These precipitated sandy particles are called residual silica gel. Residual silica gel is chemically identical to the silicate framework and differs from it only in fineness. The porous glass thus prepared is called type C glass.

In order to form smaller pores in the matrix, it is necessary to choose a lower temperature, when cooling the two-phase glass, as the phase separation temperature is slightly lower than the melting point of the sodium borate phase, but close to it (about 490 °C). As a result of keeping it at this temperature for hundreds of hours, the sodium borate phase forms rather small bubbles. After the cooling to room temperature, two-phase glass becomes small (on the order of tens of nanometers) sections of mutually interwoven silicate and sodium borate phases. After etching the unstable sodium borate phase as described above, a sufficiently finely porous silicate glass is formed, which also contains residual silica gel. The porous glass thus prepared is called type A glass.

The presence of residual silica gel in the pores may be both useful and harmful depending on the nature of the task. The presence of silica gel in the pores reduces their effective size, i.e. makes glass finer. At the same time, it reduces the free space for the formation of nanoparticles, improves the adsorption properties of glass, but worsens its mechanical properties. Indeed, silica gel swells in a humid environment, which leads to a mechanical deformation of the plate [13–15]. However, the mechanical properties of glass of any type may be improved by a special processing [16]. Silica gel can passivate the surface of the formed nanoparticles of a created ensemble, thus hindering the aggregation process [1, 17].

In the cases where the presence of silica gel is undesirable, the authors of book [18] developed a leaching technology. The made porous glass is etched in an alkaline potassium hydroxide etchant. By interacting with the surface of silica, this etchant dissolves residual silica gel, practically without affecting the walls of the frame of porous silicate glass. Type A glass thus treated is called type B glass, and type C glass is called type D.

In [19, 20] devoted to the application of the adsorption properties of porous glass for biological prosthetics, it was found that leached glasses of B and D types still contain a small amount of residual silica gel, which is called secondary. The mechanism of formation of secondary silica gel can be understood from the following model [12]. The walls of the silicate matrix do not different from residual silica gel due to their chemical composition, but they are structured differently. The alkaline etchant will not only dissolve the residual silica, but also etch the walls of the matrix itself, forming new silica, which is secondary. It cannot be distinguished neither in composition nor in structure from residual ("primary") one. Therefore the terms "residual" and "secondary" are usually interchangeable in the literature. The silica gel, which is present in types B or D matrices, is predominantly "secondary" one. The processes of etching of residual silica gel and etching of the walls of the matrix proceed at different rates due to the different surface developments of these formations. In Fig. 21.1, curve *1* describes the process of etching of silica, and curve *2* describes the process of etching of the matrix.

It can be seen from the figure that curve *2* is more gentle in the region of short etching times, since the process of etching is much slower than the dissolution of silica gel due to a more compact packaging of matrix walls. However, it has variable slope. This is due to the fact that, in the process of surface etching, the frame develops, and the rate of its destruction increases. The longer the etching time, the more the steepnesses of curves *1* and *2* become closer. If we assume that the initial amount of silica in the matrix corresponds to the line N_1, then the time for etching is t_1 (Fig. 21.1). But, during this time due to the etching of the matrix, the N_2 of new, secondary silica is formed, and etching, of it will take time t_2, during which new silica is formed again. Therefore, secondary silica can never be completely etched: it will form again and again, until the sample is completely dissolved. This mechanism is valid for both types of glasses A and C. Thus, when talking about glasses B and D, it is undermined not freeing them from silica, but only depleting them with silica.

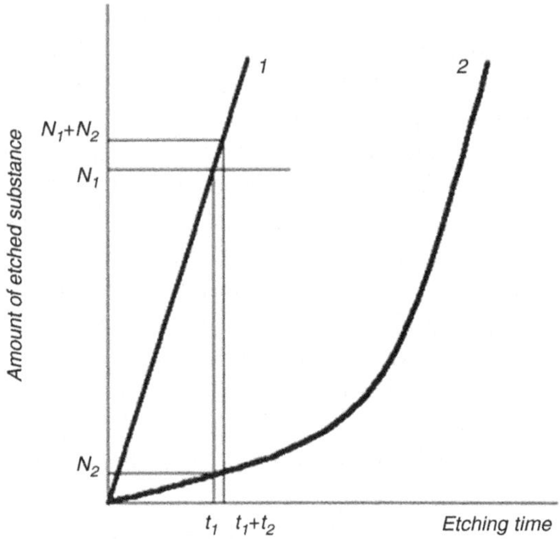

Fig. 21.1 The model of a porous glass leaching [30]: *1* – etching rate of residual silica; *2* – the etching rate of the silicate frame

The way of getting rid of silica gel is the method of the so-called carbon treatment [21, 22]. The method is based on the fact that porous glass is almost pure silicon dioxide. Since carbon is a chemically more active element than silicon, it can displace the latter from the oxide under appropriate temperature conditions. To carry out the carbon treatment, the porous glass is impregnated with an aqueous solution of glucose, which, according to the well-known equation, is

$$C_6H_{12}O_6 - (t) \rightarrow 6C + 6H_2O \tag{21.1}$$

when heated to a temperature of about 150 °C, it gives pure carbon in the form of highly dispersed graphite in the intraporous space with the release of water. The end of the reaction can be detected visually by the blackening of the sample.

Further, the obtained system is subjected to the additional annealing in the regime corresponding to the displacement of silicon from oxide without simultaneous collapse of pores (i.e., within 300–400 °C). In this case, chemically inert carbon dioxide is formed, which is heavier than air and therefore will linger in the pores for some time, preventing the flow of oxygen into them. As a result, pure silicon atoms are restored, which, being deprived of the opportunity to oxidize, will crystallize into silicon nano-sized crystallites and will settle on the pore walls in the form of "dew". The end of the reaction can again be fixed visually by the enlightenment of the sample. It is clear that the described process will be the more intense, the smaller the particles of SiO_2 will take part in it. Therefore, it is advisable to carbonize glasses type A and C containing a sufficient amount of residual silica gel. During the carbon treatment, the silica gel present in the pores is almost completely converted into silicon clusters, while the pore walls only partially lose oxygen. The resulting clusters settle on the formed dangling bonds.

Table 21.1 Conditions for the creation of porous glass of different types

	The low temperature – 490 °C of phase separation in the original glass	The high temperature – 650 °C of phase separation of the original glass
The glass type after unstable phase etching	A	C
Type of glass after leaching	B	D
Type of glass after carbon processing	β	δ

Thus, the carbonizing is an alternative leaching method for removing secondary silica gel from pores. The carbon-treated glass of type A is called glass of type β, and glass of type C is called glass of type δ. In their main parameters, these types of glasses differ significantly from glasses of type B and D.

The described technology can be briefly summarized in Table 21.1.

21.3 Porosity of Porous Glasses

Porosity and related parameters are important characteristics for any type of porous glass. Porosity is usually determined by the gravimetric method, by entrainment of the mass. The manufactured two-phase glass is weighed on a precision torsion balance before (m_0) and after (m_1) the etching of the unstable sodium borate phase (also after the leaching for glasses of types B and D and after the carbon treatment for glasses of type β and δ). Then the porosity η is determined by the formula [23, 24]

$$\eta - \left(1 - \frac{m_1}{m_0}\right) \times 100\%. \quad (21.2)$$

Due to the difference in molecular weights of the sodium borate ($Na_2O \times B_2O_3$) and silicate phases (SiO_2), the porosity thus determined for various glasses lies in the range 60–80% [25]. The porosity determined by the ablation is directly related to the total volume of voids in the glass.

In some cases, it is advisable to determine the porosity in volume percents, which is directly related to the development of the inner surface (the specific surface of pores) of porous glass. In this case, the difference not only affects the molecular masses, but also the densities of both phases. The value of the porosity determined in this way is 50–65 vol %. Here, a larger volume of the sodium borate phase molecules is compensated by a less dense packaging of silicate-phase molecules [24]. For a standard sample $20 \times 15 \times 0.5\,mm^3$ in size, it can be tens of square meters [18].

The main parameters associated with porosity for all considered types of glasses are shown in Table 21.2.

Table 21.2 Parameters of porous glasses of different types associated with porosity

Glass type	Total void volume, mm^3/g	The specific surface of pores, m^2/g
A	292	54.7
B	395	38.6
β	340	44.0
C	389	34.3
D	478	29.8
δ	431	31.3

From the table, a number of conclusions can be drawn about the structural features of porous silicate glasses. It can be seen that glasses B and D obtained after the leaching have a larger volume of voids. This can be easily explained, since, in the glasses A and C from which they were obtained, the residual silica gel occupied some place and was freed after the leaching. In terms of the total volume of voids, glasses β and δ obtained by the carbon treatment occupy an intermediate place between the initial and leached glasses. This is also understandable, since gel is removed during the leaching and, during the carbon processing, just loses oxygen and settles on the pore walls in the form of silicon clusters. The latter occupy a less space, than silica gel. Silica gel particles, being sandy formations, contributed to the surface development of initial glasses A and C. Therefore, after the leaching, the unfolding of the inner surface of the corresponding glasses decreases. For β and δ glasses, this characteristic again takes on an intermediate value, since silicon clusters which appeared on the pore surface of the initial glasses after the carbon treatment give an additional contribution to the surface development, but a smaller one than that of sandy silica gel. In addition, it can be seen that large-pore glasses correspond to a larger volume of voids and a less developed inner surface of the pores.

Another important characteristic of any type of porous glass is the pore size distribution. Traditionally, this characteristic is determined by penetrating mercury porosimetry. It is a destructive typically test method of industrial mass control over properties of products. The method is based on the injection of mercury into the pores of a sample at various pressures which correlate with the external pore diameter, shape, and complexity of the porous structure. Since the mercury penetrated into the sample cannot be completely removed from the controlled ones, they must be disposed of after the measurement is completed, and the remaining ones are to be certified as identical to them. The expensive, bulky, and unsafe devices – porosimeters – have been developed to implement the specific test control [21]. However, this method is not applicable to porous silicate glasses containing pores of about 10 nm or less, since it gives a guaranteed inflated result. Indeed, the pressures corresponding to such pore sizes are 1500–1600 kg/cm^2, which exceeds the breaking point of quartz glass (about 1100 kg/cm^2 [1]). Thus, the thin partitions between the individual pores will break, and, instead of several small pores, the method will fix one large one, which will significantly distort the pore size distribution spectrum.

An alternative method, without such drawback, is the adsorption-desorption method developed in [26, 27]. This method does not require the use of aggressive or toxic liquids or gases, such as benzene, toluene, mercury, etc., although all of them (except for mercury) can be used as a working substance. For measuring pores' size distribution in porous glass, ordinary water can be selected as the working substance. It is preferable due to the small effective size of its molecule (about 0.23 nm), which allows one to obtain pore size distribution spectra for glasses with very small pores.

The method is based on the correlation of the dielectric constant of porous glass with the concentration of inclusions in pores [28]. The investigated matrix is a three-component system of silicate porous glass, air, and water contained in pores. In this case, the contribution of the porous skeleton to the dielectric constant of the matrix does not change during measurements, and the mutual ratio of the contributions of air and water (which are the mentioned inclusions in our case) depends on thermodynamic conditions. These conditions can be changed by changing the concentration of water in pores by means of the molecular condensation. Since, the sample capacity depending on the distribution of voids in the matrix will change during the water adsorption-desorption in pores in different ways, we can estimate the sizes of pores by fixing these changes.

The sample is placed in a sealed chamber, with the humidity up to 100%, which leads to the complete filling of matrix's pores with water. Further, with a continuous very accurate (up to 0.001 pF) measurement of the capacitance C, the chamber is dried very slowly (for 48 h) to 5–6% relative humidity. In this case, the desorption of the absorbed water takes place, in which the pores of a specific diameter are released from the adsorbate. This corresponds to gently sloping parts of the sample capacity/humidity dependence (see Fig. 21.2, which shows a typical curve for a model medium with two pore fractions). The more the pores of a given diameter, the longer is the shallow section. The steepness of the curve near the flat portion makes it possible to estimate the pore scatter of this fraction [29, 30].

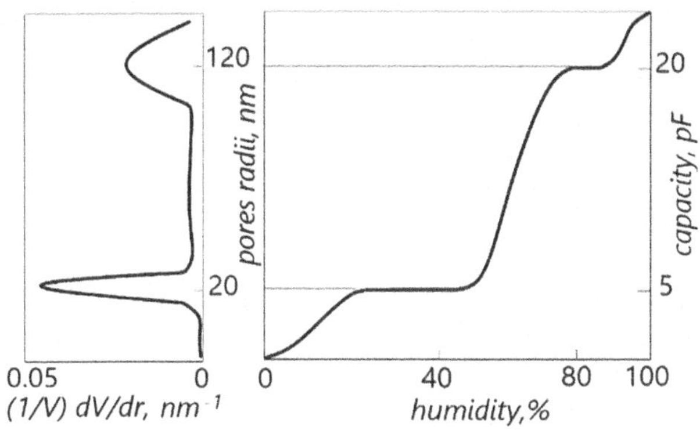

Fig. 21.2 Calculation scheme for the pore size distribution by the isotherm of water desorption from a model porous substance with two pore fractions

Table 21.3 Parameters of the pore size distributions for porous glasses of different types

Glass type	The average pore diameter, nm	Preferred pore diameter, nm
A	30	20
B	40	30
β	34	25
C	80	120
D	120	130
δ	99	125

Thus, experimentally determining the sample capacity/environmental humidity dependence, the information on the size distribution of pores in it based on Dolimor and Hill equation [31] can be estimated. The measurement process should be isothermal: the temperature deviation in a chamber during the entire measurement should not exceed 0.2 °C.

It is worth to note that, for real silicate porous glasses, the isotherms of water desorption turn out to be more complicated than for the model shown in Fig. 21.2, but the essence of the calculations is not changed. As usual, in different sets of glasses, the pore size distributions can be slightly different. Therefore, before using specific glasses as a model medium, it is recommended to construct firstly the pore size distribution spectra in them. The main parameters of the pore size distribution for all considered types of glasses are shown in Table 21.3.

We note that relatively small pores, as well as sufficiently large ones, are characteristic of all types of glasses. Table 21.3 shows that, for finely porous glasses A, B, and β, the predominant pore size is smaller than average one, and, for large-porous, C, D, and δ glasses, the situation is opposite. This regularity can easily be explained by the structural features of glasses of all types, which follow from the structural features of SiO_2. In fact, all pores that appear before the leaching with some scatter are grouped into two fractions with the most probable sizes of about 20 nm and about 120 nm [32]. In a glass A, the former prevail, while the latter, although they exist, are rare. In glass C, everything happens, on the contrary. This explains cregularity evident from the table. This regularity remains after the leaching, but the pores are depleted in silica gel, and they are slightly etched due to the removal of silica gel. During the carbon treatment, silica gel is not removed, but is converted into silicon clusters, which settle on the pore walls, while continuing to occupy some place in the interpore space. Therefore, the pores in glasses β and δ, compared with glasses A and C, are also etched, but to a less extent than during the leaching.

It is also notable that, in contrast to the leaching, the carbon treatment leads to a change in the structure of the matrix. Silicon clusters appear on the walls of its pores structurally similar to porous silicon obtained by the laser ablation [33]. Therefore, the external peaks corresponding to porous silicon appear in the photoluminescence spectra. Therefore, glasses of types β and δ are rarely used as a matrix in luminescent studies of nanoparticles ensembles, but their absorption properties are useful for a number of applied problems, such as, for example, the ophthalmic bioprosthetics [19, 20].

21.4 Formation of Nanoparticles' Ensembles

When creating semiconductor clusters in a matrix of porous glass, it is necessary to distinguish whether the element is in the chemical composition of the matrix or not. In the first case, clusters can be created by a suitable processing of the matrix itself. In the second case, the corresponding substances should be introduced into it from the outside. As was noted above, the carbon treatment leads to the formation of silicon clusters inside the pores. In fact, pores' surface in the matrix is coated with porous silicon. This process can be controlled by subjecting β or δ type glass to the repeated carbonizing. The significant difference in this case is that, in the first carbonizing, the residual silica gel reacted primarily, when the repeated carbon treatments lead to the reduction of silicon from the matrix walls themselves. Wherein, as shown in [22], clusters are formed consisting of seven or eight coordination spheres. This is due to the fact that a cluster consisting of less than seven spheres is unstable, since it contains too many dangling bonds, and a cluster exceeding eight coordination spheres just has no time to originate during the reaction: this is energetically disadvantageous.

In [19], the properties of glass during multiple carbon processing, up to 14-fold onse were considered. In all cases, either the existing silicon clusters build up the 8th coordination sphere, or new clusters are formed from seven coordination spheres. This is perhaps the only case where an ensemble of semiconductor nanoparticles can be created directly from the matrix body by the simple thermochemical processing.

However, in many applied problems, it is necessary to create an ensemble not of silicon nanoparticles, but of nanoparticles of some other substance. In this case, the necessary substance can be introduced into thematrix from the outside. The simplest is the creation of nanoparticles of dyes in a matrix. These are high-molecular substances, which, when dissolved in a standard solvent, dimethylformamide $(CH_3)_2NCO$, perfectly impregnate the matrix and, after a heat treatment, are evenly distributed in the pores. In this case, the solvent is removed and, if glass containing residual silica gel was used as a matrix, the formed particles hardly aggregate [34].

Some substances, which are to be used for the formation of nanoparticles' ensembles, appeared to be insoluble in solvents not destroying the model medium. However, these materials can be synthesized as a result of the chemical reaction of substances that are readily soluble in water. This, in particular, concerns such a widespread semiconductor material as CdS. CdS clusters are formed inside the porous matrix as follows [4, 5, 7]. $Cd(NO_3)_2$ is dissolved in deionized water. $(NaPO_3)_3$ was added to the solution as a catalyst, and the matrix was impregnated with the resulting mixture. To ensure a uniform distribution of cadmium-containing molecules in the matrix, the system is kept for 2 h at room temperature. Next, in order to remove the excess water, the sample is dried at 150 °C for half an hour. The sulfidization is carried out by passing gaseous H_2S through a porous system for 2 h at room temperature. In order to increase the concentration of CdS clusters, the procedure can be repeated several times.

When trying to create nanoparticles' ensembles used in AgBr emulsions, the molecules just cover the surface of the porous sample without penetrating into the pores. Therefore, it is necessary to use a binder, which will collect AgBr particles. When porous glass is impregnated, a binder together with AgBr molecules will enter the pores. Moreover, gelatin traditionally used in photoemulsions is unsuitable as a binder, since its molecules are too large and will not enter the pores of the matrix [8]. Therefore, polyvinyl alcohol (Polinol), whose molecules are able to penetrate into the smallest pores, can be used to transport AgBr molecules to the pores of the matrix. Typically, polyvinyl alcohol is used to introduce $AgNO_3$ molecules into the matrix. After the half-hour holding of the obtained system at about 40 °C (to ensure the uniform distribution of inclusions in the pores of the matrix), it is saturated with bromine vapor during one day, which displaces the nitric acid residue.

The additional difficulties arise in the formation of ensembles of SnO_2 nanoparticles. A practically insoluble oxide may be obtained by thermosynthesis [35] from $SnCl_4$, which is perfectly soluble in alcohol. However, such a solution in high concentrations is a effective glue for glass and thus is able to destroy the matrix before nanoparticles are formed. This property of the solution is lost, if its concentration is less than 30%. Our studies show that the best luminescent response is obtained, if the glass is impregnated with an alcohol solution of $SnCl_4$ with a concentration in the interval 5–7.5% [36]. The most effective thermosynthesis is carried out at temperatures exceeding 600 °C. However, this temperature is comparable with the temperature of the porous glass melting [18] and is able to distort its structure. Thermosynthesis is also possible at lower temperatures, but the process becomes longer. This, however, is even a positive factor, since the low synthesis temperature promotes the formation of smaller SnO_2 particles [37].

21.5 Conclusions

Due to their chemical, mechanical, absorption, and luminescent properties, porous silicate glasses can be considered as the most suitable model medium for creating the ensembles of nanoparticles of various substances used in nanoelectronics. The effectiveness of the technology for creating these ensembles is associated with the existence of several types of porous glasses that differ structurally. The chemical and structural features of the substances, whose ensemble of nanoparticles must be formed, are taken into account. Porous glass of a suitable type is used as a system of specific nanoscale tubes. The chosen model medium is saturated with solutions of the corresponding substances and subjected to a heat treatment in the suitable mode ensuring the necessary chemical reactions in the pore space, the uniform distribution of resulting nanoparticles in the matrix, and the drying of the resulting system.

References

1. Meshkovsky IK (1998) The composite optical materials based on porous matrices. St.-Petersburg. In Russian
2. Smyntyna VA, Skobeeva VM, Malushin NV (2011) The boundary influence on the optical and luminescent properties of the quantum dots of the CdS in a polymer. Fiz Khim Tverd Tela 12(2):355–358. In Ukrainian
3. Smyntyna VA, Skobeeva VM, Malushin NV (2012) Influence of the surface on the spectrum of luminescence NC CdS in gelatine matrix. Photoelectronics 21:50–56
4. Gevelyuk SA, Doycho IK, Mak VT, Zhukov SA (2007) Photoluminescence and structural properties of nano-size CdS inclusions in porous glasses. Photoelectronics 16:75–79
5. Rysiakiewicz-Pasek E, Polańska J, Gevelyuk SA, Doycho IK, Mak VT, Zhukov SA (2008) The photoluminescent properties of CdS clusters of different sizes in porous glasses. Opt Appl XXVIII(1):93–100
6. Rysiakiewicz-Pasek E, Zalewska M, Polańska J (2008) Optical properties of CdS-doped porous glasses. Opt Mater 30(5):777–779
7. Doycho IK, Gevelyuk SA, Ptashchenko OO, Rysiakiewicz-Pasek E, Zhukov SO (2008) Luminescence kinetics peculiarities of porous glasses with CdS inclusions. Photoelectronics 17:43–47
8. Doycho IK, Gevelyuk SA, Ptashchenko OO, Rysiakiewicz-Pasek E, Tolmachova TN, Tyurin OV, Zhukov SO (2010) Photoluminescence features of AgBr nanoparticles formed in porous glass matrices. Opt Appl 40(2):323–332
9. Ermakov V, Kruchinin S, Fujiwara A (2008) Electronic nanosensors based on nanotransistor with bistability behaviour. In: Bonca J, Kruchinin S (eds) Proceedings NATO ARW "Electron transport in nanosystems". Springer, pp 341–349
10. Rodionov VE, Shnidko IN, Zolotovsky A, Kruchinin SP (2013) Electroluminescence of Y_2O_3: Eu and Y_2O_3:Sm films. Mater Sci 31:232–239
11. Ermakov V, Kruchinin S, Hori H, Fujiwara A (2007) Phenomena in the resonant tunneling through degenarate energy state with electron correlation. Int J Mod Phys B 11:827–835
12. Doycho IK (2015) Study of the photoluminescent properties of nanoparticles ensembles of dyes. In: Smyntyna VA (ed) Non equilibrious processes in the sensor structures, Odessa Mechnikov National University, Odessa, pp 120–170
13. Gevelyuk SA, Doycho IK, Rysiakiewicz-Pasek E, Marczuk K (2000) Relative changes of porous glass dimensions in humid ambiance. J Porous Mater 7:465–467
14. Gevelyuk SA, Doycho IK, Lishchuk DV, Prokopovich LP, Safronsky ED, Rysiakiewicz-Pasek E, Roizin YaO (2000) Linear extension of porous glasses with modified internal surface in humid environment. Opt Appl 30(4):605–611
15. Gevelyuk SA, Doycho IK, Prokopovich LP, Rysiakiewicz-Pasek E, Safronsky ED (2002) Humidity dependencies of porous sol-gel and silica glass linear sizes. Mater Sci 20(2):23–27
16. Rysiakiewicz-Pasek E, Vorobyova VA, Gevelyuk SA, Doycho IK, Mak VT (2004) Effect of potassium nitrate treatment on the adsorption properties of silica porous glasses. J Non-Cryst Solids 345–346:260–264
17. Tyurin OV, Bercov YM, Zhukov SO, Levitskaya TF, Gevelyuk SA, Doycho IK, Rysiakiewicz-Pasek E (2010) Dye aggregation in porous glass. Opt Appl 40(2):311–321
18. Mazurin OV, Roskova GP, Averianov VI, Antropova TV (1991) Biphasic glasses: structure, properties, applications. Nauka, Leningrad. In Russian
19. Rysiakiewicz-Pasek E, Gevelyuk SA, Doycho IK, Prokopovich LP, Safronsky ED (2003) Antibiotic hentamicini sulphate effect on photoluminescent properties of the silicate porous glasses, which are suitable for ophthalmologic protesting. Opt Appl 33(1):33–39
20. Rysiakiewicz-Pasek E, Gevelyuk SA, Doycho IK, Vorobjova VA (2004) Application of porous glasses in ophthalmic prostetic repair. J Porous Mater 11:21–29

21. Gevelyuk SA, Doycho IK, Prokopovich LP, Rysiakiewicz-Pasek E, Marczuk K (1998) The influence of anneal of incorporated carbon on the photoluminescence properties of porous glass and porous silicon. In: Stoch L (ed) Polish ceramic bulletin 19, ceramics 57/porous and special glasses. Proceedings of the 4-th international seminar PGL'98. Polish Ceramic Society, Krakow, pp 59–64
22. Gevelyuk SA, Doycho IK, Prokopovich LP, Rysiakiewicz-Pasek E, Safronsky ED (2003) Influence of carbon multiple treatments on the photoelectrical properties of porous glasses. Radiat Eff Defect 158:427–432
23. Alexeev-Popov AV, Roizin YO, Rysiakiewicz-Pasek E, Marczuk K (1993) Physical properties of organosilicate porous glasses. Mol Cryst Liquid Cryst 230:197–201
24. Alexeev-Popov AV, Roizin YO, Rysiakiewicz-Pasek E, Marczuk K (1993) Porous glasses for optical application. Opt Mater 2:249–255
25. Plachenov TG, Kolosentsev SD (1988) Porometry. Khimiya. In Russian
26. Dubinin MM (1978) Capillary phenomena and the information about porous structure of adsorbers. Izvestiya AN SSSR, Ser Khim 1:101–125
27. Roizin YO, Rysiakiewicz-Pasek E, Safronskiy ED (1995) Moisture sensitivity of the porous tin dioxide. Technol Constr Electr Devices 2:53–54
28. Greg SJ, Sing KSW (1967) Absorption, surface area and porosity. Academic, London
29. Rysiakiewicz-Pasek E, Safronsky ED (1995) Wplyw wody na wiasnosci elektryczne szkiei porowatych. In: IV sympozium Naukowo-Techniczne. Bialostok, pp 94–99
30. Roizin YO, Gevelyuk SA, Prokopovich LP, Savin DP, Rysiakiewicz-Pasek E, Marczuk K (1997) Water absorption and mechanical properties of silica porous glasses. J Porous Mater 4(3):151–155
31. Rusanov AI, Goodrich FCh (1980) The modern theory of capillarity. Khimiya, Leningrad
32. Savin DP, Gevelyuk SA, Roizin YO, Mugenski E, Sokolska I (1998) Comparison of some properties of nanosized silicon clusters in porous glasses. Appl Phys Lett 72(23):3005–3007
33. Gevelyuk SA, Doycho IK, Prokopovich LP, Savin DP (1999) The structural and luminescent properties of the porous silicon, obtained by the laser ablation method. Photoelectronics 8:18–21
34. Doycho IK, Gevelyuk SA, Rysiakiewicz-Pasek E (2015) Photoluminescence of tautomeric forms of nanoparticle ensembles of dyes based on the 4-valence stannum complexes in porous silica glass. Photoelectronics 24:30–37
35. Uchiyama H, Shirai Y, Kozuka H (2011) Formation of spherical SnO_2 particles consisting of nanocrystals from aqueous solution of $SnCl_4$ containing citric acid via hydrothermal process. J Cryst Growth 319(1):70–78
36. Gevelyuk SA, Grinevych VS, Doycho IK, Lepikh YaI, Filevska LM (2019) Photoluminescence of SnO_2 nanoparticle ensemble in porous glass with column structure. In: IEEE proceedings of 8-th international conference on advanced optoelectronics and lasers (CAOL-2019) 06–08 Sept 2019, Sozopol, sub. P101. to be published
37. Ivanov VV, Sidorak IA, Shubin AA, Denisova LT (2010) SnO_2 powders production by means of decomposition of thermo not stable compounds. J Siberian Fed Univ Eng Technol 2(3):189–213

Chapter 22
The Revised SI Systems of Units of 2018 and Its Impact on Nanotechnology

W. Nawrocki

Abstract The revised SI is the system of units in which the standards of units dependent on physical constants and quantum effects. The redefinition of the SI system on the base of physical constants is its great advantage. However, the revised SI system is understood rather by narrow community of scientists and experts of metrology, not by a wide population of users. In particular this remark relates to the definitions of the kilogram and the kelvin in the REVISED SI. Critical remarks on the new SI concern sophisticated definitions, the set of base units and high costs of implementation the definition of the kilogram. The definitions only repeat the values of physical constants. New standards of the kilogram (watt balance or XRCD method) are at least 30–50 times more expensive than the "old" standard – the artefact of PtIr alloy. In result, only about 10 national metrology institutes in the world are able to install and operate new standards of the kilogram. Before and after adopting the revised SI there are the proposals to replace the ampere, the base unit, by the wolt (or the ohm), the kelvin by the joule, and the unit of mole by a new unit of amount of substance (Leonard's proposal). In the paper the experimental set-up of the standard of mass based on the watt balance is presented.

Keywords SI system · Base units · Physical constant

22.1 Introduction

The revised SI (SI of 2018) is the long-time awaited system of units in which the standards of units dependent of material properties have been replaced by the standards based on physical constants and quantum effects [1]. The main reason for a new definition of the SI was the request for the new definition of the kilogram.

W. Nawrocki (✉)
Faculty of Computing and Telecommunications, Poznan University of Technology, Poznan, Poland
e-mail: waldemar.nawrocki@put.poznan.pl

J. Bonča, S. Kruchinin (eds.), *Advanced Nanomaterials for Detection of CBRN*,
NATO Science for Peace and Security Series A: Chemistry and Biology,
https://doi.org/10.1007/978-94-024-2030-2_22

Before 2018 the kilogram was the only unit between the seven base units of the SI of 1960 which is defined using a standard of an artefact – the International Prototype of the Kilogram (IPK). The IPK is kept in the BIPM. Its official copies are distributed between about 40 – laboratories around the world. The IPK served as the international standard of mass till 20 May 2019. The redefinition of the SI system on the base of physical constants is its great advantage.

22.2 The Revised SI System of Units

The redefinition of the SI system (Fig. 22.1) was the opportunity for new definitions all seven base units in the revised SI (the SI of 2018). The base units in the SI of 2018 are defined using fundamental physical constants (c – speed of light, h – Planck's constant, e – elementary electric charge, k_B – Boltzmann's constant, N_A – Avogadro's constant) and two not fundamental constants (ν_{Cs} – transition frequency in ^{133}Cs, K_{cd} – luminous efficiency of monochromatic radiation of frequency). The Committee on Data for Science and Technology (CODATA) has prepared the special adjustment of four fundamental physical constants (CODATA 2017 adjustment) for the definition of the revised SI system [2]. The values of h, e, k_B and N_A from CODATA 2017 reflect the new results in measurements of them. Their values of this constants are included in the definitions of the base units of the revised SI of 2018.

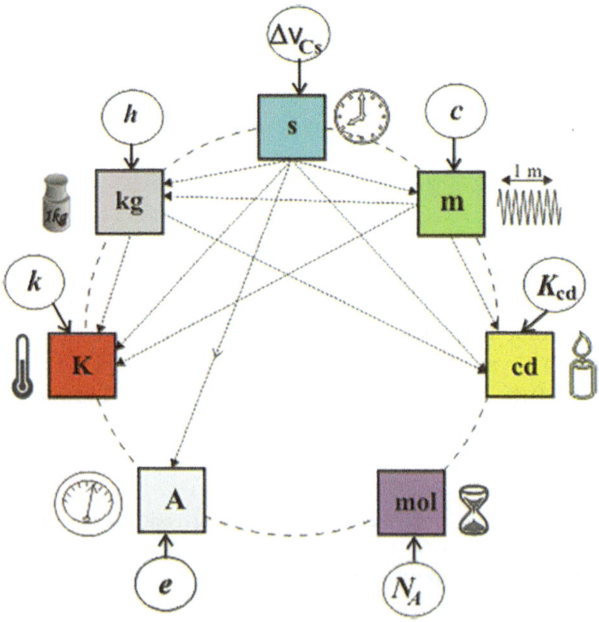

Fig. 22.1 Base units of the new SI system and their interrelations

Finally, the revised SI system of units was approved in November 2018. The revised SI system entered into force in May 2019 [1]. The definitions of units in the revised SI are listed below.

- The second is defined by taking the fixed numerical value of the caesium frequency $\Delta\nu_{Cs}$, the unperturbed ground-state hyperfine transition frequency of the caesium 133 atom, $\nu_{Cs} = 9\,192\,631\,770$ Hz.
- The metre is defined by taking the fixed numerical value of the speed of light in vacuum $c = 299\,792\,458$ m/s.
- The kilogram is defined by taking the fixed numerical value of the Planck constant h to be $6.626\,070\,15 \times 10^{-34}$ kg m^2 s^{-1}.
- The ampere is defined by taking the fixed numerical value of the elementary charge e to be $1.602\,176\,634 \times 10^{-19}$ C.
- The kelvin is defined by taking the fixed numerical value of the Boltzmann constant k to be $1.380\,649 \times 10^{-23}$ J K^{-1}.
- The mole is the SI unit of amount of substance. One mole contains exactly $6.022\,140\,76 \times 10^{23}$ elementary entities. This number is the fixed numerical value of the Avogadro constant, N_A, when expressed in mol^{-1} and is called the Avogadro number.
- The amount of substance, symbol n, of a system is a measure of the number of specified elementary entities. An elementary entity may be an atom, a molecule, an ion, an electron, any other particle or specified group of particles.
- The candela is defined by taking the fixed numerical value of the luminous efficiency of monochromatic radiation of frequency 540×10^{12} Hz, $K_{cd} =$ 683 lm W^{-1}.

In order to implement the revised SI, for each of the five newly defined base units of measures, a link between the macroscopic standard and the physical constant must be found. The problem is that the relevant physical constant have been demonstrated in weak physical effects but standard must be macroscopic. At present this problem is solved at a sufficient level.

Note that the Planck constant is expressed in units [kg m^2 s^{-1}]. Thus the kilogram depends on the h and vice versa. This mutual relation make possible to define the kilogram in a function of h and to develop the standard of the kilogram dependent on the planck constant.

22.3 Mass Standards Based on the Planck Constant

A new definition of the kilgram was crucial for acceptance the revised SI system of units. In the currently recommended direction of development of mass standards the unit of mass is related to the Planck constant h through electric energy, according to the proposed redefinition of the kilogram. Three solutions for the standard in accordance with the new definition are being developed. In the considered standards:

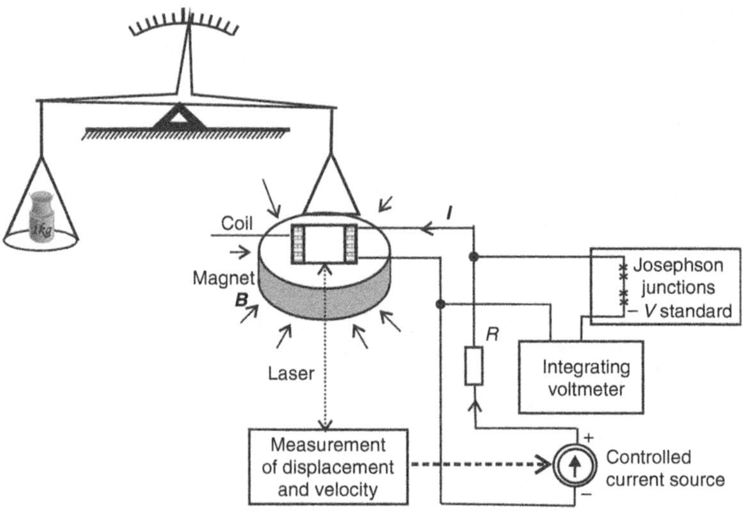

Fig. 22.2 Set-up of the standard of the kilogram based on mechanical and electrical energy balance using the watt balance

- The mass of a superconductor is levitated by a magnetic field generated by a current flowing in a coil;
- The gravitational force on the mass is balanced by the attractive electrostatic force of an electrometer plate;
- A watt balance with a moving coil measures the force acting on the coil with current in a magnetic field. The required coefficient of mechanical force exerted on the coil by the magnetic flux is determined directly by the measurement of the voltage generated when the coil is in motion. It is also necessary to measure the velocity of the coil; an interferometer and a stopwatch are used for this purpose. The realization of the unit of mass by the watt balance standard is based on a virtual comparison of the mechanical power with the electrical power. The standard holds promise to realize the unit of mass with an uncertainty of the order of one part per 10^8.

Figure 22.2 shows the block diagram of the watt balance – standard of the unit of mass with mechanical and electrical energy balance – installed at the National Physical Laboratory in the UK [3, 4]. The watt balance is also called the Kible balance from the name of its inventor. In the NPL system a very large permanent magnet generates a magnetic field with a 'horizontal' induction vector $\boldsymbol{B}$ (Fig. 22.3).

The magnetic flux acts on a horizontal circular coil suspended from one arm of the balance – Fig. 22.2. The current I in the coil with the wire length L generates a 'vertical' force, which is balanced by the gravitational force $F_{\text{grav}} = g \times M$ on the mass M. After the weighing experiment the coil is set in vertical motion (along the z axis) by rotating the arm of the balance, and passes through the initial position with a velocity v, as a result of which a voltage V is generated in the coil – Fig. 22.4. Since

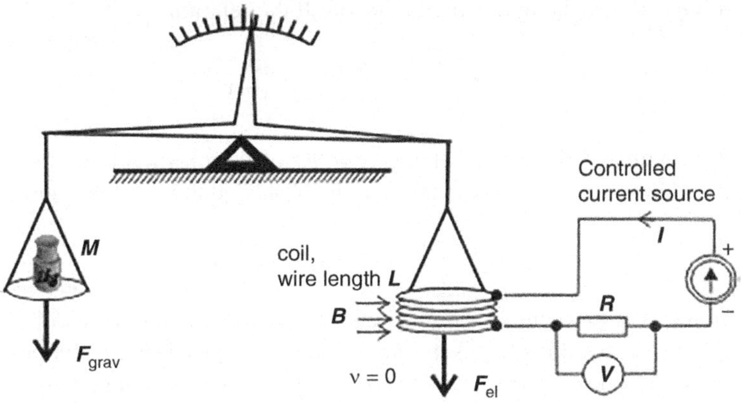

Fig. 22.3 Static phase in the watt balance

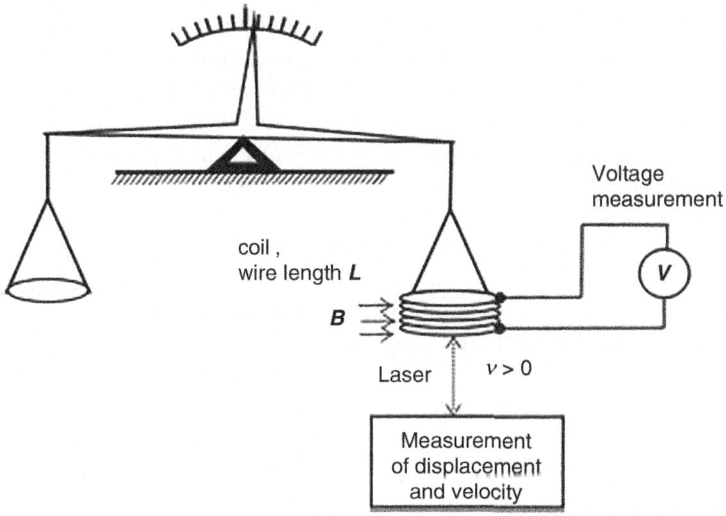

Fig. 22.4 Moving phase in the watt balance experiment

the energy E of the interaction between the current I and the magnetic flux Φ is $I\Phi$, the measured vertical component of the interaction force is given in Eq. (22.1)

$$F_{\mathrm{el}} = F_{\mathrm{grav}},$$

$$\frac{\partial E}{\partial z} = I\frac{\partial \Phi}{\partial z} = BIL = M \times g, \tag{22.1}$$

where g is the acceleration of gravity experienced by the mass M.

The voltage V and the magnetic flux Φ fulfill the relation:

$$V = \frac{\partial \Phi}{\partial t} = \frac{\partial \Phi}{\partial z}\frac{\mathrm{d}z}{\mathrm{d}t} = \frac{\partial \Phi}{\partial z}\nu. \tag{22.2}$$

The elimination of the rate of change $\partial\Phi/\partial z$ of the magnetic flux between the weighing and moving experiments leads to the basic equation of the standard:

$$M \times g \times v = I \times V. \tag{22.3}$$

Variation of the gradient of the magnetic flux and the components of the linear and angular velocities must be taken into account or reduced by careful positioning (alignment in the axis) of the coil and the other parts of the equipment with respect to the vertical axis so that the effect of these additional factors is negligible. In practice this means that the individual angular deviation must be below 10^{-4} radian.

The electric current I in the coil is determined from Ohm's law, $I = V/R$, with the voltage measured with a quantum standard based on the Josephson effect, and the resistance with a standard using the quantum Hall effect. Constant potential difference ΔV of up to a few volts can be measured with an uncertainty of 1 part per 10^9. The voltage V realized by the Josephson standard is a function of the frequency f of irradiation of the Josephson junctions, the physical constants e and h, and a natural number n_1:

$$V = n_1 f(h/2e). \tag{22.4}$$

Resistance can be also measured with an uncertainty of 1 part per 10^9 as a function of the quantum Hall resistance R_{H}:

$$R_{\mathrm{H}} = (h/e^2)/n_2. \tag{22.5}$$

Denoting the plateau number in the characteristic of the quantum Hall effect sample, the integer n_2 ranges from 2 to 4. If $n_2 = 2$, then $R_{\mathrm{H}} \approx 12906\,\Omega$; this value can be scaled by a factor k in order to measure any resistance in the range from $1\,\Omega$ to $100\,\mathrm{k}\Omega$ with a Harvey cryogenic current comparator. With a Josephson effect standard and a quantum Hall effect standard the current I can be determined in the range from a few milliamperes to a few amperes, by the formula:

$$I = \frac{V}{R} = \frac{n_1 n_2 f e}{2k}. \tag{22.6}$$

The equation (22.3) for the standard of the unit of mass can be converted to the equations (22.7), (22.8), and (22.9) below:

$$Mgv = \frac{V}{R}V, \tag{22.7}$$

$$Mgv = \left(\frac{n_1 h}{2e} f\right)^2 \frac{n_2 e^2}{h}, \tag{22.8}$$

$$M = k_W \times h, \tag{22.9}$$

where h is the Planck constant and the coefficient k_W includes g, v and f, quantities determined by the measurement of length and time and defined by physical constants and the integers n_1 and n_2.

The relation (22.9) indicates that in measurements using the standard with mechanical and electrical energy balance the measured value of mass M is mainly determined by the Planck constant, and to a lesser degree by the parameters of the equipment.

The formula (22.1) compares the *virtual* electrical and mechanical power. No *actual* energy dissipation, such as that due to friction in the bearings of the balance or electrical heating of the coil, enters this equation. This provides grounds for adopting the equation (22.1) with an uncertainty required in the metrology of standards.

The measurement technique used in the NPL [3] differs slightly from the general concept presented at the beginning of this section. The balance is biased with a mass of 500 g. The current I in the coil generates a force that balances the gravitational force on this mass, $g \times 500$ g. Then, a reference mass (e.g. a mass of 1 kg) is placed on the same scale as the coil, as a result of which the current in the coil changes direction to maintain the equilibrium. The watt balance is kept in equilibrium by an adjusted current I, the intensity of which is controlled by a computer. Thus, the gravitational force $g \times M$ corresponds to the total change in current in the coil. This substitution method eliminates many systematic errors of electrical, magnetic and mechanical origin. The effect of the hysteresis of the balance pointer (due to its inelastic deformation) is minimized by reducing the deflection of the balance in the weighing experiment. Also, before each observation the arm of the balance is set in oscillation with decaying amplitude to reduce and randomize the offset related to friction.

The mass samples used at the NPL are made of gold-plated copper to reduce the interaction by the dispersed magnetic flux from the magnet. The samples are stored in vacuum, with the exception of the short time interval necessary for their transfer to and from vacuum containers. The samples are compared with the BIPM prototype of the kilogram by the agency of its NPL copy.

The coil is moved in the vertical direction by rotating the arm of the balance to induce a voltage V in the coil. This causes also slight horizontal movements of the coil and, as a consequence, its swinging as well as a slight extra voltage in the coil. Contributing to the measurement error, this extra voltage is minimized by careful alignment of the coil in the axis. The effect of the horizontal swinging is reduced by averaging the voltage generated as a result of this motion over one period of oscillation or a longer time. Substrate vibrations present another problem, destabilizing the motion of the coil (causing velocity noise) and generating as much as ca. 0.2% of correlated voltage. These effects are reduced to a minor level also by averaging both the velocity of the coil and the voltage it generates.

The density of the magnetic flux from the permanent magnet varies with temperature with a coefficient of 400 parts per 10^6 K^{-1}, and with the magnetic field in the environment. Even though the parameters of the environment are stabilized in order to reduce the fluctuations in the magnetic flux density, the magnetic flux can still change slightly in the time necessary for the weighing and moving experiments. The effect of changing magnetic flux can be taken into account on the assumption that the changes in the flux are continuous and slow. The results of the weighing and moving experiments are used for the determination of the constant reference mass, which is the final result of the experiment. The expected uncertainty of realization of the unit of mass by the system used at the NPL is below 1 part per 10^8 in a single series lasting the whole night. Similar performance is expected at the NIST and METAS (Bundesamt für Metrologie, Switzerland). The understanding of the sources of the type B uncertainty in this standard of the unit of mass and the correction of their effects will certainly take some time.

22.4 Remarks on the Revised SI

Measures are created and established for people, first of all for use of common people. Systems of units should be defined in the way understood for common people. However, the revised SI system is understood rather by narrow community of scientists and experts of metrology, not by a wide population of users. In particular this remark relates to the definitions of the kilogram and the kelvin in the SI.

The BIPM informs on its website "the recommended practical system of units of measurement is the international system of units with the international abbreviation SI". However, the revised SI can hardly be called a *practical* system of units.

Any system should include interrelations between its components otherwise we are dealing with a set of components, not with the system. The definition of the revised SI looks like a set of physical constants. Interrelations between the components in the revised SI system are hidden.

The definitions of units in revised SI don't define them clearly (Appendix 3 of the Resolution 1). Five of them include the sentence "it is defined by taking the fixed numerical value of the *particular physical* constant..." instead of a practical definition of an unit. For instance, the definition of the metre from the "old" SI "the metre is the length of the path travelled by light in vacuum during a time interval of 1/299792.458 of a second" is much better understood than the definition of the metre from the revised SI.

Intuitively understandable is the greater importance of base units than derivatives. It is the use of base units for definition derived units. However, in the new SI, standards of the ampere, the kelvin and the candela are realized using derived units. The ampere, the base units of the SI, is realized using only two derived units (the volt and the ohm) and Ohm's law – the recommended realization. It is a pity that the ampere was not replaced either by the volt or the ohm in the set of base units of the new SI. Each of these two derive units can be realized at least 100 times more accurately than the ampere.

Kilinin and Kononogov [5] see that the new definition of the kelvin using the Boltzmann constant "is not advisable in view of the present-day level of accuracy and reliability of determination of the value of the Boltzmann constant".

Leonard published some critical remarks concerning the definitions of the mole presented above [6]. He introduced his remarks assuming the total amount of substance $n(S)$ and the corresponding number of the entities $N(S)$. The number-specific amount of substance is $n(S)/N(S)$. He has proposed "using the name entity and symbol ent for this fundamental physical invariant and formally adopting it as an atomic-scale unit in use with SI..."

The price of implementation the both methods (watt balance and XRCD) for the new standard of the kilogram is very high. A setup of the watt balance includes measuring the voltage through Josephson junctions, resistance measurements using quantum Hall effect, length measurements with lasers and cryogenic installations (liquid He). All this components of the setup are expensive. Highly qualified staff is necessary to operate the watt balance. In the case of the X-ray crystal density technique (XRCD) a sphere made of ^{28}SI is used. The monocrystalline silicon sphere, used for the first XRCD experiments, and measurements on dimensions of the sphere cost more than 2 million euros. In result, only about 10 national metrology institutes in the world are able to install and operate the new standards of the kilogram. For comparison, 900 g of platinum and 100 g of iridium, used for a prototype of one kilogram, cost together about 30,000 USD.

22.5 Proposals to Change the Revised SI

The SI system of units, with of seven base units (m, kg, s, A, K, cd and mol) is not the only possible system. The following is a hypothetical system of units in which the 5 base units of measures may be defined by the fundamental physical constants (c, e and h) and frequency, ν, and the other two (candela and mole) using the already defined other base units – Fig. 22.5.

In this system the unit of temperature, the kelvin, is replaced by the unit of energy, the joule [7, 8], and the ampere is replaced by the volt. Numerous arguments are presented in favour of such change. The set of base units should comprise the measurement units of physical quantities, which are the most important for human life and activity, e.g. applied in trade. Therefore, among base units there are the units for length (metre), mass (kilogram), and time (second). Because energy is also the subject of common trade exchange on a large scale, the unit of energy (joule) and the accuracy of its standardization have a great effect on trading and commercial activity. Energy is perhaps the most universal physical quantity in nature. One of the formulations of the first thermodynamics principle reads: "The quantity of

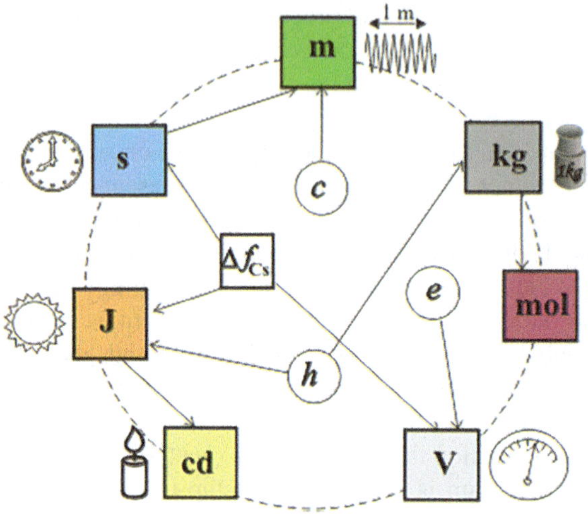

Fig. 22.5 Base units of a hypothetical system of units (with the joule instead of the kelvin and the volt instead of the ampere) and their mutual interrelations

energy in the universe is always constant". Different forms of energy: mechanical energy or work, thermal energy, electric energy and nuclear energy, enable mutual comparisons of the standards of mechanical quantities, thermal quantity standards, electric quantity standards and mass standards, as well as comparing them to the standard of energy unit [8–10].

$$E = h \times f; \quad E = m \times c^2; \quad E = V \times I \times t; \quad E = F \times l;$$

$$E = m \times v^2/2; \quad E = k_B \times T; \quad Q = c_h m \times \Delta T,$$

where: E – energy, Q – heat, c – speed of light in vacuum, c_h – specific heat, h – Planck's constant, k_B – Boltzmann's constant, f – frequency, F – force, I – electric current, l – length, m – mass, t – time, T – temperature, V – electric voltage, v – linear velosity.

As far as a limit resolution of measurements is concerned., we usually have to take into consideration energy changes affecting (influencing) the measuring sensor. Energy is one of four physical quantities occurring in inequalities describing quantum limits of measurement resolution, according to the Heisenberg uncertainty principle. Three others quantities are: time, length (for position) and momentum:

$$\Delta x \times \Delta p \geqslant \hbar/2; \quad \Delta E \cdot \Delta t \geqslant \hbar/2. \tag{22.10}$$

So far the idea of replacing the kelvin by the joule has not been supported by other scientists.

22.6 Conclusions

Adoption the revised SI system of units in 2018 is a good step in the development of measurements. This system could be better after consideration some remarks on it and implementation them.

Acknowledgments The author is grateful for funding the scientific cooperation with the Bogoliubov – Infeld Programme fund.

References

1. 9th edition of the SI Brochure (2019) https://www.bipm.org/en/measurement-units/rev-si
2. Newell DB et al (2018) The CODATA 2017 values of h, e, k, and N_A for the revision of the SI. Metrologia 55:L13–L16
3. Kibble BP, Robinson IA (2003) Replacing the kilogram. Meas Sci Technol 14:1243–1248 .
4. Stock M (2011) The watt balance: determination of the Planck constant and redefinition of the kilogram. Philos Trans R Soc A 369:3936–3953
5. Kalinin MI, Kononogov SA (2010) Redefinition of the unit of thermodynamic temperature in the international system of units. High Temp 48:23–28
6. Leonard BP (2007) The atomic-scale unit, entity: key to a direct and easily understood definition of the SI base unit for amount of substance. Metrologia 44:402–406
7. Nawrocki W (2019) Introduction to quantum metrolog. The revised SI system and quantum standards, 2nd edn. Springer, Heidelberg
8. Nawrocki W (2006) Revising the SI: the joule to replace the Kelvin as a base unit. Metrol Meas Syst 13:171–181
9. Kruchinin S, Nagao H, Aono S (2010) Modern aspect of superconductivity: theory of superconductivity. World Scientific, p 232. ISBN-9814261602
10. Kruchinin S, Pruschke T (2014) Thermopower for a molecule with vibrational degrees of freedom. Phys Lett A 378:157–161

Chapter 23
CBRNe as Conceptual Frame of an All Hazards Approach of Safety and Security: The Creation of Organic Networks of Military, Civil, Academic/Research and Private Entities at National and International Level to Generate Solutions for Risk Reduction – A European and Italian Perspective

M. Carestia, F. Troiani, R. Caldari, M. Civica, F. Bruno, C. Vicini, D. Di Giovanni, A. Iannotti, C. Russo, M. Thornton, L. Palombi, F. d'Errico, C. Bellecci, P. Gaudio, and A. Malizia

Abstract The world has changed. The global crisis related to the reduction of energy fossil resources, the reduction of potable water resources and the war for the control of energy sources are part of the causes which can lead to an intentional

---

F. Troiani · R. Caldari · M. Civica · F. Bruno · C. Vicini · A. Malizia
Sogin s.p.a, Rome, Italy

M. Carestia (✉)
Department of Industrial Engineering, University of Rome Tor Vergata, Rome, Italy

International Master Courses in Protection Against CBRNe Events, University of Rome Tor Vergata, Rome, Italy
e-mail: mariachiara.carestia@uniroma2.it

D. Di Giovanni · C. Russo
Department of Industrial Engineering, University of Rome Tor Vergata, Rome, Italy

International Master Courses in Protection Against CBRNe Events, University of Rome Tor Vergata, Rome, Italy

Health Safety Environmental Research Association Rome, Rome, Italy

A. Iannotti
Department of Industrial Engineering, University of Rome Tor Vergata, Rome, Italy

International Master Courses in Protection Against CBRNe Events, University of Rome Tor Vergata, Rome, Italy

International Alliance CBRN, HESAR, Rome, Italy

J. Bonča, S. Kruchinin (eds.), *Advanced Nanomaterials for Detection of CBRN*, NATO Science for Peace and Security Series A: Chemistry and Biology,
https://doi.org/10.1007/978-94-024-2030-2_23

CBRNe (Chemical, Biological, Radiological, Nuclear, and explosive) event together with the potential misuse of CBRNe agents, once considered to be solely the concern of the military, now needs to be considered also in the public domain. This kind of events could also be the consequence of an unintentional release of substances (i.e., an accident of a truck containing a Toxic Industrial Chemical), or of natural events like a tsunami or an earthquake. Thus, the high percentage of risk connected to their occurrence is clear. The aim of this paper is to analyze the CBRNe risk scenario and propose a standardized CBRNe joint approach for the establishment of "CBRNe technical tables" composed by cluster Entities from around Europe to manage the production of technologies, research, education and training activities in order to: (1) establish and maintain a roster of capabilities and facilities, and (2) organize to share expertise, and (3) plan to pool resources. The authors will show the results in the "Italian case study" as a potential pilot project analyzing the strengths and weaknesses to propose new ideas to improve it.

Keywords CBRN · All hazard approach · Education · Training · Resources pooling

M. Thornton
International Master Courses in Protection Against CBRNe Events, University of Rome Tor Vergata, Rome, Italy

International Alliance CBRN, HESAR, Rome, Italy

L. Palombi
International Master Courses in Protection Against CBRNe Events, University of Rome Tor Vergata, Rome, Italy

Department of Biomedicine and Prevention, University of Rome Tor Vergata, Rome, Italy

F. d'Errico
School of Engineering, University of PisaUniversity of Pisa, Pisa, Italy

C. Bellecci
International Master Courses in Protection Against CBRNe Events, University of Rome Tor Vergata, Rome, Italy

P. Gaudio
Department of Industrial Engineering, University of Rome Tor Vergata, Rome, Italy

International Master Courses in Protection Against CBRNe Events, University of Rome Tor Vergata, Rome, Italy

23.1 Introduction

The accidental or deliberate release of chemical agents, pandemic and epizootic biological diseases, and radiological and nuclear substances is a threat to health, environment and economy.

Concern arise either from huge events like the Chernobyl and the Fukushima accidents, despite being considered "black swan" events, from the use of unconventional weapons during conflicts worldwide like the recent use of sarin and chlorine gas in Iraq and Syria and of the nerve agent VX at Kuala Lumpur airport in February 2017.

Besides that, the biological threat is posed by emerging and re-emerging diseases like the Ebola virus disease outbreak in West Africa in 2014–2016 and recently in 2019, as well as by the uncertain use of the scientific knowledge and the availability of laboratory equipment that is now sold on the most well-known e-commerce platforms [1].

All together, these concerns and threats, represent a chemical, biological, radiological and nuclear (CBRN) global risks scenario that requests a strong collaborative effort by academia, practitioners, policy makers and industry, to be tackled.

23.1.1 CBRN Threat: An Evolving Scenario

Events of dispersion of Chemical, Biological, Radiological and Nuclear (CBRN) materials due to natural, accidental or intentional events are considered risk situations and nowadays represent one of the most critical concerns for safety and security [2].

The full extent of the potential CBRN threats cannot be predicted because they can evolve in non-linear ways and can be affected by several outside factors, including economy, flow of goods and people, meteorological conditions, etc. [3]

Threats related to CBRN agents use are evolving rapidly alongside changes in the political environment and developments in technology. The continued use of chemical weapons in armed conflict has shown the fragile nature of existing arms control regimes. In addition, several recent attacks in Asia and Europe using toxic chemicals and radioactive materials suggest that a new concern – state-sponsored assassination or attempted assassination – must now be incorporated into national security policy. Such confirmed use of CBRN materials by both state and non-state actors in these contexts highlights substantial challenges that the world is facing [4].

23.2 CBRN in the European Union

The European Union and the Euro-Atlantic community in general are currently facing two main security threats related to terrorism, represented by foreign terrorist fighters trained in Syria and able to execute complex attacks (also known as called "returnees issue"), and the risk represented by the influence and jihadization capability of ISIS and other salafijihadist groups which can result in a potential escalation of attacks all over Europe [3].

According to the European Union Terrorism Situation and Trend Report report, CBRN materials remain highly attractive to terrorists and several incidents in 2015 involved actual or attempted malevolent use of CBRN materials with criminal or unknown intentions. Incidents involving attempted sale of radioactive materials by organized crime groups occurred in Moldova [5].

There is a clear connection between transportation and terrorism is clear. In 9/11 attacks to the Twin Towers and the Pentagon, or during the bombing of the Brussels Airport in 2016, transportation (airplanes in the first case and a truck in the second one) has been the means to accomplish terrorists' objectives. Nevertheless, because transportation is the mechanism that allows the global supply chain to operate, and world economies to function and grow, it is paramount too establish and maintaining, through varying threat levels, a continuum of throughput and security [6].

To add more complexity to a "conventional" CBRN terroristic scenario, we can also think of wider CBRN potential scenarios like, for instance, that of food security. The food chain can be considered either a target for a CBRN attack, or a critical infrastructure jeopardized by the consequences of a CBRN event, in wealthy countries but most and foremost, in low and middle income countries, where limited access to food and water resources can either be the consequence or the cause of a CBRN event [7].

In the wake of the Salisbury nerve agent attack, the European Council stated in March 2018 that "the EU must strengthen its resilience to Chemical, Biological, Radiological and Nuclear-related risks ("CBRN risks") [8], moreover, there are indications that the scale of global cybercrime business is now greater than that of drug trafficking. There is a continued interest in dangerous materials from terrorist groups such as Daesh (IS) which certainly has the capacity to make chemical weapons. The Molenbeek (IS) network showed interest in obtaining nuclear materials. The growing capabilities of terrorist groups with respect to CBRN materials present a serious and growing threat [9].

23.2.1 Actions to Increase CBRN Preparedness and Reponse

Being a union of Member States, the European Union recognized the urgency to define a more harmonized approach to CBRN, as well as the need to pool and share resources to increase prevention, preparedness and response to CBRN events.

The overall approach of the EU concerns both the domestic and the external dimension.

The CBRN EU External Dimension

The European Union Chemical Biological Radiological and Nuclear Risk Mitigation Centres of Excellence Initiative (or EU CBRN CoE) [10] was launched in response to the need to strengthen the institutional capacity of countries outside the European Union to mitigate CBRN risks. The EU CBRN CoE initiative is led, financed and implemented by the European Commission, The funding for CoE projects comes from the EU Instrument Contributing to Security and Peace (IcSP) [11, 12], in close coordination with the European External Action Service (EEAS) and with the support of the UN (UNICRI) and other International Organisations and local experts.

The several projects (eighty, to today), financed under this instrument [13], are structured around the approach reported below:

- Based on the EU CBRN CoE prescribed methodology, CBRN CoE National Focal Points and their CBRN National Teams are responsible for assessing their respective national needs.
- Following the assessment, National Teams develop their own National Action Plans with the ultimate goal of developing an integrated and effective CBRN policy that is in line with internationally agreed standards.
- Where gaps are identified the CoE aims to work with the countries to address any possible shortcomings by means of tailored regional projects.

The CBRN EU Domestic Dimension

The conclusions of the EU CBRN Action Plan state that "in light of the evolving threats, Europe needs to pool resources and expertise to develop innovative, sustainable and effective solutions. Cooperation efforts across the EU along the lines set out in this Action Plan can result in significant security gains and lead to tangible results [13]."

To this extent, several CBRN related projects have been funded during the past two decades (108 with the search criteria CBRN on the CORDIS portal and just mentioning the projects under framework programmes). The topics of the different projects span from CBRN sensors to CBRN standard operating procedures, cross border cooperation, to training and education.

Among all these projects the eNOTICE (European Network Of CBRN TraIning CEntres) [14] project is worthen a special mention as it aims at creating a sustainable network of CBRN Training Centers [15], which will play the role of catalyst for a strict collaboration between practitioners, technology providers and other relevant CBRN stakeholder, using "Joint Activities" (JA). JA are defined as activities organized by a CBRN Training Center and made available for other participants than those who regularly take part to it. Following the eNOTICE approach, not only this means pooling resources for better investments and avoid the duplication of efforts, but also that joint testing, training, demonstration and validation are key for increasing preparedness and response capability between the EU Member states with a bottom-up approach.

23.3 The Italian Case Study: The Academia as Hub for CBRN Collaboration

JA, are a good instrument for building a smooth and effective CBRN response capability starting from the "on-field" experience, but also another environment proved to be of great value to promote CBRN education, exchange of expertise and start new collaborations, like the one that brought together Sogin S.p.A., the Italian State owned company responsible for the decommissioning of Italian nuclear plants and the management of radioactive waste, and the CBRNe Master courses organized by the University of Rome Tor Vergata, who worked together on the research project "SOS-Alert Solution – Cross-border cooperation project for enhanced detection and interception of illicit CBRN materials on the Slovakian-Ukrainian border" [16, 17].

Since 2009, with the launching of the first post graduating course in "Protection against CBRN events", the Italian academic institution embraced the idea of CBRN as an academic discipline.

The first edition of the course, in national language, collected the need of a high level CBRN academic education for practitioners (military, first responders, law enforcement agency…), and gave the opportunity to open such course to researches, people from the industry and fresh graduate students.

From that moment, the University of Rome Tor Vergata took on the role of facilitator of communication and engine for cooperation among all these different actors.

In ten years, not only the course became international and officially taught in English (since 2014), but it was necessary to split the original course, in a first level course for first responders and first responders' trainers, and in a second level course for advisors of decision makers, to meet the expectations and training needs of these two different target audiences.

Ten years from the first course delivered, the trend concerning new collaborations, activities, courses attendance is increasingly positive with 186 students for the II level Master course since 2009 to 2019 and 43 students for the I level Master course from 2013 to 2019, coming from Europe, Asia and Africa and America.

The simultaneous presence of several expertise in the same academic environment, allows the students not only to get CBRN education and training, but most and foremost, to create their own network, share their knowledge and get to see things from different perspectives. This improve their ability to communicate and relate with the different actors involved in planning, preventing, preparing, for and responding to CBRN events.

Such a prolific environment, pushed many of the students to elaborate innovative thesis that become:

- scientific publications, (more than 180 [18]),
- patents,
- ground for research projects and new collaborations between public or private entities,

with the support of a didactic board of more than 250 lecturers coming from all over the world [19], and

In 2016, the Master Courses provided fertile ground for the launching of the first CBRN book series [20] that was officially recognized also by the Italian Ministry of Education, University and Research. Moreover, since 2009, several scientific events have been organized under the umbrella of the Master courses, in collaboration with public and private entities:

- 7 international workshops [21],
- 1 Scientific international conference on CBRN (SICC 2017) [22], with peer to peer reviewed proceedings edited by Springer [23].

At the same time, the growing interest related to CBRN education and training at the academic level, brought many public Italian, foreign and international entities to sign cooperation agreements with the Master Courses. In many cases, this allowed the Master students to get access to working and training facilities, like those managed by Sogin S.p.A. and have a "on-the-field" experience, including the possibility to learn how to operate CBRN detection equipment, as well as to learn how to use personal and collective protective equipment [24].

23.4 Conclusion

When speaking about CBRN, the term "contamination" has usually a negative connotation. However, in this short review of national and international initiatives devoted to CBRN preparedness, training and education, a common factor to increase the ability to prevent and reply effectively to these events, appears to be the "cross-contamination" between experts, practitioners, technology providers, academia and policy makers. It is therefore paramount, to seek for every initiative and opportunity that can promote the exchange of expertise, knowledge and best practice among all those actors.

References

1. Zilinskas RA, Mauger R (2015) Biotechnology E-commerce: a disruptive challenge to biological arms control. CNS occasional paper No. 21 middlebury institute of international studies. Available at: https://www.nonproliferation.org/wp-content/uploads/2015/05/biotech_ecommerce.pdf
2. Bruno F et al (2017) CBRN risk scenarios. In: Proceedings of the NATO advanced research workshop on nanostructured materials for the detection of CBRN Kiev, Ukraine 14–17 Aug. https://doi.org/10.1007/978-94-024-1304-5_23
3. Final Report Chemical, Biological, Radiological, and Nuclear Threats (2017) Atlantic Treaty Association
4. Reassessing CBRN Threats in a Changing Global Environment (2019) Su F, Anthony I (eds) SIPRI
5. Ukraine, Turkey, and Poland (2016) Europol, "European Union Terrorism Situation and Trend Report 2016". European Police Office
6. CBRN Risks in Maritime and Land Containers Transport Collected Essays (2017) Cartocci V, Gerlini M, Giovanetti A (eds) ENEA National Agency for new technologies, energy and sustainable economic development. ISBN: 978-88-8286-347-0
7. Moramarco S (2018) Food security and proper nutrition: a public health and humanitarian priority in pre- and post-cbrn events. Def S&T Tech Bull 11(2):299–309
8. Fifteenth Progress Report towards an effective and genuine Security Union Brussels, 13.6.2018 COM (2018), p 470
9. van der Meer A, Aspidi A (2017) Security, development, and governance CBRN and cyber in Africa. In: Martellini M, Malizia A (eds) Cyber and chemical, biological, radiological, nuclear, explosives challenges. Terrorism, security, and computation. Springer:Dordrecht, The Netherlands
10. EU CBRN CoE. More information available at: https://ec.europa.eu/jrc/en/research-topic/chemical-biological-radiological-and-nuclear-hazards/cbrn-risk-mitigation-centres-of-excellence/countries
11. Regulation (EU) 2017/2306 of the European parliament and of the council of 12 December 2017 amending regulation (EU) No. 230/2014 establishing an instrument contributing to stability and peace. Available online at: https://eur-lex.europa.eu/legal-content/EN/TXT/PDF/?uri=CELEX:32017R2306&from=EN
12. More information on the EU CBRN CoE financed projects are available at: https://ec.europa.eu/jrc/en/research-topic/chemical-biological-radiological-and-nuclear-hazards/cbrn-risk-mitigation-centres-of-excellence/projects
13. Communication from the commission to the European parliament, the council, the European economic and social committee and the committee of the regions action plan to enhance preparedness against chemical, biological, radiological and nuclear security risks. Brussels, 18.10.2017 COM (2017)
14. Enotice – European Network of CBRN training centres. More information available at: https://www.h2020-enotice.eu/
15. D2.5 framework and sustainability plan forthe European CBRNTraining Centresnetwork (2019). Available at: https://cloud.h2020-enotice.eu/index.php/s/jRrkPSTCkWXnF7P#pdfviewer
16. SOS-Alert Solution – Cross-border cooperation project for enhanced detection and interception of illicit CBRN materials on the Slovakian-Ukrainian border. More information available at: https://eeagrants.org/archive/2009-2014/projects/SK08-0001
17. Bruno F et al (2017) CBRN risk scenarios. In: Proceedings of the NATO advanced research workshop on nanostructured materials for the detection of CBRN Kiev, Ukraine 14–17 Aug 2017. https://doi.org/10.1007/978-94-024-1304-5_23
18. List of publications. Available at: http://www.mastercbrn.com/page/326/cbrne-scientific-papers/

19. List of lecturers. Available at: http://www.mastercbrn.com/page/41/didactic-board/
20. List of CBRN book series. Available at: http://www.aracneeditrice.it/aracneweb/index.php/collana.html?col=CBRNe
21. More information. Available at: http://www.mastercbrn.com/page/1340/international-and-national-workshops-in-cbrne/
22. More information about SICC 2017. Available at: https://www.sicc-series.com/old-conference-sicc2017/
23. Enhancing CBRNE safety & security: proceedings of the SICC 2017 conference. Available at: https://www.springer.com/gp/book/9783319917900
24. Palombi L et al (2014) Building a chemical, biological, radiological, nuclear, explosive events tech advisor and first responders to support top decision makers during emergencies. In: Speciale: ENEA technologies for security 1, 138. https://doi.org/10.12910/EAI2014-101

Chapter 24
Decrease in the Concentration of Hazardous Components of Exhaust Gases from a Combustion Chamber of a Heat Engine

V. Morozov and I. Morozova

Abstract We consider the application of a means of electrophysical influence on a hydrocarbon fuel with the purpose to decrease the concentration of the hazardous components of exhaust gases, by affecting the working process in the combustion chamber of a heat engine.

Keywords Environmental indices · Working process in a combustion chamber · Heat engine · Hydrocarbon fuel · Activation energy · Magnetic field · Magnetic moment · Nucleus spin · Resonance frequency · Nuclear polarization

24.1 General Statement of the Problem

At the present time, one of the environmental problems is the pollution of air by hazardous ejections of the motor transport. The main priority problems are oriented to the search for alternative types of a fuel which should improve the environmental situation and decrease the toxicity of exhaust gases ejected by the motor transport.

Atmospheric air is the defining factor for the life of people, animals, and phytoworld, as well as for the soil fertility. The ejections of hazardous substances affect significantly the greenhouse effect and the climate change, which influences, in turn, the deterioration of the health of all alive creatures and organisms. The exhaust gases of an internal-combustion engine contain about 280 different components. Most of them are toxic by their chemical properties and render the corresponding influence on human organism.

V. Morozov · I. Morozova (✉)
National Aviation University, Kyiv, Ukraine
e-mail: iramoro@ukr.net

J. Bonča, S. Kruchinin (eds.), *Advanced Nanomaterials for Detection of CBRN*, NATO Science for Peace and Security Series A: Chemistry and Biology, https://doi.org/10.1007/978-94-024-2030-2_24

The priority directions aimed at a decrease in the toxicity of exhaust gases of engines are the following ones: the further improvement of their constructions; use of alternative types of a fuel; and design of engines and units with a low content of toxic components in the exhaust gases of internal-combustion engines.

The majority of trucks, industrial heavy equipment, and agricultural technique are equipped with Diesel engines. Therefore, it is necessary to improve their environmental, economic, and efficiency parameters.

The analysis of the results of scientific studies indicates that the amount of hazardous substances in exhaust gases depends on: the adjustment of a fuel-supplying equipment, type of an engine, its technical state, working mode, and various operation factors. All this should be considered at the choice and a treatment of various types of fuels.

In the recent decades, the problems of the struggle against the contamination of the environment attract the enhanced attention. The toughening of the norms concerning the content of hazardous components in the exhaust gases of internal-combustion engines led, as expected, to the impossibility to solve the following one-in-two problem by means of traditional methods: to decrease the level of ejections of toxic components and to conserve or to increase the economic efficiency of a fuel. The analysis of the domestic and foreign scientific-technical works indicates that the problem of a decrease in the toxicity and an improvement of the efficiency of engines is considered in a lot of scientific-research works [1–3].

24.2 Analysis of Unsolved Problems

One of the sources of the formation of hazardous substances is the process of combustion of a fuel. The units, where the process of combustion is realized with the purpose to obtain the mechanical energy, are called heat machines or engines. In this case, they are divided into two classes: with external supply of heat (the combustion products and the working medium are separated) and with internal combustion (combustion products are the working medium). The former includes, for example, steam boilers, Stirling engines, and the latter is presented by gas-turbine, piston, and rotor-piston engines. In the present work, we will consider the problems of the formation of hazardous substances by internal-combustion engines (ICEs). The main distinction of similar engines from one another is the type of a fuel-air mixture intended for the combustion which is preliminarily prepared or is produced during the working process. The ignition of a mixture can be realized by various means: forcibly by a spark or heated surface and due to the self-ignition caused by the compression. The character of the ignition plays a minor role, though it also affects significantly the running of the process of combustion. The mentioned ICEs differ, in addition, from one another by the design of a combustion chamber (nonseparated, semiseparated, and separated ones), aggregation state of a used fuel (liquid or gas-like), composition of a fuel-air mixture (by the oxidizer – fuel ratio), temperature and pressure of supplied air, etc. All those various factors influence the specific features of the course of the process of combustion and, respectively, the processes of formation of hazardous substances.

The combustion chambers of heat engines are related to the process of combustion of liquid hydrocarbon fuels in a turbulent flow of moving air. If a fuel and an oxidizer are introduced into the reaction zone from different sides, then the combustion duration is determined by the mixing time and the diffusion. The diffusion mechanism of combustion is characteristic of drops of a liquid fuel and works in the burning of a candle or a kerosene lamp.

Another limitedly typical case is the kinetic combustion of a homogeneous (mixed) blend of a fuel and an oxidizer. In this case, the prepared blend enters the reaction zone, and the combustion duration will be mainly determined by the time necessary for the running of a chemical reaction. A similar process is called kinetic, since the combustion duration is determined by the mechanism and the rate of a chemical reaction.

The combustion chamber is the most significant part of heat engines. It must ensure the supply of heat to a working medium and should be designed so that the process of combustion be supported by itself. Moreover, the field of temperatures of combustion products at the input to a turbine should be maximally uniform with the mean temperature lower than the maximal temperature in the flame core.

In the traditional combustion chambers, the process of combustion is running in the primary and secondary zones. In the primary zone, the fuel-air mixture is formed, and almost all amount of a fuel is burned. In the secondary zone, the residual fuel burns down, and the combustion products are cooled to the required temperature at the input to a turbine. The ecological characteristics of combustion chambers depend completely on the structure of the process of combustion in the primary and secondary zones.

The ecological characteristics are evaluated by the indices of emission of hazardous gas-like substances (except for carbon black). The index is the ratio of the mass of a substance in g to 1 kg of a consumed fuel. Respectively, the indices EI_{CO}, $EI_{C_nH_m}$, EI_{NO_x} are in use.

It is worth to note the growing requirements to heat engines such as an increase in the completeness of the combustion of a fuel, a decrease in hazardous ejections, and the use of alternative fuels (heavy fractions, synthetic gas, hydrogen) with the supply of water or water vapor in a combustion chamber. Therefore, the questions of the preparation and supply of a fuel attracts a more and more attention.

The complicated running of the processes of combustion puts significant obstacles to the design of combustion chambers which should satisfy the contradictory requirements. On the one hand, the combustion chambers must have: a high completeness of combustion of a fuel ($\eta_c \geq 0.99$); the steadiness of combustion in broad intervals of the coefficients of excess of air (from $\alpha_{\min} = 1.0 \ldots 1.5$ to $\alpha_{\max} = 20 \ldots 40$), pressures, and velocities; low losses of the total pressure ($\sigma_{cp} = 0.97 \ldots 0.99$); and the support of given circumferential and radial epures of the gas temperature at the input to a turbine. On the other hand, it is necessary to ensure: a low level of ejections of hazardous substances (CO, C_nH_m, NO_x, smoke), low deposits of a carbon in a fire tube; a low prime cost of the products, simplicity of the maintenance in operation; high reliability and lifetime; the ensuring of the workability of a combustion chamber without repair during a definite assured term; small mass and overall dimensions.

In order to satisfy such severe requirements, we need a profound comprehension of physico-chemical processes running in a combustion chamber. Its principal distinctive feature, as compared with other units of an engine, is the absence of permanent reliable methods of design and the use of empiric approaches for this purpose.

Among the modern methods aimed at an enhancement of the efficiency of combustion chambers, we distinguish the method of electromagnetic influence on hydrocarbon fuels. The electromagnetic influence is realized on the stage of preparation of a fuel to the combustion.

24.3 Statement of the Problem

To decrease the concentration of hazardous components of exhaust gases from the combustion chambers of heat engines by means of an improvement of physico-chemical and operational properties of a fuel by the electrophysical method.

24.4 Basic Part

The intense development of the power engineering, industry, and transport causes inevitably a growth in the consumption of a hydrocarbon fuel, which increases, in turn, the amount of products of its combustion ejected in the atmosphere.

By data of the many-year monitoring, the amount of environmentally hazardous chemical compounds, substances, and elements of the fuel combustion products ejected in the atmosphere is doubled each 12–14 years. Therefore, the problem of contamination of the atmosphere by fuel combustion products is referred to one of the global problems of the contemporaneity.

Toxicity of exhaust gases ejected in the air environment depends mainly on the quality, grade, and type of combustible hydrocarbon fuel, conditions of implementation of the process of its combustion, and the technical state of heat engines and fuel-burning setups.

For example, the application of low-grade fuels favors, on the one hand, a decrease in current operational expenditures for the purchase of a fuel and increases, on the other hand, the amount of environmentally hazardous contaminants ejected in the atmosphere.

The mechanism of formation of environmentally hazardous contaminants of the atmosphere such as the combustion products of a hydrocarbon fuel involves the mechanisms of simultaneously running physico-chemical, kinetic, thermodynamical, heat-exchange, and other processes, is very complex, and is slightly studied.

Each mechanism of formation of hazardous ejections is characterized by the presence of a number of driving forces called the factors which determine the qualitative and quantitative compositions of contaminants of the air basin. Such factors include the discharge, structure, and chemical composition of a fuel; qualitative and quantitative compositions of combustible components of a fuel; the quality of the dispersion of a fuel; physico-chemical indicators of the quality of a fuel; the quality of a mixing, the quantitative ratio of a fuel and air; temperature and the degree of nonuniformity of its distribution over the combustion volume; commensurability of the combustion volume and the volume of a combustion chamber (furnace), etc.

To decrease the hazardous influence of ejected exhaust gases on the air environment, the following measures are developed and used [4]: the mounting of filters and catalysts on smoke stacks and eduction pipes; application of burners producing the low levels of nitrogen oxides (NO_x); two-stage combustion of a fuel; recirculation of smoke-containing (exhaust) gases; introduction of water or water vapor into the combustion zone; introduction of admixtures in a fuel or in the combustion zone; chemical methods of purification of exhaust gases. Unfortunately, the used above-indicated measures affect mainly the consequences and do not remove the causes of the creation of contaminants.

One of the promising directions aimed at a decrease in the gas and thermal contaminations of the air basin, is the removal of causes for the appearance of hazardous contaminants by means of the electrophysical influence (EPI) on a hydrocarbon fuel.

The EPI is based on the interaction of a molecular liquid system with external magnetic and electromagnetic fields. It is known that the atomic particles with a magnetic moment, if fall in a magnetic field, orient in it and acquire the Zeeman additional energy [5–19]. For a proton, such energy

$$E = -\mu_n H_0.$$

Under ordinary conditions, the nuclear polarization is insignificant. But, as the strength of an external magnetic field increases, the difference in the energies for a spin directed against the field and a spin oriented along it increases as well, which leads to an increase in the polarization. In the bulk of a substance, there is the total magnetic moment of nuclei M (magnetization) directed along the external field:

$$M = \frac{\mu_n^2 H_0 n_0}{kT}.$$

A sharp decrease in the strength of the magnetic field leads to that the additional energy of a proton is absorbed, due to the superfine interaction, by hydrogen and the atoms of elements which are located nearby and possess a nonzero electron spin. Such atoms can belong to paramagnetic oxygen, free radicals, etc. In other words, the paramagnetic centers interact with the basic chemical compound. An increase in the polarization of hydrogen nuclei at a change in the magnetic field strength causes inevitably the destruction of electron shells in molecules due to the superfine interaction. This causes, in turn, the induction currents affecting the physical properties of fluids.

To increase the efficiency, stability, and universality of the magnetic influence on liquid systems, we propose the electrophysical method based on the resonance absorption of the energy by the proton system of a fluid. The method consists in the simultaneous influence of an inhomogeneous constant magnetic field and a resonance high-frequency electromagnetic field on the fluid. In this case, the frequency of oscillations coincides with the frequency of a precession of nuclei in the given magnetic field, and we can observe the selective absorption of the energy of a generator at the frequency

$$\omega = \gamma H_0,$$

where γ is the gyromagnetic ratio of a nucleus.

The alternating electromagnetic field H_1 induces a perturbation of the system of spins, which is described by the Hamiltonian

$$\mathcal{H}'(t) = \mathcal{H} \cos \omega t = q_n \beta (J_x H_{1x} + J_y H_{1y} + J_z H_{1z}) \cos \omega t.$$

The influence of the electromagnetic field causes the intense transitions "from top to bottom" and vice versa between the energy levels, and the Boltzmann distribution is violated. In this case, the electromagnetic energy of the field is partially absorbed by the proton system of a fluid, the absorption signal is measured, and the energy of the medium increases.

The power P of the alternating electromagnetic field which is absorbed by spins is as follows:

$$P = \hbar \omega W_{\hbar} = \hbar \omega n_0 \frac{W}{1 + 2WT_1}.$$

Thus, the resonance absorption of the electromagnetic energy of an oscillatory circuit by the proton system of a fluid and the subsequent exchange of this energy with the molecular system of a fluid cause a change in the physical and physico-chemical parameters of a fuel (Fig. 24.1). Finally, the electrophysical influence causes a change in the rate constant of a chemical reaction K_f due to, by the Arrhenius law, a change in the activation energy E_a:

$$K_f = k_0 \sqrt{T} e^{\frac{-E_a}{RT}},$$

where k_0 is the coefficient depending on the molar masses of a substance and the sizes of reacting molecules, and $e^{\frac{-E_a}{RT}}$ is the Arrhenius exponential function. This affects, in turn, the rate and completeness of the combustion.

Based on the proposed method, we designed and produced a unit for the electrophysical influence on a fuel [14]. It allows one to decrease the amount

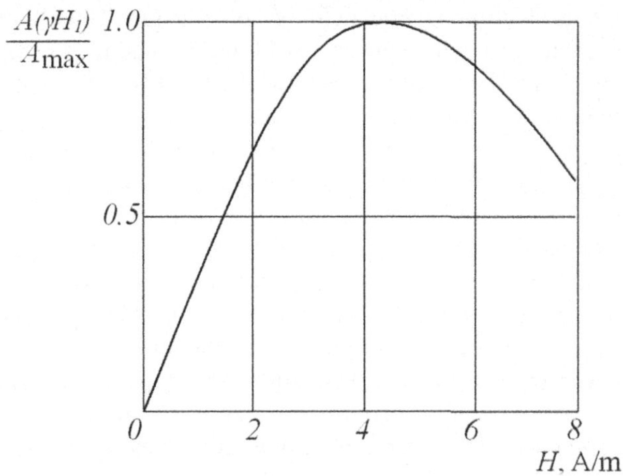

Fig. 24.1 Dependence of the absorption signal amplitude on the alternating electromagnetic field

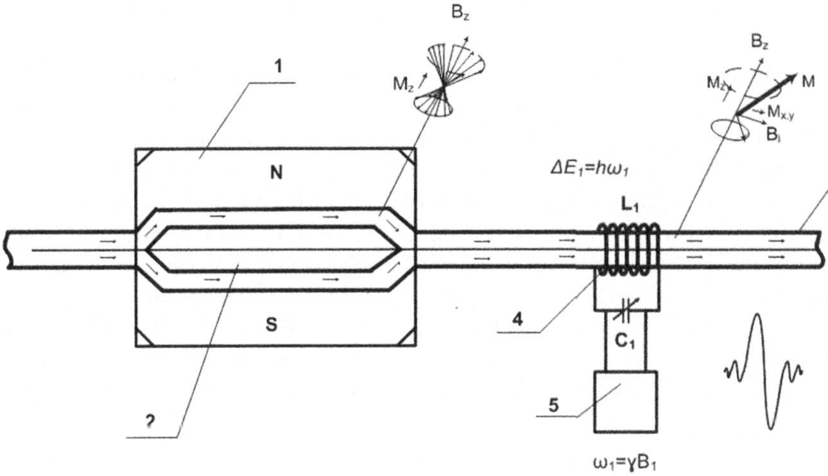

Fig. 24.2 Scheme of a unit for the electrophysical influence on a hydrocarbon fuel

of environmentally hazardous chemical compounds and substances of the fuel combustion products ejected into the atmosphere (Fig. 24.2).

The fluid driven in motion by a pump enters chamber (2) of a polarizer placed between the poles of a constant magnet (1) with the inhomogeneous strength of the magnetic field and stays there for some time $t > 3T_1$. From the polarizer, along a pipe with small cross-section (3), the fluid enters the coil of an oscillatory circuit (4) of a high-frequency generator (5). This coil is located in the inhomogeneous magnetic field of a constant magnet (1). Thus, the fluid undergoes the action of two fields: the inhomogeneous constant magnetic field H_0 of magnet (1) and the alternating electromagnetic field H_1 of coil (4). Thus, there occurs the electrophysical influence at which the energy of the alternating electromagnetic field of an oscillatory circuit of generator (5) is absorbed by hydrogen nuclei of the basic chemical compound, i.e., of a hydrocarbon fluid.

Polarizing magnet (1) ensures the highest value of the magnetization of nuclei in a fluid flowing out of it and creates an inhomogeneous magnetic field in the zone of coil (4), where the resonance absorption of the energy by the proton system of a fluid at the electrophysical influence is realized.

The essential indicator of the quality of Diesel fuels is the tendency to the ignition at the contact with air. The combustibility of fuels is evaluated by the cetane number. Its value determines, in turn, the duration of the delay of ignition τ_i of a fuel.

The executed tests showed [15] that, after the electrophysical influence on a Diesel fuel, the cetane number increases by 2.5 … 3.5 units. This testifies to an increase in the combustibility of a hydrocarbon fuel, i.e., to a decrease in the period of delay of the ignition τ_i which is determined by the Semenov formula

$$\tau_i = \frac{\text{const}}{A_m} P^{-\nu} e^{\frac{E_a}{RT}},$$

where A_m is a factor depending on the composition of a reacting mixture; ν is the summary order of the reaction of branching; P is the pressure of air in a cylinder; E_a is the activation energy; and T is the temperature of a reacting mixture.

A change in the cetane number characterizes only one of the components of the total duration of delay of the ignition, namely, τ_{chem}, whereas τ_{phys} in the test on the same engine remains practically invariable. The duration of staying of a fuel in the dropwise state is as low as fractions of one second, whereas the rate of oxidation with the formation of peroxides and other easily combustible products of partial oxidation is significant and depends on the activation energy E_a. The rate of the chemical reaction of combustion of a fuel at the given time moment is proportional to the product of the concentrations of reacting substances at the same time moment:

$$W_f = \frac{dC_f}{d\tau} = k_f C_f^{\nu_f} C_k^{\nu_k},$$

where k_f is the reaction rate constant; C_f and C_k are the concentrations of a fuel and oxygen, respectively; ν_f and ν_k are the reaction orders, respectively, by a fuel and oxygen; and τ is the time.

The formula for the rate constant of a chemical reaction is given by the Arrhenius law:

$$K_f = k_0 \sqrt{T} e^{\frac{-E_a}{RT}},$$

where k_0 is a coefficient depending on the molar masses of substances and the sizes of reacting molecules. Hence, as the activation energy E_a decreases due to the electrophysical influence on the energy of interaction between molecules of a fuel and oxygen, whereas the rate of combustion of a fuel W_f increases. Due to a decrease in τ_i, the duration of the first phase is reduced. Hence, a less fraction of a fuel enters a cylinder of the engine for this time. This leads to a shift in the start of the second phase (fast combustion) to the left from the upper dead point. A decrease in the amount of a fuel, which allows a stable ignition of a mixture during the period of delay of the ignition τ_i leads to the best indices of stiffness and efficiency of the working process of a Diesel engine.

An increase in the oxidation reaction rate W_f in the main phase of combustion is one of the requirements aimed at an increase in the efficiency and favors an increase in the completeness of the combustion. The improvement of the processes of combustion of a fuel in the initial phases leads to a decrease in the phase of torching, which induces a decrease in the temperature of exhaust gases, concentration of carbon oxide by 0.64 ... 0.7%, concentration of hydrogencarbons by 25 ... 35%, and concentration of nitrogen oxide by 12 ... 16%.

24.5 Conclusions

The established effect of a higher rate of oxidation of a hydrocarbon fuel after the electrophysical influence decreases the burning temperature, which decreases the ejections of CO, C_nH_m, and NO_x. The burning temperature is decreased due to an increase in the specific heat capacity of gases in a cylinder and a decrease in the total concentration of oxygen. A decrease in the concentration of hazardous substances in exhaust gases due to the established effect allows one to sharply reduce the ejections of hazardous substances into the atmosphere by means of an increase in the power efficiency of an engine.

References

1. Blümich B (2000) NMR imaging of materials. Oxford University Press, Oxford
2. Internal-Combustion Engines (2007). In: Lukanin VN, Shatrov MG (eds) 3 Pts. Pt. 1. Theory of working processes. Vysshaya Shkola, Moscow. In Russian
3. Dorfman YaG (2010) Magnetic properties and structure of substances. URSS, Moscow. In Russian
4. Erokhov VI (2013) Toxicity of modern automobiles. Methods and means to decrease the hazardous ejections into the atmosphere. Forum, Moscow. In Russian
5. Kaganov MI, Tsukernik VM (2016) Nature of magnetism. URSS, Moscow. In Russian
6. Kruchinin S, Nagao H, Aono S (2010) Modern aspect of superconductivity: theory of superconductivity. World Scientific, Singapore, p 232
7. Kruchinin S, Dzezherya Yu, Annett J (2006) Imteractions of nanoscale ferromagnetic granules in a London superconductors. Supercond Sci Technol 19:381–384
8. Dzhezherya Yu, Novak IYu, Kruchinin S (2010) Orientational phase transitions of lattice of magnetic dots embedded in a London type superconductors. Supercond Sci Technol 23:1050111–105015

9. Kruchinin S, Nagao H (2012) Nanoscale superconductivity. Int J Mod Phys B 26:1230013
10. Kruchinin S (2016) Energy spectrum and wave function of electron in hybrid superconducting nanowires. Int J Mod Phys B 30(13):1042008
11. Ermakov V, Kruchinin S, Pruschke T, Freericks J (2015) Thermoelectricity in tunneling nanostructures. Phys Rev B 92:115531
12. Kruchinin S, Pruschke T (2014) Thermopower for a molecule with vibrational degrees of freedom. Phys Lett A 378:157–161
13. Kulchitskii AR (2004) Toxicity of automobile and tractor engines. Akadem. Proekt, Moscow. In Russian
14. Morozova IV, Morozov VI (2010) Method and a unit for an increase in the economic and ecological indicators of the operation of an internal-combustion engine. Avtoshl Ukr 13:58–60
15. Morozova IV, Morozov VI, Tereshchenko YuM (2016) Improvement of the operational indicators of heat engines with the help of the electrophysical influence on a fuel. Naukoem Tekhnol 1(29):102–106
16. Sergeev NA, Ryabushkin DS (2013) Foundations of the quantum theory of niclear magnetic resonance. Logos, Moscow. In Russian
17. Baumgarten C (2006) Mixture formation in internal combustion engines. Springer, Berlin
18. Garg R, Agarwal AK (2013) Fuel energizer: the magnetizer (a concept of liquid engineering). Int J Innov Res Dev 2(4):389–403
19. R. Kunz, Chemist. Magnetic treatment of fuel. http://www.probonoscience.org/pennysolutions/recipes/automobile/buyer_beware/magnetic_fuel_treatment.htm

Chapter 25
Visual Analytics in Machine Training Systems for Effective Decision

Iu. Krak, K. Kruchynin, O. Barmak, E. Manziuk, and S. P. Kruchinin

Abstract The approaches to the formation, development of a formal and mental model based on the use of visual analytics are proposed. It is based on the description of model building technologies. An example of information technology that allows getting a formal model based on the transformation of the mental model through the space of formalized universal forms is given. This allows the model to be used in a different usage and execution environment. Model development is carried out using loops the improvement of the base model or transforming the use of the model from another runtime. An example of equipment and tools for the construction and transformation of models is demonstrated.

Keywords Visual analytics · Classification · Mental model · Formal model · Visualization · Decision making

25.1 Introduction

One of the directions in the development of machine learning is the use of human analytical capabilities to build models. This area of improvement consists of the development of techniques, approaches and information technologies in which a human, namely his intellectual capabilities, are effectively used. According to this, machine learning workflows with interactive interaction are formed. Such a direction as visual analytics is actively developing with the formation of software systems for human-machine interaction. Machine learning is widely used to search

Iu. Krak · K. Kruchynin
V.M. Glushkov Institute of Cybernetics, Taras Shevchenko National University, Kyiv, Ukraine

O. Barmak · E. Manziuk
Khmelnitsky National University, Khmelnitsky, Ukraine

S. P. Kruchinin (✉)
Bogolyubov Institute for Theoretical Physics, Kyiv, Ukraine

J. Bonča, S. Kruchinin (eds.), *Advanced Nanomaterials for Detection of CBRN*,
NATO Science for Peace and Security Series A: Chemistry and Biology,
https://doi.org/10.1007/978-94-024-2030-2_25

for elements of data similarity and to identify data grouping by integral signs of similarity. This allows you to identify data groups in order to systematize and conclude the belonging of new data to one or another group. These approaches are applicable to data of various natures, such as images, text, feature tables, ungrouped data, and others. The prospects of this area are based on the fact that systems of complementary and integral areas are developed for the effective use of the capabilities of both the machine and the human to achieve the desired result of machine learning.

25.2 The Role of a Human in the Visual Analytical Process

The interaction of a machine and a human is very important and sometimes crucial. Interaction involves the exchange of certain information, while the presentation of this information should have the property of understanding. That is, information must be transformed and presented in such a way that the recipient and user of this information can understand and interpret it. The purpose of obtaining information may be ultimate for the consumer. An example of this use is the various types of tables, graphs, charts that are prepared for a human. A human most fully and comprehensively uses the visual representation of information. Therefore, various types of visualization are important for a human. Often graphics are the final product for the human who prepares the machine. For a machine, a human presents information in the form of numbers forming data volumes. However, the human, providing the machine with data, wants to get the result of working on the data in the form of an easy-to-use information presentation. Based on the computation results, a human can change or perform any actions on data or algorithms, providing feedback for the machine and thereby improving the results of machine work. Thus, a human becomes involved in the process of obtaining the necessary computed results. A human is loops involved in the computing process and becomes its necessary part. This direction in the development of human-machine interaction was called "human-in-the-loop" [1, 2]. A direction of visual analytics was formed according to [3] analytical reasoning is based on interactive visual interfaces. Visual analytics allows implementing techniques to deepen knowledge and understanding, evaluate, decision making. Visual representation allows shortening the path to understanding data and information by a human. It also allows for improving the process of research and training. Data transformation for visual presentation can improve communicativeness of representative informativeness. The process of insight can be represented as a "sensemaking loop" [4] (see Fig. 25.1).

Data researching, relationships and dependencies, as well as improving and deepening knowledge in accordance with the definition of areas of visual analytics [2] focused on the human, namely on the process of insight by a human.

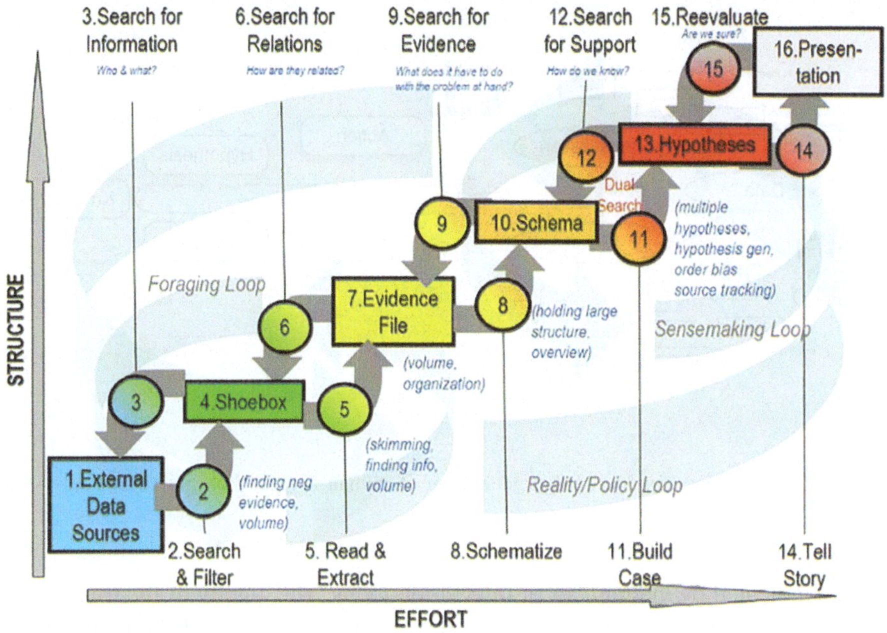

Fig. 25.1 Sensemaking loop for intelligence analysis P. Pirolli et al. [4]

25.3 The Effectiveness of the Interface Interaction Process

Improving the process of visual analysis allows you to shorten the path of insight. In this direction, the process of interaction between machine and human is being improved. Interface interaction is the main part of visual analytics, as it is the only connecting link. This link plays a key role in the transfer of information between human and machine. An important issue is the quality of interaction. Since the interaction is process-oriented, it is difficult to assess how much knowledge has expanded and, as a result, to assess the quality of the interaction itself. In the general case, in our opinion, a qualitative interaction should reduce the number of loops interactions, for example, the number of “sensemaking loops”. If a human uses a large number of interaction loops to obtain the understanding he needs, this indicates that a small amount of information is transmitted in one loop. Therefore, it is necessary to improve visual presentation to increase the amount of information content transmitted. A human uses visual analytics for a specific purpose. The visualization application should not contain many visual elements, for example, graphs of tables and so on, since in this case, the system transmits a lot of information and a human receives a lot of information. But not all the information transmitted corresponds to the purpose of using the visual application, therefore a lot of information is transmitted, however, the information content is deficient, which

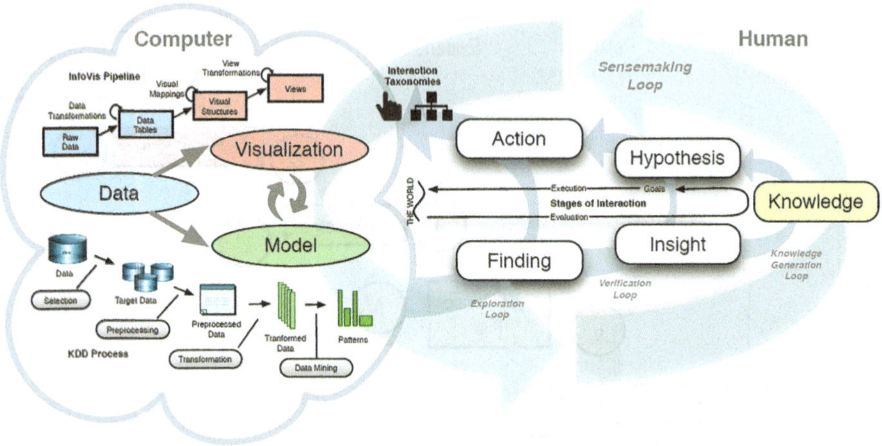

Fig. 25.2 The process for knowledge generation of human and machine interaction D. Sacha et al. [5]

can be indicated by the number of interaction loops. The quality of the transmitted information content may also be insufficient due to inadequate information content regarding the purpose of using visualization. The development of the human-machine interaction process is the knowledge generation model for visual analytic [5]. The processes occurring both on the machine side and on the human side are considered in more detail and sequentially. The proposed interaction scheme presented in the Fig. 25.2 separates the processes and is focused on obtaining a specific result, which is knowledge.

This process is also human-oriented but allows specifying the result of using visualization. The knowledge acquired by a human is the result of the use of visual analytics and lends itself to more formalization than focusing on the process of insight. The final product, in this case, is knowledge.

Effective visualization is to show as much information as possible in the simplest form. This aspect is very important because it allows expanding the use of visual analytics for analysts with different qualifications. Thus, the trends in the use of visual analytics indicate an increasing involvement of a human in the process of extracting knowledge from data and developing visual analytics workflow.

25.4 Model as the Final Product of Using Visual Analysis

Like the previous approach, visual analysis is used for the end-user – a human. However, it should be noted that the machine may also be the final consumer of the product of visual analytics. In this case, the human acts as a necessary integrated part of the system for obtaining the final product that will be used by the machine. Such areas as Interactive machine learning [6–10] allows you to use the intellectual capabilities of a human. The machine produces the final product while including the human in the loops process of improving the result. Visual analytics, or rather intellectual abilities of a human, are used to build the final product of machine learning VIS4ML proposed in paper [11].

Machine learning uses such approaches as SVM, neural networks, random forest, and others to form models. These models are fully created by the machine, and the results of their work are evaluated by a human using quality indicators and visualization. The developed models, as well as the input data, are refined and improved on the basis of an iterative process. Human participation in this process is now necessary to improve and achieve the necessary results. Thus, a human makes a valuable contribution and helps the machine to achieve the development goal, which is the building of a model. This model is defined by paper [12] is formal and is used by the machine. Thus, we get the advantage and benefit using the intellectual capabilities of a human. Human is increasingly integrating into the process of building models by machine as part of the direction of visual analytics. Therefore, there is a need to develop analytical systems for the interaction of a machine and a human-based on visualization. Visualization is by far the most informative for humans.

In the paper [14] proposed a graphical representation of visual analytics as a process in which knowledge was extracted from data through visual representation, research, and data processing (see Fig. 25.3) and given approach was subsequently used in [5, 12] and others. Data transformation is an important step for visual analytics. Data transformation should provide visual representativeness of information according to the purpose of the study.

When orienting the working process of interaction between a machine and a human to obtain a final product, namely a model, as a decision-making mechanism, it is necessary to determine the consumer of this product. We use the definition of the model as an information processor that converts the input data into the required result. The most promising goal of using visual analytics is not to obtain a specific solution, but to build a mechanism for obtaining solutions. In the general case, this is the goal of machine learning, namely the construction and training of a model that subsequently provides the necessary solutions within the framework of the problem used. Two forms of the model are determined – formal, for use by the machine and mental [12, 13] for human use. A formal model is defined as a computational model presented in the form of a used machine. This form of the model is used in machine learning. The mental model is defined as a certain complete form of

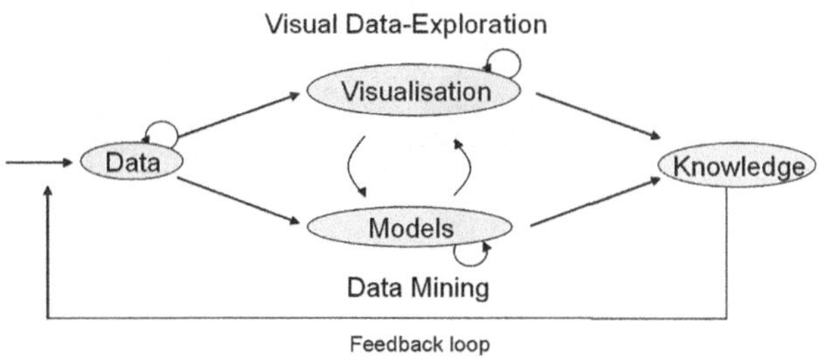

Fig. 25.3 Visual analytics activities D. Keim et al. [14]

knowledge that can be defined for decision-making without using the processes of extension of insight. This form of the model is used by a human to make decisions. The model is formed as a result of training a human in the process of visual analytics, and in the general case, the result of use does not differ from the formal model of machine learning. The machine helps a human to form a mental model as a result of cognitive activity and fixing knowledge. The mental model is completely contained in the human mind and is used exclusively by a human. The goal is to build a mental model as a human's understanding of the logical relationships between data, their properties, formed in the form of features, patterns, components, and so on [14, 15]. According to the parameters of the goal and the results of use, the formal model differs from the mental model only in end-users, which are either a machine or a human.

It should be noted that in interactive machine learning a human is used to improve the final product – a formal model. A human does not directly participate in the building of a formal model. The model is built exclusively by the machine. A human can evaluate the results of the constructed model, change or improve the algorithm used, modify the data or change the parameters. To obtain a formal model, the approach [11], where the initial model is formed, which is iteratively improved by a human by means of visual analytics.

Let's transform the common view of visual analytics by [16] into a functional representation in order to determine the final result in the form of a model as demonstrated at the Fig. 25.4.

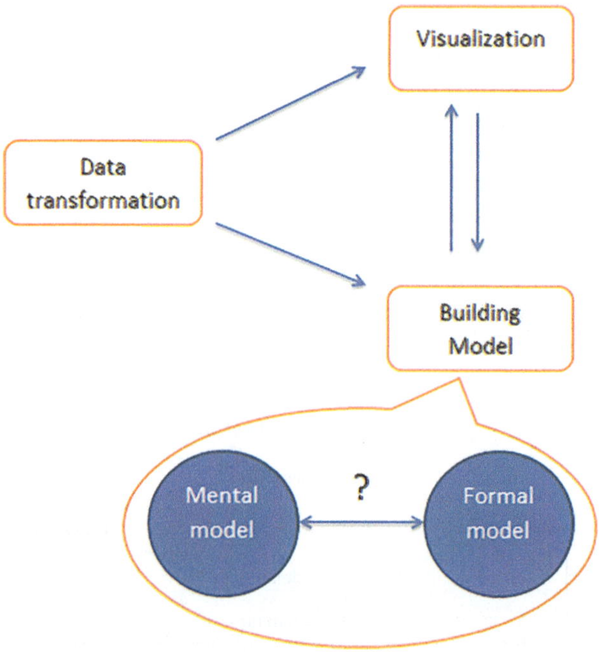

Fig. 25.4 Functional adapted representation of the visual analytics D. Keim et al. [16] focused on getting the model

The functional presentation allows determining the use of visual analytics to obtain the final result – the model. In the general case, this is a single model, but depending on the end-user, it has two representations: formal or mental. The formation of these representations is fundamentally different because it has a different usage environment.

25.5 Generalization Approaches for Using a Model

Considering the use of visual analytics, the most convenient approach is not focused on the process of human-machine interaction and the effectiveness of this process, but on obtaining the final product, which is the model. This approach allows specifying the purpose of using visual interaction. At the same time, the effectiveness of interface interaction allows speeding and shorting the path to obtaining the final result.

The resulting product is presented in the form of a model and is focused on the end-user. A user is a machine or human. Orientation to the end-user is due to a radically different environment for using the model. The formal model is used by the machine. An example is a trained model in machine learning. The mental model is used by human and is contained in his mind. The generated mental model can be transmitted and used in the appropriate environment, namely, other humans. It should be noted that a well-formed mental model can be used by humans regardless of the user's qualifications. This is the main advantage of visual analytics orientation to obtain a model. The user qualification requirement must be set at the stage of model building. Qualification is an important parameter of the user for the building of the model with the necessary parameters of quality and efficiency. The mental model at the stage of use is described by instructions for formal execution and needs only an appropriate execution environment. The execution environment of the mental model is the human mind. The machine cannot use the mental model; for the machine, it must use the formal model. The difference between the formal and mental models in general consists of the fundamental difference between the execution environment and the use of the model. This difference is so significant that it is necessary to use a workflow-oriented to the execution environment (a human or a machine) in order to obtain an appropriate model.

In this regard, of great interest are techniques that allow establishing relationships between models, the mutual transformation of models, porting, building of a generalized form of the model.

Consider the approaches that allow building a model using the advantages of another model, forming informative relationships between them.

The first approach is to synchronize models. Models may differ in their execution mechanisms but have similar results. In this case, the models are also similar and can be replaced. Thus, we can replace the model execution environment; notice a machine on a human or vice versa.

This approach is also used during the construction phase of the model. Based on the visualized information, human forms concepts about regarding result. Initially, a mental model is formed. Further, the iteratively formal model is synchronized with the mental one in accordance with the performance indicators. At the same time, feedback is also very often present to improve the mental model. An example of such use is VIS4ML [5] also Interactive Machine Learning [6].

The process of forming models is possible in two ways.

1. The basic model is formed completely with the necessary parameters. Further, the developed model is fully synchronized with the basic model for the results of work. With this form, the main goal is to maximize the similarity of the results of the models.
2. The initial approximation of the base model is initialized. Then, using another model, the base model is iteratively improved. With this use, for example, the mental model is adjusted in the cognitive process, using the results of the formal model as an improvement tool. The basic model with this form of education is ahead of development, and the other model is "pulled" to the results of the basic model. In complex tasks, this form is most often used and requires a highly qualified user.

Another approach is to use the opposite (with regarding the environment) model. A model can be formed in the ways described in the first approach. A different environment may use the model further. This approach requires adaptation of the built model to a different environment. The approach is the most attractive because it allows expanding the boundaries of the use of the model. For this, it is necessary to develop methods for projecting the model into another runtime environment or other methods for using the model in another environment.

Consider an example in which a mental model is formed. Further, the formal model is formalized in a common space. Common space is the space of forms of representation of universal models. This space of presentation forms allows creating models that are universal in relation to implementation in the runtime environment. If the metal model has the property of projection into the space of universal forms, it can be implemented in a different runtime environment. At the same time, it is necessary to project not the model completely, but its crucial part, that part which is formed for making decisions.

25.6 Transformation of the Model Through the Space of Formalized Forms

If a human becomes an integral part of the analytical system, it is necessary to provide effective feedback from the human to the system. Since the mental model is built in the human mind, the projection of the mental model into the analytical system is possible through the space of formalized forms of representation. Inside the analytical system, there should be a representation of the mental model. A human needs to provide an effective tool for interacting with the analytical system. Here is an example of data classification. Visually, the data is arranged in such a way that it is obvious to a human that the data are divided into two classes. This concept, which has developed in his mind, is a mental model. It is necessary to ensure the mapping of the mental model into formal forms of representation in the analytical system. A human should be able to identify the boundaries of classes. The boundaries of classes should be interpreted by the internal model of the system as the rules for dividing data into classes. Further, when classifying new objects, the system determines the position of the object relative to the boundaries of the classes and indicates to which class the object belongs. This example is a practical implementation of a human-centered analytical classification system.

In the papers [17, 18] proposed an approach according to which a mental model is formed, which is further used by the machine as a different runtime environment. An approach and tools for projecting the use of a mental model by a machine are proposed. This approach is implemented by the example of data classification based on clustering. Grouped data, using methods to reduce the dimension of a set of features, are transformed into a visually accessible set. A mental model is created that is used by a human to form hypervolumes and class boundaries. A decision tree is formed for data belonging to certain classes. The class boundaries and the tree obtained on the basis of the mental model of class membership represent the mental model in the space of universal forms. These presentation parameters are well-formalized in a universal form, which is the basis of the formal model. Thus, the training of the formal model is carried out on the basis of the mental, which is its basis.

25.7 The Use of Human in Visual Analytics

Visual analytics allows combining in a single system the use of the capabilities of the machine and the human while effectively using the strengths of this association. This association allows not only to use the intellectual capabilities of a human but also to be useful for a human.

1. Human intellectual capabilities are an important resource that must be integrated into the model building system. The development of an effective interaction interface is an important link in the productive integration of humans. A human becomes not only a user but an important structural element. At the same time, for the absence of a human, the system becomes unproductive. Such systems can be called a hybrid, based on the heterogeneity of the distributed functionality between human and machine. The task that is posed does not consist in the deep integration of a human into the process of interaction, but in the effective use of the features of the intellectual capabilities of a human, minimizing quantitative estimate of interface interaction. This allows improving and shortening the path

to obtaining the desired result by maximizing the information content of the interaction and its optimization in the direction of the goals.

2. The process of interaction in relation to a human in most cases has a cognitive aspect. This is most evident in the process of forming the model, especially the mental one. The process of building a mental model is cognitive for a human. A human gains new knowledge about the subject area and self-learns.
3. The orientation of the use of visual analytics to obtaining the final product of the model does not exclude the use of the process component. Product orientation can simplify the formalization of the validity of using visual analytics. The form of building a model is a process. And this shows all the advantages of visual analytics in relation to a human. It can be like gaining new knowledge, experience of use, deep insight, broadening one's horizons, and so on.
4. Visual analytics allows a human to use research approaches and "understand" the data. A human understands the aspects of data, features of signs, internal relationships, structure, generalizing signs and so on. This allows making the necessary data transformations, their transformation and finding hidden features. Identification of hidden features and relationships of data makes adjustments to the construction of models and improve their quality characteristics.
5. Advanced training of a human inevitably accompanies the process of visual analysis. This is due to the research component, which leads to human self-development. The analytical work of visual analytics improves a human's qualification abilities in the field of data analysis, formats the aspect of data mining. Human skills are improved by immersion in visual data analytics. This aspect is most effective when building a model. This stage is largely creative in which a human can show his creative abilities. Self-expression and the ability to evaluate the results of their work is an important emotional aspect that affects the quality of the final result.
6. Visual analytics in practical use is a dynamically improving environment. The intellectual development of humans is accompanied by the process of using visual analytics. A user with acquired new knowledge, improved skills sees ways to improve visual analytics workflow in domain areas and improve visual analytics tools. This is an evolutionary aspect of improving human-machine interaction through visual analytics.

25.8 Conclusion

In this paper, we identified areas and aspects of using visual analytics. At the same time, the main emphasis was aimed at formalizing the ways of using visual analysis to specify the final product. This made it possible to determine the directions and aspects of the considered analytical process.

1. The visual analytical process must be used with the focus on obtaining the final product in the form of a formal or mental model.

2. Obtaining a model is the result of effective interaction between humans and machines during the productive exploitation of the advantages of the interacting parties.
3. The effectiveness of the analysis is determined by minimizing the interface interaction and increasing the information content of the visual representation in accordance with the purpose of its use.
4. Visual analytics workflow is determined by the end-use environment of the building model and focuses on a human or a machine.
5. The construction of a working model is determined by the directions of using model synchronization with priority development of the base model or using another model in relation to its execution environment.
6. Changes execution environment of the model is caused by the ability of the model to project into the space of formalized universal forms of representation.
7. The deep integration of a human into the model building system with the aim of exploiting his intellectual analytical abilities is accompanied by an integral component – advanced training, the acquisition of new knowledge, research and intellectual development.

A practical example of using a mental model in a machine environment based on projection techniques through the space of universal formalized forms is considered.

The direction of developing techniques for using models in other runtime environments and developing methods for building universal models is the most attractive and promising.

References

1. Endert A, Hossain MS, Ramakrishnan N, North C, Fiaux P, Andrews C (2014) The human is the loop: new directions for visual analytics. J Intell Inf Syst 43(3):411–435
2. Endert A, Ribarsky W, Turkay C, Wong BLW, Nabney I, Díaz Blanco I, Rossi F (2017) The state of the art in integrating machine learning into visual analytics. Comput Graph Forum 36(8):458–486
3. Thomas J, Cook K (eds) (2005) Illuminating the path: research and development agenda for visual analytics. IEEE-Press, New Jersey, p 455
4. Pirolli P, Card S (2005) The sensemaking process and leverage points for analyst technology as identified through cognitive task analysis. In: Proceedings of international conference on intelligence analysis, McLean, vol 6, pp 1–6
5. Sacha D, Stoffel A, Stoffel F, Kwon Bum Chul K, Ellis G, Keim DA (2014) Knowledge generation model for visual analytics. IEEE Trans Visual Comput Graph 20(12):1604–1613
6. Lee T, Johnson J, Cheng S (2016) An interactive machine learning framework. Arxiv: abs/1610.05463
7. Holzinger A, Plass M, Kickmeier-Rust M, Holzinger K, Crisan GC, Pintea CM, Palade V (2019) Interactive machine learning: experimental evidence for the human in the algorithmic loop. Appl Intell 49(7):2401–2414
8. Kruchinin S, Nagao H, Aono S (2010) Modern aspect of superconductivity: theory of superconductivity. World Scientific, Singapore, p 232
9. Sugahara M, Kruchinin SP (2001) Controlled not gate based on a two-layer system of the fractional quantum Hall effect. Mod Phys Lett B 15:473–477

10. Kruchinin S, Klepikov V, Novikov VE, Kruchinin D (2005) Nonlinear current oscillations in a fractal Josephson junction. Mater Sci 23(4):1009–1013
11. Sacha D, Kraus M, Keim DA, Chen M (2019) Vis4ml: an ontology for visual analytics assisted machine learning. IEEE Trans Visual Comput Graph 25(1):385–395
12. Andrienko N, Lammarsch T, Andrienko G, Fuchs G, Keim D, Miksch S, Rind A (2018) Viewing visual analytics as model building. Comput Graph Forum 37(6):275–299
13. Liu Z, Staskoj T (2010) Mental models, visual reasoning and interaction in information visualization: a top-down perspective. IEEE Trans Visual Comput Graph 16(6):999–1008
14. Kryvonos IG, Krak IV (2011) Modeling human hand movements, facial expressions, and articulation to synthesize and visualize gesture information. Cybern Syst Anal 47(4):501–505
15. Kryvonos IG, Krak IV, Barmak OV, Kulias AI (2017) Methods to create systems for the analysis and synthesis of communicative information. Cybern Syst Anal 53(6):847–856
16. Keim D, Andrienko G, Fekete J-D, Görg C, Kohlhammer J, Melançon G (2008) Visual analytics: definition, process, and challenges. In: Kerren A, Stasko JT, Fekete J-D, North C (eds) Information visualization: human-centered issues and perspectives. Springer, Berlin, pp 154–175
17. Manziuk EA, Barmak AV, Krak YV, Kasianiuk VS (2018) Definition of information core for documents classification. J Autom Inf Sci 50(4):25–34
18. Barmak A, Krak Y, Manziuk E, Kasianiuk V (2019) Information technology of separating hyperplanes synthesis for linear classifiers. J Autom Inf Sci 51(5):54–64

Chapter 26
Apoptosis in Atherosclerosis and the Ways of Its Regression

A. Ahsan and A. T. Mansharipova

Abstract Apoptotic cell death played an important role in the pathogenesis of atherosclerosis.

Several studies indicate the involvement of endothelial cell apoptosis in acute coronary syndromes, because apoptotic endothelial cells have pro-coagulant [1] and pro-inflammatory activities.

Atherosclerotic plague rapture causes myocardial infarction. Plague raptures are associated with increased fibrous cap macrophages, reduced fibrous cap vascular smooth muscle cells (VSMCs), and increased VSMC apoptosis. Because VSMCs are the only cells within plagues capable of synthesizing structurally important collagen isoforms, VSMC apoptosis might promote plague rupture [2].

Keywords Apoptosis · Atherosclerosis · NO

26.1 Introduction

Impaired production or activity of NO leads to vasoconstriction, platelet aggregation, SMCs proliferation and migration, leukocyte adhesion, and oxidative stress [3]. NO replacement therapy by NO donors restored nitric oxide deficits associated with dysfunctional endothelium [4]. Also NO donors serve as an anti apoptotic regulator of caspase activity via S-nitrosation of the Cys-163 residue of caspase-3 [5].

High-density lipoproteins (HDLs) are considered anti-atherogenic because of their key role as a carrier of excess cellular cholesterol in the reverse cholesterol transport (RCT) pathway [6–10].

A. Ahsan
LLC Lynx Eurasia, Almaty, Kazakhstan

A. T. Mansharipova (✉)
Kazakh Russian Medical University, Almaty, Kazakhstan
e-mail: dralma@mail.ru

J. Bonča, S. Kruchinin (eds.), *Advanced Nanomaterials for Detection of CBRN*,
NATO Science for Peace and Security Series A: Chemistry and Biology,
https://doi.org/10.1007/978-94-024-2030-2_26

Besides providing of cholesterol efflux from peripheral tissues, HDLs provid anti-inflammatory, anti-oxidative, anti-aggregatory, anti-coagulant, and pro-fibrinolytic activities which exerted by different components of HDL, namely apolipoproteins, enzymes, and even specific phospholipids. This complexity further emphasizes that changes in the functionality of HDL rather than changes of plasma HDL-cholesterol levels determine the anti-atherogenicity of therapeutic alterations of HDL metabolism [11].

Enrichment of plasma lipoproteins (LP) with anionic phospholipid phosphatidilinositol (PI) increased the negative surface potential of all classes of LP including HDL [12].

PI enrichment of HDL probably enhances the HDL activation the PI3K/Akt pathway via lysophospholipid receptor S1P3/EDG3 (sphingosine-1 phosphate subtype 3/endothelial differentiation gene 3). This HDL/receptor interaction results in the phosphorylation of Akt, suppressed activity of caspase-3/7 and protect endothelial cells from apoptosis [13].

26.2 Aim of the Study

Our aim was to determine the effect of NO donor encapsulated in phospholipid Phosphatidylinositol nano capsules (Alpha X1) at the apoptosis processes in atherosclerotic lesions.

26.3 Materials and Methods

26.3.1 Animals and Study Design

Thirty five 6 month old male Chinchilla rabbits (3500 g average initial weight) were housed individually in stainless steel cages in temperature- and humidity-controlled rooms, with 12 h of light and 12 h of darkness daily. The rabbits were observed daily and weighed once a month.

5 rabbits with regular diet (Purina rabbit chow (Dytes Inc.,USA)) served as control (group 1).

30 rabbits were fed with 2% cholesterol-enriched diet (Dytes Inc., USA) for 6 months for experimental hyperlipidemia and atherosclerotic lesions development. After 6 months of feeding, animals average weighs were 5000 g. These 30 rabbits were randomly subdivided into 3 groups with continued feeding of cholesterol – enriched diet.

Group 2 consisted of 10 animals continued the experiment without treatment.

Group 3 consisted of 10 animals were treated with Alpha X1 for 10 days and Group 4 was treated with Alpha X1 for a period of 20 days.

Hair of the back area of each rabbit from group 3 and 4 were carefully removed (1,5 cm in diameter) with an electric shaver. Alpha X1 was transdermally rubbed at the shaved area of animal body twice a day during 10 days for group 3 and for 20 days for group 4. Drug doses were calculated according to the interspecies coefficient of doses recalculation.

Experimental protocol was approved by the Ethics Review Committee for Animal Experimentation of Kazakh Russian Medical University.

26.3.2 Sacrifice Schedule

Sacrifices of the animals were conducted under anesthesia with sodium pentobarbital (40 mg/kg) by cervical dislocation. After treatment with Alpha X1 for 10 days first sacrifice was conducted for 10 animals from groups 2 and 3 (20 animals total). Group 4 continued treatment for another 10 days. Second sacrifice was conducted for 10 animals from group 4 after 20 days of treatment. Control animals were sacrificed at the end of whole experiment.

26.3.3 Necropsy Procedures

After each termination the chest and abdomen cavities were rapidly opened, the abdominal aortas, were carefully dissected free from surrounding tissues, rinsed of blood in distilled water, cut longitudinally and small pieces of each aorta with atherosclerotic damages were taken for each pre-arranged investigation.

26.3.4 Tissue Processing

Part of excised aorta pieces were fixed in 10% buffered formalin solution for 48 h. Afterwards the fixed tissues were washed with distilled water from excess fixative and subjected to series of spirit solutions with growing concentrations (70%–80%–90%–96%–96%–100%–100%) for 1 h in every solution. Dehydrated tissues are defatted in xylene and embedded into paraffin wax at the apparatus Leica TP 1020 (Germany).

Paraffin-embedded sections of the above tissues were sectioned at 4 μm (microtome Leica SM 2000R, Germany).

26.3.5 In Situ Detection of Apoptotic Cells

Apoptosis in aorta paraffin – embedded sections was evaluated with use of ApopTag Peroxidase In Situ Oligo Ligation (ISOL) technique (Chemicon International, Inc) according to the manufacturer's instructions.

Examined under light optical microscope (LEICA DM 4000B, Leica Microsystems CMS GmbH, Germany) connected with digital camera (Leica DFC 320 Leica Microsystems Ltd.) under magnification ×200.

26.3.6 *Immunohistochemistry*

P53 and BAX expression in aorta paraffin-embedded sections were conducted with P53 (FL-393): sc-6243 (Santa Cruz Biotechnology, Inc.) and BAX (P-19): sc-526 (Santa Cruz Biotechnology, Inc.) according to the manufacturer's instructions.

Sections were counterstained with hematoxylin and examined under light optical microscope (LEICA DM 4000B, Leica Microsystems CMS GmbH, Germany) connected with digital camera (Leica DFC 320 Leica Microsystems Ltd.) under magnification ×200. Cytoplasmic staining for P53 and bax was considered a positive result.

26.3.7 *Morphometric Analysis*

Counting of ISOL-Positive Nuclei

All specimens per rabbit (on an average one to three pieces of tissue) were analyzed, yielding 20–30 sections each.

Cells were considered apoptotic when cell nuclei demonstrated positive ISOL staining and apoptotic morphology. For quantification of ISOL-positive cells, four fields per section were examined at 200-fold magnification. The apoptotic index was calculated using the formula: 100 × (number of ISOL + cell nuclei per field/total number of cell nuclei per field) [14].

Counting of p53 and BAX Expression

For quantitative analysis of the percentage of the cells that displayed positive immunohistochemical stain for P53 and bax proteins (cytoplasmic staining), light optical microscope was used. From these, color images were digitized on the computer. Each time before measurement of the image, the contrast, brightness were standardized and calibration of the measurement system with appropriate slides were done. The edges of the sections were excluded from the evaluation. For quantitation of immunostaining intensities, we measured the inverse mean density based on the RGB color parameter were measured. The circle profile tool of Image-Pro Plus program (Media Cybernetics) was used for measurements of cytoplasmic P53 and bax immunostaining, respectively. According to previous reports in each case, 20 randomly selected color video images were done. These represented 6800 measurements for each protein that was analyzed in each biopsy. For each set of measurements, a curve was made according to immunostaining intensity. The units arbitrarily ranged from 0 (intensity absent) to 230 (top intensity). In each case, the percentage of ISOL (+) cells gave the apoptotic body index (ABI) for each case.

26.3.8 Statistical Analysis

Results were reported as mean ± SD, median, and range. Intergroup comparisons – with regard to correlation of thymus pathology and MG stage with staining results – were performed using one-way analysis of variance (ANOVA). When the equal variance test or normality test failed, the Kruskall–Wallis non-parametric test was applied. In order to address the problem of multiple comparisons, these tests (ANOVA, Kruskall–Wallis) were followed by a posthoc Bonferroni test. Spearman's rank correlation was used to examine the possible associations between the immunohistochemical or ISH results for bcl-2 and bax, the immunohistochemical results for Ki67, and the results for TUNEL stain, on the one hand, and the thymus weight and size, on the other. Spearman's rank correlation was also used to detect any relationships between bcl-2, bax, Ki67, and ABI. Data were analyzed using the SigmaStat (Jandel Scientific, USA). Significance was defined as $P < 0.05$ (Figs. 26.1 and 26.2).

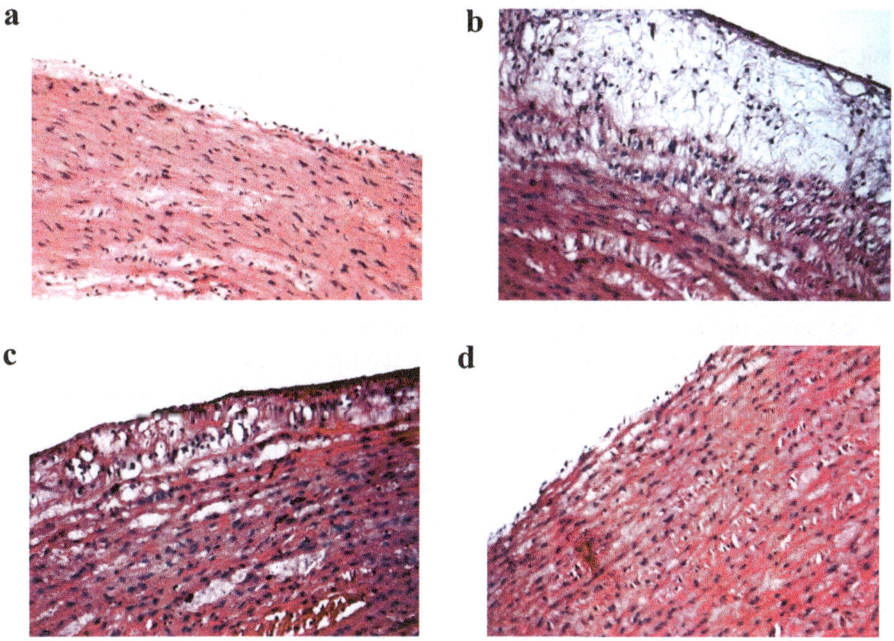

Fig. 26.1 Shows H&E staining of aorta (original magnification ×200). (**a**) group 1 – control. Absence of signs of atherosclerotic lesions. Normal morphology of aorta wall. (**b**) group 2-cholesterol-fed rabbit. Numerous subendothelial transparent vacuoles, foam cells with oval vacuoles, some of foam cells penetrated in initial region of medial aorta layer. Smooth muscle cells (SMCs) proliferation. Some of SMCs contain small vacuoles. (**c**) group 3 – cholesterol fed rabbit after 10 days of treatment with Alpha X1. Solitary subendothelial vacuoles and foam cells. (**d**) group 4 – cholesterol fed rabbit after 20 days of treatment with Alpha X1. Flat endotheliocytes with homogenously pink cytomlasm and solitary cytoplasmatic vacuoles. Clear basal membrane

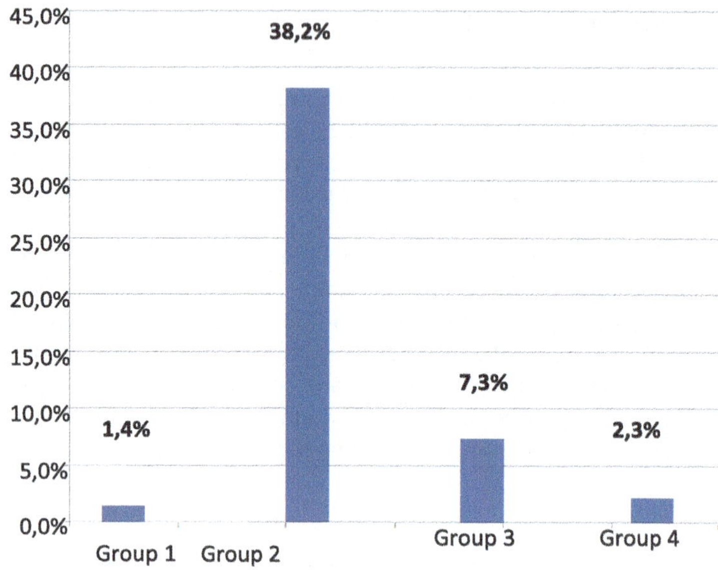

Fig. 26.2 When analyzing aortic tissue and counting areas stained with Oil Red O, it was revealed that 1.4 ± 0.2% of aortic tissue was stained in the control group, while in group 1 in the tissues of animals with atherosclerosis, the percentage of staining of aortic tissue was 38.2 ± 3.6%. After treatment with Alpha X1 for 10 days (group 3), lipid staining of the aortic tissue is observed at 7.3 ± 0.8%, and after treatment for 20 days (group 4), this indicator amounted to 2.3 ± 0.6%

26.4 Results

When analyzing aortic tissue and counting areas stained with Oil Red O, it was revealed that 1.4 ± 0.2% of aortic tissue was stained in the control group, while in group 1 in the tissues of animals with atherosclerosis, the percentage of staining of aortic tissue was 38.2 ± 3.6%.

After treatment with Alpha X1 for 10 days (group 3), lipid staining of the aortic tissue is observed at 7.3 ± 0.8%, and after treatment for 20 days (group 4), this indicator amounted to 2.3 ± 0.6%.

In situ detection of apoptotic cells.

In situ determination of apoptotic cells in the aortic tissues of rabbits from group 1, group 2, group 3 and group 4 revealed the presence of massive apoptotic processes in group 2, compared with group 1. The number of apoptotic bodies in aortic tissue in groups 3 and 4 was significantly reduced compared to group 2. After application of Alpha X1 for 20 days in the aortic tissues of group 4 rabbits, apoptotic processes were less pronounced than in group 3. In situ data for the determination of apoptotic bodies in aortic tissue are presented in Figs. 26.3 and 26.4.

The diagram shows the percentage of apoptotic bodies in aortic tissues in all groups of rabbits (mean ± s.d.).

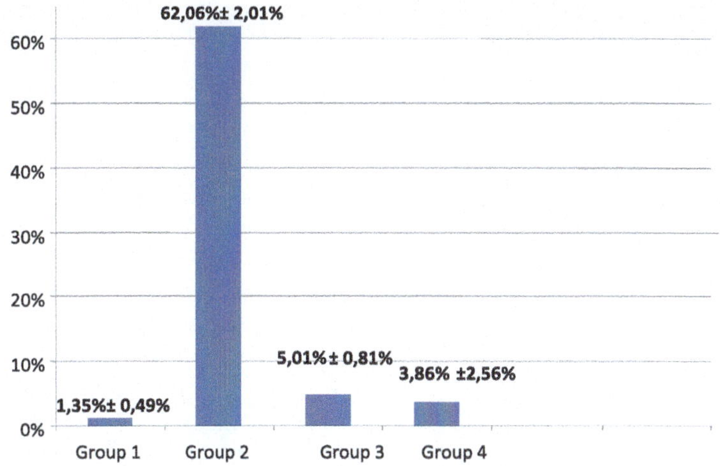

Fig. 26.3 The diagram shows the percentage of apoptotic bodies in aortic tissues in all groups of rabbits (mean ± s.d.)

26.5 Detection of P53 Expression

The determination of P53 expression in aortic tissues of rabbits from group 1, group 2, group 3, and group 4 revealed the presence of a massive immunohistochemical reaction in group 2 (p), compared with group 1. The expression of P53 in aortic tissues in group 3 and group 4 was significantly reduced compared to group 2. In the aortic tissues of group 4 rabbits, after application of Alpha X1 for 20 days, P53 expression was less pronounced than after application of this combination for 10 days in group 3. Data on the determination of P53 expression in aortic tissues are presented in Fig. 26.5.

26.6 Detection of BAX Expression

The determination of BAX expression in the aortic tissues of rabbits from group 1, group 2, group 3, and group 4 revealed the presence of a massive immunohistochemical reaction in group 2, compared with group 1. The expression of BAX in aortic tissues in group 3 and group 4 was significantly reduced compared to group 2. In the aortic tissues of group 4 rabbits, after application of Alpha X1 for 20 days, BAX expression was less pronounced than after application of this combination for 10 days in group 3. Data for the determination of BAX expression in aortic tissues are presented in Fig. 26.6.

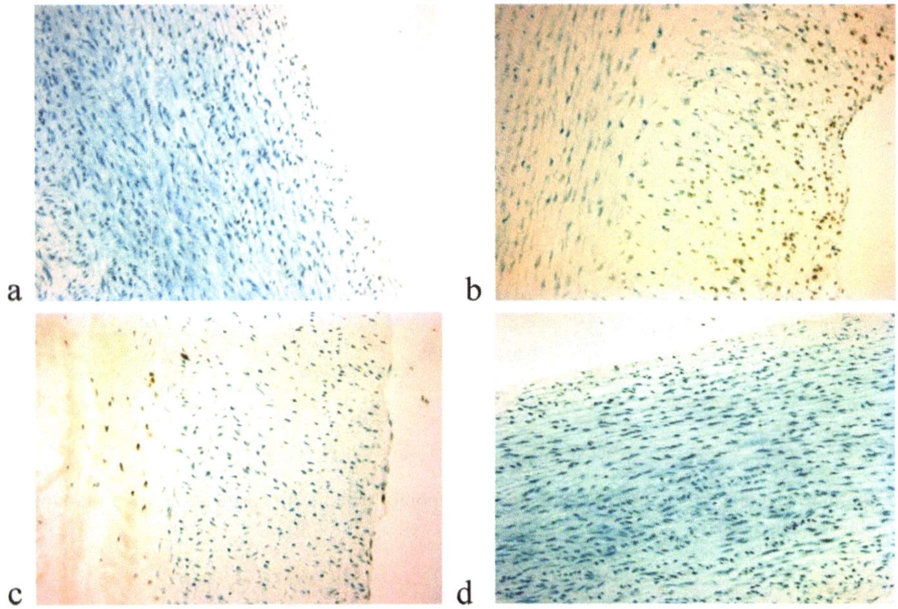

Fig. 26.4 In situ detection of apoptotic cells in aortic tissues (ApopTag Peroxidase In Situ Oligo Ligation (ISOL) technique, 4 μm thick, original magnification × 200). (**a**) group 1 – aorta of control rabbit. No data were found for apoptosis. (**b**) group 2 – aorta of cholesterol-fed rabbit. Multiple small brown-colored bodies in the endothelium and subendothelial space are determined. (**c**) group 3 – aorta of cholesterol fed rabbit after 10 days of treatment. The presence of single, brown-stained cores. (**d**) group 4 – aorta of cholesterol fed rabbit after 20 days of treatment A meager cluster of brown-stained core elements

26.7 Discussion

26.7.1 Discussion of the Results: Quantitative Analysis of Myointimal Thickening of the Aortic Wall

When analyzing the results, it was found that compared with the control group, in group 2, after experimentally induced atherosclerosis, the thickness of the aortic wall increased by 12%. Moreover, after treatment with Alpha X1 myo-optimal thickening decreased after 10 days of treatment by 7.3%, and after 20 days of treatment – by 31%.

The data obtained indicate that in atherosclerosis the processes of vascular wall infiltration by cellular and non-cellular elements, such as monocytes, neutrophils, and LDL, occur, which leads to a thickening of the myo-optimal zone and the formation of atherosclerotic damage.

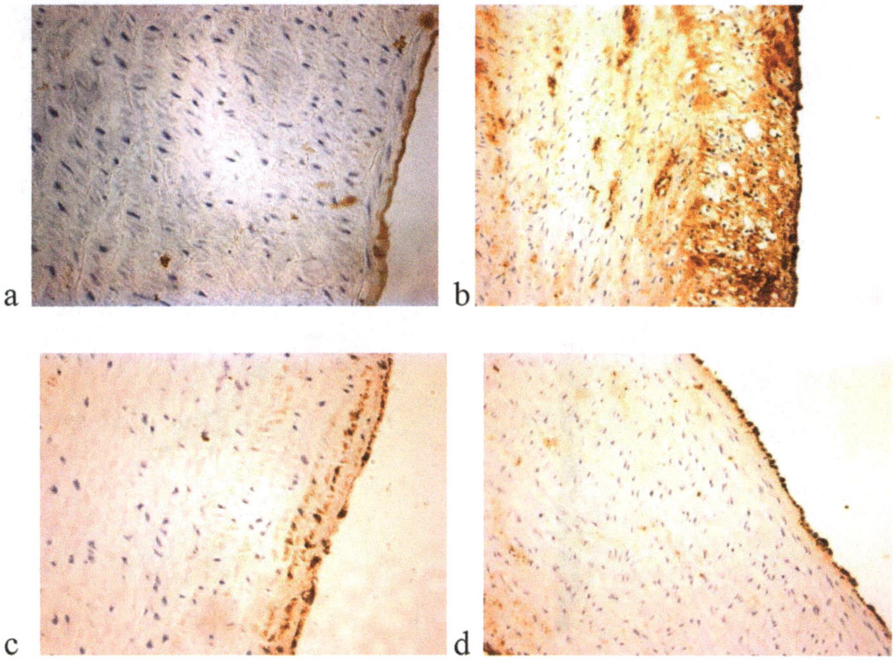

Fig. 26.5 The pictures show the expression of P53 in aortic tissues (P53 (FL-393): sc-6243 (Santa Cruz Biotechnology, Inc., 4 μm thick, original magnification × 200). (**a**) group 1 – aorta of control rabbit. When immunohistochemical studies of the expression of P53 data for apoptosis were not found. (**b**) group 2 – aorta of cholesterol-fed rabbit. The presence of granules stained in intense brown color, located in the endothelium, basement membrane and sections of the middle aortic membrane, is noted. The staining intensity decreased towards the adventitia membrane. (**c**) group 3 – aorta of cholesterol fed rabbit after 10 days of treatment with Alpha X1. The immunohistochemical reaction P53 revealed the presence of brown granules in the endothelium and subendothelial space of the aorta. (**d**) group 4 – aorta of cholesterol fed rabbit after 20 days of treatment. A meager accumulation of brown-stained core elements, endothelial cells, and individual small foci of the subintimal space was determined

There was abundant subendothelial lipid infiltration in atherosclerotic vascular lesions. The percentage of staining of tissues of group 2 was 27 times higher compared with the control group.

However, during therapy with Alpha X1 for 10 days, lipid infiltration of the vascular wall decreases by 5.2 times, and with 20 days of treatment – by 16.6 times, compared with group 2, which indicates a pronounced antiatherosclerotic effect of this preperation.

At the same time treatment with Alpha X1. The positive effect of reducing myointimal aortic wall infiltration is based on the effect of phosphatidylinositol on cholesterol reverse transport. Phosphatidylinositol, integrating into lipoproteins stimulates and enhances the reverse transport of cholesterol from peripheral tissues, which is due to a pronounced significant decrease in myointimal aortic wall infiltration.

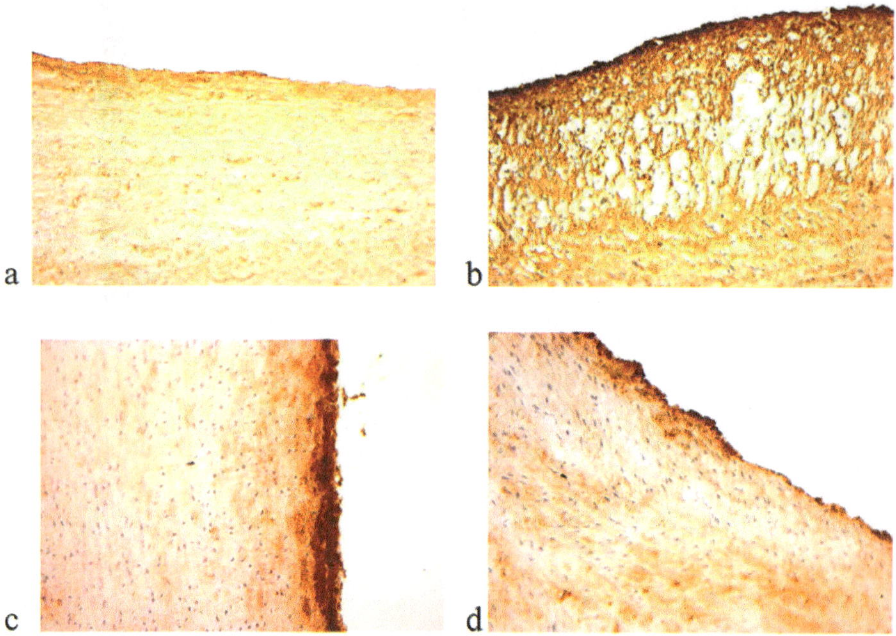

Fig. 26.6 The pictures above show BAX expression in aortic tissues (P-19): sc-526 (Santa Cruz Biotechnology, Inc., 4 μm thick, original magnification × 200). (**a**) group 1 – aorta of control rabbit. In immunohistochemical studies of BAX expression, data for apoptosis were not found. (**b**) group 2 – aorta of cholesterol-fed rabbit. The presence of granules stained in intense brown color, located in the endothelium, basement membrane and sections of the middle aortic membrane, is noted. The staining intensity decreases towards the adventitia membrane. (**c**) group 3 – aorta of cholesterol fed rabbit after 10 days of treatment with Alpha X1. The BAX immunohistochemical reaction revealed the presence of brown granules in the endothelium and subendothelial space of the aorta. (**d**) group 4 – aorta of cholesterol fed rabbit after 20 days of treatment A meager accumulation of brown-stained elements of the nucleus, endothelial cells, and individual small foci of the subintimal space is determined

26.7.2 Discussion of the Results: Quantitative Analysis of Apoptotic Death of Aortic Wall Cells

The obtained data of quantitative analysis of cells that entered the apoptotic phase indicate that during the development and progression of atherosclerotic processes, massive apoptotic processes takes place, which leads to aggravation of pathological processes in the lesion, the spread and progression of atherosclerotic processes. In group 2, apoptosis-positive nuclei were 45.9 times more than in control group.

The data obtained during experimental treatment indicate a positive anti-apoptotic effect of Alpha X1, which is expressed in a significant decrease in apoptosis of vascular wall cells in the focus of atherosclerotic lesion. The number of apoptotic nuclei of the vascular wall was 12.3 times in group 3 (10 days of treatment) and 16 times less in group 4 (20 days of treatment) than in control group.

The data obtained indicate a sufficiently pronounced cytoprotective anti-apoptotic effect of Alpha X1, which probably can serve as the basis for its use not only as a drug with a major vasodilating effects and shows significant decrease in apoptotic cell due to the double apoptosis inhibitory effect of both components. The data obtained indicate the predominant therapeutic effect of Alpha X1.

References

1. Nofer J-R, Levkau B, Wolinska I, Junker R, Fobker M, von Eckardstein A, Seedorf U, Assmann G (2001) Suppression of endothelial cell apoptosis by high density lipoproteins (HDL) and HDL-associated lysosphingolipids. J Biolog Chem 276(37):34480–34485
2. Boyle JJ, Weissberg PL, Bennett MR (2003) Tumor necrosis factor-promotes macrophage-induced vascular smooth muscle cell apoptosis by direct and autocrine mechanisms. Arterioscler Thromb Vasc Biol 23:1553–1558
3. Herman AG, Moncada S (2005) Therapeutic potential of nitric oxide donors in the prevention and treatment of atherosclerosis. Eur Heart J 26:1945–1955
4. Russo G, Leopold JA, Loscalzo J (2002) Vasoactive substances: nitric oxide and endothelial dysfunction in atherosclerosis. Vascul Pharmacol 38(5):259–269
5. Rössig L, Fichtlscherer B, Breitschopf K, Haendeler J, Zeiher AM, Mülsch A, Dimmeler S (1999) Nitric oxide inhibits caspase-3 by S-nitrosation in vivo. J Biol Chem 274:6823–6
6. Brewer HB (2004) Increasing HDL cholesterol levels. N Engl J Med 350(15):1491–1494
7. Repetsky P, Vyshyvana IG, Nakazawa Y, Kruchinin SP, Bellucci S (2019) Electron transport in carbon nanotubes with adsorbed chromium impurities. Materials 12:524
8. Ermakov V, Kruchinin S, Pruschke T, Freericks J (2015) Thermoelectricity in tunneling nanostructures. Phys Rev B 92:115531
9. Kruchinin S, Pruschke T (2014) Thermopower for a molecule with vibrational degrees of freedom. Phys Lett A 378:157–161
10. Filikhin I, Peterson TH, Vlahovic B, Kruchinin SP, Kuzmichev YuB, Mitic V (2019) Electron transfer from the barrier in InAs/GaAs quantum dot-well structure. Phys E Low-dimensional Syst Nanostruct 114:113629
11. Nofer JR, Kehrel B, Fobker M, Levkau B, Assmann G, von Eckardstein A (2002) HDL and arteriosclerosis: beyond reverse cholesterol transport. Atherosclerosis 161(1):1–16
12. Stamler CJ, Breznan D, Neville TAM, Viau FJ, Camlioglu E, Sparks DL (2000) Phosphatidylinositol promotes cholesterol transport in vivo. J Lipid Res 41(8):1214–1221
13. DeKroon R, Robinette JB, Hjelmeland AB, Wiggins E, Blackwell M, Mihovilovic M, Fujii M, York J, Hart J, Kontos C, Rich J, Strittmatter WJ (2006) APOE4-VLDL inhibits the HDL-activated phosphatidylinositol 3-kinase/akt pathway via the phosphoinositol phosphatase SHIP2. Circ Res 99(8):829–836
14. Hassan AHM, Lang IM, Ignatescu M, Ullrich R, Bonderman D, Wexberg P, Weidinger F, Glogar HD (2001) Increased intimal apoptosis in coronary atherosclerotic vessel segments lacking compensatory enlargement. J Am Coll Cardiol 38:1333–1339

Author Index

J. Bonča, S. Kruchinin (eds.), *Advanced Nanomaterials for Detection of CBRN*,
NATO Science for Peace and Security Series A: Chemistry and Biology,
https://doi.org/10.1007/978-94-024-2030-2

Subject Index

J. Bonča, S. Kruchinin (eds.), *Advanced Nanomaterials for Detection of CBRN*,
NATO Science for Peace and Security Series A: Chemistry and Biology,
https://doi.org/10.1007/978-94-024-2030-2

The manufacturer's authorised representative in the EU is Springer Nature Customer Service Centre GmbH, Europaplatz 3, 69115 Heidelberg, Germany. If you have any concerns regarding our products, please contact ProductSafety@springernature.com

Printed and bound by CPI Group (UK) Ltd, Croydon, CR0 4YY
30/09/2026
02225333-0012